Statistical Inference

The Minimum Distance Approach

MONOGRAPHS ON STATISTICS AND APPLIED PROBABILITY

General Editors

F. Bunea, V. Isham, N. Keiding, T. Louis, R. L. Smith, and H. Tong

Monographs on Statistics and Applied Probability 120

Statistical Inference
The Minimum Distance Approach

Ayanendranath Basu

Indian Statistical Institute
Kolkata, India

Hiroyuki Shioya

Muroran Institute of Technology
Muroran, Japan

Chanseok Park

Clemson University
Clemson, South Carolina, USA

CRC Press
Taylor & Francis Group
Boca Raton   London   New York

CRC Press is an imprint of the
Taylor & Francis Group an **informa** business

A CHAPMAN & HALL BOOK

Chapman & Hall/CRC
Taylor & Francis Group
6000 Broken Sound Parkway NW, Suite 300
Boca Raton, FL 33487-2742

First issued in paperback 2022

© 2011 by Taylor and Francis Group, LLC
CRC Press is an imprint of Taylor & Francis Group, an Informa business

No claim to original U.S. Government works

ISBN 13: 978-1-03-247763-3 (pbk)
ISBN 13: 978-1-4200-9965-2 (hbk)

DOI: 10.1201/b10956

Visit the Taylor & Francis Web site at
http://www.taylorandfrancis.com

and the CRC Press Web site at
http://www.crcpress.com

To Manjula, Srabashi, and Padmini   – AB –

To my wife, my father, and mother   – HS –

To Late Professor Byung Ho Lee   – CP –

Contents

Preface

In many ways, estimation by an appropriate minimum distance method is one of the most natural ideas in statistics. A parametric model imposes a certain structure on the class of probability distributions that may be used to describe real life data generated from a process under study. There hardly appears to be a better way to deal with such a problem than to choose the parametric model that minimizes an appropriately defined distance between the data and the model.

The issue is an important and complex one. There are many different ways of constructing an appropriate "distance" between the "data" and the "model." One could, for example, construct a distance between the empirical distribution function and the model distribution function by a suitable measure of distance. Alternatively, one could minimize the distance between the estimated data density (obtained, if necessary, by using a nonparametric smoothing technique such as kernel density estimation) and the parametric model density. And when the particular nature of the distances has been settled (based on distribution functions, based on densities, etc.), there may be innumerable options for the distance to be used within the particular type of distances. So the scope of study referred to by "Minimum Distance Estimation" is literally huge.

Statistics is a modern science. In the early part of its history, minimum distance estimation was not a research topic of significant interest compared to some other topics. There may be several reasons for this. The neat theoretical development of the idea of maximum likelihood and its superior performance under model conditions meant that any other competing procedure would have had to make a real case for itself before being proposed as a viable alternative to maximum likelihood. Until other considerations such as robustness over appropriate neighborhoods of the parametric model came along, there was hardly any reason to venture outside the fold of maximum likelihood, particularly given the computational simplicity of the maximum likelihood method in most common parametric models, which minimum distance methods in general do not share.

The growth of the area of research covered by the present book can be attributed to several factors. Two of them require special mention. The first one is the growth of computing power. As in all other areas of science, research in statistical science got a major boost with the advent of computers. Previously intractable problems became numerically accessible. Approximate methods could be applied with enhanced degrees of precision. Computational complexity of the procedure became a matter of minor concern, rather than the major deciding factor. This made the construction of distances and the computation of the estimators computationally feasible. The second major reason is the emergence of the area of robust statistical inference. It was no longer sufficient to have a technique which was optimal under model conditions but had weak robustness properties. Several minimum distance techniques have natural robustness properties under model misspecifications. Thus, the computational advances and the practical requirements converged to facilitate the growth of research in minimum distance methods.

Among the class of minimum distance methods we have focused, in this book, on density-based minimum distance methods. Carrying this specialization further, our emphasis, within the class of density-based distances has been on the chi-square type distances. Counting from Beran's path breaking 1977 paper, this area has seen a major spurt of research activity during the last three decades. In fact, the general development of the chi-square type distances began in the 1960s with Csiszár (1963) and Ali and Silvey (1966), but the robustness angle in this area probably surfaced with Beran. The procedures within the class of "ϕ-divergences" or "disparities" are popular because many of them combine strong robustness features with full asymptotic model efficiency.

There is no single book which tries to provide a comprehensive documentation of the development of this theory over the last 30 years or so. Our primary intention here has been to fill in this gap. Our development has mainly focused on the problem for independently and identically distributed data. But we have tried to be as comprehensive as possible in this regard in establishing the basic structure of this inference procedure so that the reader is sufficiently prepared to grasp the applications of this technique to more specialized scenarios. We have discussed the estimation and hypothesis testing problems for both discrete and continuous models, extensively described the robustness properties of the minimum distance methods, discussed the inlier problem and its possible solutions, described weighted likelihood estimators and considered several other related topics. We trust that this book will be a useful resource for any researcher who takes up density-based minimum distance estimation in the future.

Apart from minimum distance estimation based on chi-square type distances, on which we have spent the major part of this book, we have briefly looked at three other topics. These may be described as (i) minimum distance estimation based on the density power divergence; (ii) some recent developments on goodness-of-fit tests based on disparities and their modifications, and (iii) a discussion of the applications of these minimum distance methods in information theory and engineering. We believe that the last item will make the book useful to scientists outside the mainstream statistics area.

In this connection it is appropriate to mention some closely related books that are available in the literature. The book by Pardo (2006) gives an excellent description of minimum ϕ-divergence procedures and is a natural resource for this area. However, Pardo deals almost exclusively, although thoroughly, with discrete models. In our book we have also provided an extensive description of continuous models. Besides, the robustness angle is a driving theme of our book, unlike Pardo's case.

Our discussion of the multinomial goodness-of-fit testing problem has been highly influenced by the classic by Read and Cressie (1988). However, we have made every effort not to be repetitive, and only described such topics not covered extensively by Cressie and Read (or extended their findings beyond the power divergence family). Unlike the minimum distance inference case where we have tried to be comprehensive, in the goodness-of-fit testing problem we have been deliberately selective.

We have also kept the description to a level where it will be easily accessible to students who have been exposed to first-year graduate courses in statistics. Our presentation, although sufficiently technical, does not assume a measure theoretic background for the reader and, except in Chapter 11, the rare references to measures do not arrest the flow of the book. The book can very well serve as the text for a one-semester graduate course in minimum distance methods.

We take this opportunity to acknowledge the help we have received from many colleagues, teachers, and students while completing the book. We should begin by acknowledging our intellectual debt to Professor Bruce G. Lindsay, two of the three authors of the current book being his Ph.D. advisees. Discussions with Professors Leandro Pardo, Marianthi Markatou and Claudio Agostinelli have been very helpful. Discussion with Professor Subir Bhandari has helped to make many of our mathematical derivations more rigorous. Many other colleagues, too innumerable to mention here, have helped us by drawing our attention to related works. We also thank Professor Wen-Tao Huang, who was instrumental in bringing this group of authors together.

Special thanks must be given to Dr. Abhijit Mandal; his assistance in working out many of the examples in the book and constructing the figures has been invaluable. Dr. Rohit Patra and Professor Biman Chakraborty also deserve thanks in this connection.

Finally, we wish to thank all our friends and family members who stood by us during the sometimes difficult phase of manuscript writing.

Ayanendranath Basu
Indian Statistical Institute
India

Hiroyuki Shioya
Muroran Instiute of Technology
Japan

Chanseok Park
Clemson University
USA

Acknowledgments

Some of the numerical examples and figures represented here are from articles copyrighted to different journals or organizations. They have been reproduced with the permission of the appropriate authorities. A list is presented below, and their assistance in permitting these reproductions is gratefully acknowledged.

The simulated results in Example 5.5 have been reproduced from *Sankhya*, Series B, Volume 64, Basu, A. (author), Outlier resistant minimum divergence methods in discrete parametric models, pp. 128–140 (2002), with kind permission from *Sankhya*.

The simulated results in Example 6.1, together with Figures 6.1, 6.2 and Table 6.1, are reproduced from *Statistica Sinica*, Volume 8, Basu, A. and Basu, S. (authors), Penalized minimum disparity methods for multinomial models, pp. 841–860 (1998), with kind permission from *Statistica Sinica*.

The real data example in Section 9.4.1, together with Figures 9.2 and 9.3 have been reproduced from *The Annals of the Institute of Statistical Mathematics*, Volume 58, Basu, S., Basu, A. and Jones, M. C. (authors), Robust and efficient estimation for censored survival data, pp. 341–355 (2006), with kind permission from the *Institute of Statistical Mathematics*.

1

Introduction

Statistical distances have two very important uses in statistical analysis. Firstly, they can be applied naturally to the case of parametric statistical inference. The idea of minimum distance estimation has been around for a while and there are many nice properties that the minimum distance estimators enjoy. Minimum distance estimation was pioneered by Wolfowitz in the 1950s (1952, 1953, 1954, 1957). He studied minimum distance estimators as a class, looked at the large sample results, established their strong consistency under general conditions and considered the minimized distances in testing goodness-of-fit. Parr (1981) gives a comprehensive review of minimum distance methods up to that point. Vajda (1989) and Pardo (2006) have provided useful treatments in statistical inference based on divergence measures.

The most important idea in parametric minimum distance estimation is the quantification of the degree of closeness between the sample data and the parametric model as a function of an unknown parameter; the degree of closeness is described by some notion of affinity, or inversely, by some notion of distance, between the data and the model. Thus, for example, the estimate of the unknown parameter will be obtained by minimizing a suitable distance over the parameter space.

In statistical inference, two broad types of distances are usually used. They represent

(i) distances between distribution functions (such as a distance between the empirical distribution function and the model distribution function) and

(ii) distances between probability density functions (such as the distance between some nonparametric density estimate obtained from the data, and the model probability density function).

Examples of distances of the first type include the Kolmogorov–Smirnov distance and the Cramér–von Mises distance; this class also includes some weighted versions of these distances, such as the Anderson–Darling distance. On the other hand, the second class includes density-based distances such as the Pearson's chi-square, the Hellinger distance, the Kullback–Leibler divergence, and the family of chi-square distances as a whole. It also includes some other families like the Bregman divergence (Bregman, 1967) and the Burbea–Rao divergence (Burbea and Rao, 1982).

The use of minimum distance methods in parametric estimation has been found to be particularly useful in situations where the data contain a significant proportion of outliers. Early references which formalized the theory of robustness in the middle of the last century include the works of Box (1953), Tukey (1960), Huber (1964) and Hampel (1968). Huber (1981) and Hampel et al. (1986) represent canonical references for robust statistical inference in a general sense. However, there is also a rich literature dealing specifically with the robustness of different minimum distance functionals. See, for example, Parr and Schucany (1980, 1982), Boos (1981, 1982), Parr and De Wet (1981) and Wiens (1987) who demonstrate the robustness of minimum distance functionals based on distribution functions. Many density-based minimum distance estimators have also been observed to have strong robustness properties. In addition, many members of the latter class of estimators are known to have the remarkable feature of combining strong robustness properties with full asymptotic efficiency; see, for example, Beran (1977), Tamura and Boos (1986), Simpson (1987, 1989b) and Lindsay (1994). This is in contrast with the previously held belief by the statistical community that robustness and efficiency are conflicting concepts and cannot be achieved simultaneously. However, unlike the density-based minimum distance estimators, the feature of full asymptotic efficiency does not appear to be shared by minimum distance methods based on the distribution function approach.

Similar properties of the density-based minimum distance procedures continue to hold for the parametric hypothesis testing problem.

While the primary use of minimum distance methods is in parametric inference, the second important use of statistical distances is in testing goodness-of-fit. In this case also one quantifies, inversely, the closeness between the data and the hypothesized model through a distance measure, and uses the actual value of the minimized distance to determine whether the match is "close enough." Several of the distances that are under consideration in this book are used for both purposes. However, there is a fundamental difference in the role of the distances in these two problems. In parametric estimation, where robustness is a major concern, the aim is to downweight the effect of observations inconsistent with the model. On the other hand, for the goodness-of-fit testing problem one hopes to magnify the effect of small deviations from the hypothesized model to achieve high power for the testing problem. Thus, the distances that are more desirable from the viewpoint of robust parametric estimation need not automatically be the better ones for the goodness-of-fit testing problem.

In this book we will give an expanded description of minimum distance inference using density-based distances. The treatment will focus primarily on the distances of the chi-square type, but some characteristics of other minimum distance procedures will also be discussed. In the process we will describe in detail the estimation procedures in discrete and continuous models, provide a full account of the robustness of these estimators, consider the hypothesis testing problem, consider modifications of the standard distances and many other related methods including the weighted likelihood derivative of the above. In particular, it will be our aim to establish that the procedures described here are genuine competitors to the classical methods – both in terms of efficiency and robustness – in many common scenarios. In addition, we study the utility of these distances in multinomial goodness-of-fit testing.

A book such as the present one invariably has to borrow material from a large pool of research work related to the theme of this book. Yet, it is inevitable that some of these works are more relevant than others and are referenced more frequently. Here we point out some of these for the benefit of the future researchers.

The earliest known references which expressed the class of chi-square distances through a single, generalized expression and studied their properties appear to be those by Csiszár (1963, 1967a,b) and Ali and Silvey (1966). These works will be referenced throughout the book. Other prominent works in this area that will be frequently referenced include Beran (1977), Tamura and Boos (1986), Simpson (1987, 1989b), and Lindsay (1994), as well as several works of the authors of this book. The works done by Leandro Pardo and his associates in this area, described adequately in the book by Pardo (2006), also represent an important resource for the present book. Cressie and Read (1984) and Read and Cressie (1988) represent classic references for our material, both in terms of developing the popular family of Cressie–Read power divergences, as well as in terms of extending the sphere of the multinomial goodness-of-fit testing problem to the distances within this general class. In other cases specific works will represent important resources for particular chapters, such as the work by Markatou, Basu and Lindsay (1997, 1998) and Agostinelli and Markatou (2001) for Chapter 7, the hypothesis testing problem for the kappa statistic presented by Donner and Eliasziw (1992) for Chapter 8, and the work done by Basu et al. (1998) and Jones et al. (2001) for Chapter 11.

For the consistency and asymptotic normality of our estimators, we have followed, wherever applicable, the multivariate proof for the consistency and asymptotic normality of the maximum likelihood estimator as given in Lehmann (1983, pp. 427–434).

1.1 General Notation

Here we describe some notation that will be used throughout the book.

1. Unless mentioned otherwise, the term "log" will represent natural logarithm.

2. Upper case letters will denote distribution functions, and lower case letters will denote the corresponding densities. Thus, g, f and f_θ will denote the density functions corresponding to the distributions G, F and F_θ.

3. The true distribution will be represented by G, while the empirical distribution function based on n independently and identically distributed observations will be denoted by G_n.

4. The notation $\chi_y(x)$ will represent the indicator function at y, while $\chi(A)$ will represent the indicator for the set A. Also $\wedge_y(x)$ will stand for the degenerate distribution which puts all its mass at y.

5. Unless mentioned otherwise, we will assume that the unknown parameter θ is a p-dimensional vector, and we will denote the parametric model by the notation $\mathcal{F} = \{F_\theta : \theta \in \Theta \subseteq \mathbb{R}^p\}$.

6. The term ∇ will represent the gradient with respect to the parameter, so that

$$u_\theta(x) = \nabla \log f_\theta(x)$$

is the score function of the model. The second derivative with respect to θ will be represented by ∇_2. Higher order derivatives will be similarly defined. Also ∇_j will represent the partial derivative with respect to θ_j and ∇_{jk} will represent the joint partial derivative with respect to θ_j and θ_k.

1.2 Illustrative Examples

In this section we provide two very simple but motivating examples, which help to showcase the situations where the minimum distance methods presented in this book will be useful. For illustration we have chosen one example in a discrete model, and another one in a continuous model. The distances used in these examples will be formally defined in Chapter 2, and the continuous extension will be discussed in Chapter 3. Here we just provide a sneak preview of their performances in these examples, without getting into the technical details of the minimum distance estimation process or the computation of these estimates.

Example 1.1. We consider a Poisson model. A random sample of size $n = 20$ was generated from a Poisson distribution with mean $\theta = 2$. The sampled data are $(x_i, f_i) = (0, 2), (1, 6), (2, 5), (3, 4), (4, 2), (5, 1)$, where x_i is the sampled value and f_i is the frequency of the value x_i. To investigate the stability of the estimators under consideration, we added one contaminating value at $x = 20$.

Figure 1.1 shows the dot-plot of the random sample and the superimposed bar diagram of the Poisson distribution with mean $\theta = 2$. Three minimum distance estimators of θ are reported in Table 1.1; these are the minimum

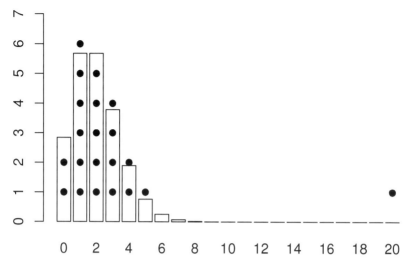

FIGURE 1.1
Superimposed bar diagram for the Poisson distribution and the dot-plot for the data.

TABLE 1.1
Parameter estimates for the contaminated Poisson data.

ML	ML+D	NED
2.905	2.050	2.074

likelihood disparity estimator, the outlier deleted minimum likelihood disparity estimator, and the minimum negative exponential disparity estimator. For discrete models the minimum likelihood disparity estimator also turns out to be the maximum likelihood estimator. The second of the three estimators in this example is simply the maximum likelihood estimator obtained from the cleaned data after the outlier has been subjectively deleted. The three estimators are labeled by ML, ML+D and NED in Table 1.1.

Notice that the full data maximum likelihood estimator and the outlier deleted maximum likelihood estimator are quite far apart; the outlier has practically shifted the maximum likelihood estimate from the neighborhood of 2 to the neighborhood of 3. On the other hand, the minimum negative exponential disparity estimator appears to effectively ignore the large observation. For the outlier deleted data the latter estimator takes the value 2.073, so that the presence or absence of the outlier makes practically no difference to the negative exponential disparity. ‖

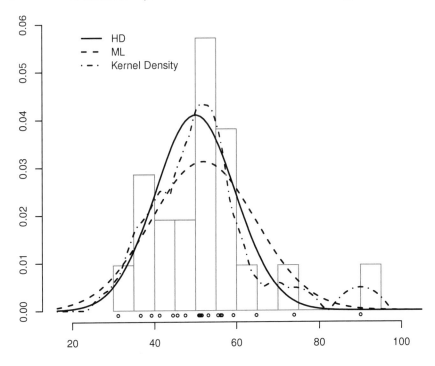

FIGURE 1.2
Histogram and normal density fits for the clinical chemistry quality control
example.

TABLE 1.2
Estimates of location and scale parameters for the clinical chemistry quality
control example.

	ML	ML+D	HD
Location	51.905	50.000	50.003
Scale	12.769	9.747	9.73

Example 1.2. The second example involves data from a clinical chemistry
quality control study. The raw data are explicitly given in Table 2 of Healy
(1979). We fit a normal model to this data using different minimum distance
estimation methods. The histogram of the data is presented in Figure 1.2. A
kernel density estimate and the normal fits provided by the maximum likelihood estimate and the minimum Hellinger distance estimate are superimposed
on the histogram. The actual observations are also plotted in the base of the
figure. The estimates of the parameters of the normal distribution are reported
in Table 1.2.

Notice that there is a substantial difference between the maximum likelihood estimates of scale for the full data, and the reduced data after removing the large outlier at 90.0. However, the minimum Hellinger distance estimate for the full data closely matches the maximum likelihood estimate for the outlier deleted data, illustrating that the minimum Hellinger distance estimate automatically downweights the effect of the large outlier. ‖

In either example, we have considered a robust alternative to the maximum likelihood estimator within the class of the minimum distance estimators that we will study in this book. As we will see, not only do these estimators have strong robustness properties, but are also asymptotically fully efficient, making them useful practical tools.

1.3 Some Background and Relevant Definitions

In this and the subsequent sections of this chapter we discuss several necessary tools that we will use repeatedly throughout this book. All of these are well known concepts of statistical theory, and we simply present them here in the format that will be directly applicable to the discussion in our context.

1.3.1 Fisher Information

Consider a random variable X having a probability density function f_θ with respect to a σ-finite measure ν on $\mathbb{R}$ where θ is the parameter of interest. (The results can be routinely extended when X represents a random vector having a probability density function with respect to a σ-finite measure on $\mathbb{R}^k$, $k > 1$). In favor of a unified notation we will continue to use the term probability density function for both discrete and continuous distributions rather than making a distinction between the probability mass function (for discrete models) and probability density function (for continuous models). Let Θ be the parameter space; for the time being we will consider the parameter θ to be unidimensional, so that $\Theta \subseteq \mathbb{R}$. We assume that the appropriate dominating conditions apply so that for any measurable set B in $\mathbb{R}$, the following relation holds:

$$\nabla \int_B f_\theta(x) d\nu(x) = \int_B \nabla f_\theta(x) d\nu(x), \tag{1.1}$$

where ∇ is the gradient with respect to θ as defined in Section 1.1. See Loeve (1977, page 129) for a discussion of the type of regularity conditions under which the above is satisfied. It is typically satisfied for regular exponential families (e.g., Lehmann and Romano, 2008, page 59). Define

$$u_\theta(x) = \nabla \log f_\theta(x) = \frac{\nabla f_\theta(x)}{f_\theta(x)}$$

to be the score function for the model represented by $f_\theta(\cdot)$. From Equation (1.1) it follows that

$$E_\theta[u_\theta(X)] = 0.$$

Define the Fisher information in the density f_θ to be

$$I(\theta) = E_\theta[u_\theta^2(X)] = \text{Var}_\theta[u_\theta(X)].$$

The Fisher Information and the score function are central to the concept of efficiency of estimators.

If $\theta = (\theta_1, \ldots, \theta_p)^T$ is a p-dimensional vector, define the j-th score function as

$$u_{j\theta}(x) = \nabla_j \log f_\theta(x) = \frac{\nabla_j f_\theta(x)}{f_\theta(x)}, \tag{1.2}$$

where ∇_j represents the gradient with respect to θ_j. In this case $u_\theta = (u_{1\theta}, u_{2\theta}, \ldots, u_{p\theta})^T$ will represent the vector of the partials. Assume that Equation (1.1) holds for the partials ∇_j, $j = 1, \ldots, p$. We define the Fisher information matrix $I(\theta)$ to be the $p \times p$ matrix which has

$$I_{jk}(\theta) = E_\theta[u_{j\theta}(X)u_{k\theta}(X)] \tag{1.3}$$

as its (j, k)-th entry. By definition the Fisher information matrix $I(\theta)$ is nonnegative definite.

Suppose that we have independent and identically distributed observations $X_1, \ldots, X_n$ from a density $f_\theta(x)$, $\theta \in \Theta \subseteq \mathbb{R}^p$, where f_θ has Fisher information $I(\theta)$. Suppose we are trying to estimate $m(\theta)$, a real valued function of θ. Let T_n be any unbiased estimator of $m(\theta)$. First assume that θ is unidimensional, the parameter space Θ is an open subset of $\mathbb{R}$, and $m(\theta)$ is continuously differentiable. Then, under standard regularity conditions the following inequality gives a lower bound for the variance of an unbiased estimator T_n of $m(\theta)$:

$$\text{Var}_\theta(T_n) \geq \frac{(m'(\theta))^2}{nI(\theta)}, \tag{1.4}$$

where $m'(\cdot)$ represents the derivative of $m(\cdot)$. This is the famous Cramér-Rao inequality, and the quantity on the right-hand side of Equation (1.4) is the Cramér-Rao lower bound (see Rao, 1973, Section 5c.2).

Now let $\theta = (\theta_1, \ldots, \theta_p)^T$ be p-dimensional, and let $m(\cdot)$ be a real valued function with domain $\Theta \subseteq \mathbb{R}^p$. Then, for $j = 1, 2, \ldots, p$, let

$$h_j(\theta) = \nabla_j m(\theta).$$

Let $h = (h_1, h_2, \ldots, h_p)^T$. Then, under regularity conditions on the density $f_\theta(x)$, the multidimensional analog of (1.4) is given by the bound

$$\text{Var}_\theta(T_n) \geq \frac{1}{n} h^T I^{-1}(\theta) h,$$

where T_n is any unbiased estimator of $m(\theta)$.

1.3.2 First-Order Efficiency

Quite often, the central limit theorem allows us to restrict attention to only those estimators T_n of $m(\theta)$ which are consistent and asymptotically normal (CAN). For such estimators there exists a positive asymptotic variance $v(\theta)$ such that

$$\frac{n^{1/2}\{T_n - m(\theta)\}}{v(\theta)^{1/2}} \xrightarrow{\mathcal{D}} Z \sim N(0, 1)$$

as $n \to \infty$; here $\xrightarrow{\mathcal{D}}$ indicates convergence in distribution. It is important to know whether there is a lower bound to the variance $v(\theta)$. The Cramér-Rao inequality for finite samples in (1.4) was thought to provide the lower bound for this variance at one time. But there exists a class of estimators, called superefficient estimators, such that for all parameter values a superefficient estimator is asymptotically normal with an asymptotic variance never exceeding and sometimes lower than the Cramér-Rao lower bound. See Le Cam (1953) and Cox and Hinkley (1974) for discussions on superefficiency. The latter work suggests that superefficiency is not a statistically important idea. We will consider alternative versions of efficiency which overcome the problem caused by superefficient estimators.

In this approach to the problem of superefficiency, we further restrict the class of CAN estimators to eliminate those which are superefficient. Let us restrict ourselves to $m(\theta) = \theta$ and assume that θ is unidimensional. The consistent and uniformly asymptotically normal (CUAN) estimators are defined to be CAN estimators for which the approach to normality of $n^{1/2}(T_n - \theta)$ is uniform in compact intervals of θ. The uniformity restriction suffices to eliminate any points of superefficiency. For CUAN estimators T_n of θ, the lower bound for the asymptotic variance of $n^{1/2}(T_n - \theta)$ is $1/I(\theta)$, where θ is the true parameter.

In the class of CUAN estimators, we will define an estimator T_n to be first-order efficient for estimating θ if the asymptotic variance of $n^{1/2}(T_n - \theta)$ equals $1/I(\theta)$ under f_θ. For the multiparameter case, the asymptotic variance covariance matrix of $n^{1/2}(T_n - \theta)$, where T_n is a first-order efficient estimator, is given by $I^{-1}(\theta)$.

Throughout this book,

$$n^{1/2}(T_n - \theta) \xrightarrow{\mathcal{D}} Z^* \sim N_p(0, I^{-1}(\theta))$$

will be our working definition of first-order efficiency or asymptotic model efficiency of T_n.

1.3.3 Second-Order Efficiency

In the class of CUAN estimators, it is easy to study the efficiency of estimators because we have a valid tight lower bound for the variance which is attained by the maximum likelihood estimator (MLE) and some other estimators under certain regularity conditions. However, there may be several estimators which are all first-order efficient in the CUAN sense. How do we distinguish between them? To make this distinction, Rao (1961) has proposed another measure called second-order efficiency. The concept is not quite as simple as that of first-order efficiency. The main point is that the second-order efficiency E_2 of an estimator T_n of θ is measured by finding the minimum asymptotic variance of the quantity $\{U_n(\theta) - \alpha(\theta)(T_n - \theta) - \lambda(\theta)(T_n - \theta)^2\}$ over $\alpha(\cdot)$ and $\lambda(\cdot)$ where $U_n(\cdot)$ is the score function for the independently and identically distributed sample defined by the gradient of $\log \prod_{i=1}^{n} f_\theta(X_i)$, based on the sample $X_1, \ldots, X_n$.

The smaller the value of E_2, the greater this efficiency. Rao (1961, 1963) and Efron (1975, 1978) are sources of more detailed discussions on the concept of second-order efficiency.

1.4   Parametric Inference Based on the Maximum Likelihood Method

The technique of maximum likelihood is the cornerstone of classical parametric inference. There is evidence in the literature of prior use of this method by mathematicians like Gauss, but the systematic development of the method was provided only by Sir R. A. Fisher in the early part of the 20th century (Fisher, 1912, 1922, 1925, 1934, 1935). The early progress of the method of maximum likelihood and the associated research of Sir R. A. Fisher has been described by many statisticians. See Savage (1976), Stigler (1986, 1999), Aldrich (1997) and Hald (1998, 1999), among others.

Let $X_1, \ldots, X_n$ be independently and identically distributed observations from a distribution modeled by the parametric family $\mathcal{F}$ as defined in Section 1.1. Let f_θ be the density function corresponding to F_θ. The likelihood function for the above sample is given by

$$L(\theta) = L(\theta | X_1, \ldots, X_n) = \prod_{i=1}^{n} f_\theta(X_i),$$

which is the joint density of the observations, treated as a function of θ. The maximum likelihood estimator of θ is any value that maximizes the above likelihood over the parameter space Θ. In practice the maximum likelihood estimator is often obtained by solving the system of *likelihood equations*, which equates the derivative of $\log L(\theta)$ (the log likelihood) to zero. Notice that $\log L(\theta)$ is an increasing function of $L(\theta)$, so that if θ is a maximizer of the log likelihood it maximizes the likelihood as well. The likelihood score equation is given by

$$\frac{\partial \log L(\theta)}{\partial \theta} = \sum_{i=1}^{n} u_\theta(X_i) = 0,$$

where u_θ is the score function defined in Section 1.3.1. In Chapter 2 we will view the maximum likelihood estimator as a special member of our class of minimum distance estimators.

Standard regularity assumptions suffice to show that the maximum likelihood estimator of θ is first-order efficient in the sense described in Section 1.3.2. The essential step in different versions of the proof of the above result available in the literature is to show that the maximum likelihood estimator $\hat{\theta}_n$ of θ admits the relation

$$n^{1/2}(\hat{\theta}_n - \theta_0) = I^{-1}(\theta_0)Z_n(\theta_0) + o_p(1), \tag{1.5}$$

where

$$Z_n(\theta) = n^{1/2}\left[\frac{1}{n}\sum_{i=1}^{n} u_\theta(X_i)\right]$$

and θ_0 is the true value of the parameter. Alternatively,

$$n^{1/2}I(\theta_0)(\hat{\theta}_n - \theta_0) = Z_n(\theta_0) + o_p(1). \tag{1.6}$$

Note that $Z_n(\theta)$ has an asymptotic normal distribution with mean zero and covariance matrix $I(\theta)$. Later on, we will derive results analogous to Equation (1.5) to establish the asymptotic efficiency of the other minimum distance estimators considered here. The subsequent chapters will show that not only do the latter class of estimators reach the asymptotic efficiency level of the maximum likelihood estimator, many members of this class also have significantly improved robustness properties.

The connection of the maximum likelihood estimator with the concept of second-order efficiency discussed in Section 1.1.3 is relatively less straightforward. However, the results of Rao (1961) show that the maximum likelihood estimator is second-order efficient at the multinomial model.

In the next section we consider the classical tests of hypothesis based on likelihood methods under the same parametric setup.

1.4.1 Hypothesis Testing by Likelihood Methods

Because of the asymptotic efficiency of the maximum likelihood estimator, statisticians have long been interested in extending this technique to the hypothesis testing problem (e.g., Neyman and Pearson, 1928; Wilks, 1938). Here we briefly describe parametric hypothesis testing problems which try to exploit the asymptotic efficiency of the maximum likelihood estimator. Details can be found in many standard texts on statistical inference and asymptotic theory (e.g., Serfling, 1980). In later chapters we will discuss other testing procedures based on the minimum distance methodology.

We consider the simple null hypothesis $H_0 : \theta = \theta_0$ against the two sided alternative. The null hypothesis may be expressed in terms of p restrictions of the form

$$R_i(\theta) = \theta_i - \theta_{0i} = 0, \quad i = 1, \ldots, p,$$

where θ_i is the i-th component of θ. A composite null hypothesis can also be expressed in terms of similar restrictions; however, the composite null will be represented by r restrictions where r is necessarily smaller than p. Equations (1.5) and (1.6) immediately show that the statistics

$$W_n = n(\hat{\theta}_n - \theta_0)^T I(\theta_0)(\hat{\theta}_n - \theta_0) \tag{1.7}$$

and

$$S_n = Z_n^T(\theta_0) I^{-1}(\theta_0) Z_n(\theta_0) \tag{1.8}$$

both have asymptotic χ_p^2 limits under the simple null, and are separated only by an $o_p(1)$ term.

Let $l(\theta) = \log L(\theta)$, and define

$$\lambda_n = 2[l(\hat{\theta}_n) - l(\theta_0)] \tag{1.9}$$

to be the log likelihood ratio test statistic where $\hat{\theta}_n$ is the maximum likelihood estimator. A straightforward Taylor series expansion under standard regularity condition shows that the statistic in (1.9) also has an asymptotic χ_p^2 limit under the simple null; in fact the expansion shows that

$$\lambda_n = W_n + o_p(1).$$

See Serfling (1980, Section 4.4) for example.

The log likelihood ratio statistic λ_n originated from Neyman and Pearson (1928), and was consolidated by the subsequent research of other authors. The statistic W_n was introduced by Wald (1943) and is generally known as the Wald statistic. The statistic S_n is the score statistic (or the Rao statistic), which was introduced by Rao (1948). In case of the simple null hypothesis, the computation of the score statistic does not involve any parameter estimation.

The above discussion shows that the three statistics λ_n, W_n and S_n are equivalent under the null hypothesis. Under some additional regularity conditions, the asymptotic equivalence of the three test statistics continue to hold under local alternatives converging sufficiently fast, although the statistics behave differently for fixed, nonlocal alternatives.

Even though it is more complicated than the simple null case, the above results may be extended to the case of the composite null. Here our null hypothesis is of the form $H_0 : \theta \in \Theta_0$, and this is to be tested against the alternative $H_1 : \theta \in \Theta \setminus \Theta_0$, where Θ_0 is a proper subset of Θ. Let the restricted parameter space Θ_0 be defined by the r restrictions $R_i(\theta) = 0$, $i = 1, \ldots, r$, where the functions $\xi_\theta = (R_1(\theta), \ldots, R_r(\theta))^T$ formalize the r restrictions imposed by the null hypothesis. The parameter space under H_0 can therefore be described by $p - r$ independent parameters $\gamma = (\gamma_1, \ldots, \gamma_{p-r})^T$, i.e., H_0 specifies that there exists a function

$$b : \mathbb{R}^{p-r} \to \mathbb{R}^p$$

where $\theta = b(\gamma)$, and γ varies over an open subset $\Gamma \subseteq \mathbb{R}^{p-r}$. The function $b(\cdot)$ is assumed to have continuous partial derivatives

$$\dot{b}_\gamma = \left[\frac{\partial b_i}{\partial \gamma_j} \right]_{p \times (p-r)}$$

with rank $p - r$; here b_i and γ_j represent the indicated components. Similarly, we assume that

$$B_\theta = \left[\frac{\partial R_i}{\partial \theta_j} \right]_{r \times p}$$

is of rank r. Denote $J(\gamma)$ to be the information matrix for the γ formulation of the model.

Define

$$\Lambda_n = \frac{\sup_{\theta \in \Theta_0} L(\theta)}{\sup_{\theta \in \Theta} L(\theta)}.$$

Then $\lambda_n = -2 \log \Lambda_n$ will represent the likelihood ratio test statistic for the composite null scenario. Similarly

$$W_n = n \xi_{\hat{\theta}_n}^T (B_{\hat{\theta}_n} I^{-1}(\hat{\theta}_n) B_{\hat{\theta}_n}^T)^{-1} \xi_{\hat{\theta}_n}$$

represents the Wald statistic where $\hat{\theta}_n$ is the unrestricted maximizer of the likelihood. Note that $n^{1/2}(\xi_{\hat{\theta}_n} - \xi_\theta)$ has an asymptotic normal distribution with mean vector zero and covariance matrix $B_\theta I^{-1}(\theta) B_\theta^T$ under the true density f_θ. If the null hypothesis holds, $\xi_\theta = 0$ under the true density, so that the asymptotic convergence of $W_n = n \xi_{\hat{\theta}_n}^T (B_{\hat{\theta}_n} I^{-1}(\hat{\theta}_n) B_{\hat{\theta}_n}^T)^{-1} \xi_{\hat{\theta}_n}$ to a χ^2 with r degrees of freedom follows immediately from the theory of quadratic forms in asymptotically multivariate normal vectors.

Let θ_n^* be the restricted maximum likelihood estimator of θ under the null hypothesis. This estimator may also be represented as $\theta_n^* = b(\hat{\gamma}_n)$, where $\hat{\gamma}_n$ is the maximum likelihood estimator of γ under the reparametrization specified by the null hypothesis. Some simple algebra shows that

$$W_n = n(\hat{\theta}_n - \theta_n^*)^T B_\theta^T (B_\theta I^{-1}(\theta) B_\theta^T)^{-1} B_\theta (\hat{\theta}_n - \theta_n^*) + o_p(1),$$

and

$$\lambda_n = n(\hat{\theta}_n - \theta_n^*)^T I(\theta)(\hat{\theta}_n - \theta_n^*) + o_p(1).$$

But $I(\theta) = B_\theta^T (B_\theta I^{-1}(\theta) B_\theta^T)^{-1} B_\theta$ (Serfling, 1980, Section 4.4.4), so that the asymptotic equivalence of the likelihood ratio test statistic and the Wald statistic holds under the null hypothesis.

See Rao (1973) for the equivalence of the score statistic with the likelihood ratio and the Wald statistics under the null.

1.5   Statistical Functionals and Influence Function

A statistic can frequently be viewed as a functional on a space of distributions. When these functionals possess certain differentiability properties, the derivatives provide information about their asymptotic behavior. In addition, these derivatives are also connected with one of the most important heuristic tools of robust statistics, the influence function. The latter measure is often used as a quantification of the amount of limiting influence on the value of an estimate or a test statistic when an additional observation is added to a very large sample. In the following, we give a brief introduction to statistical functionals and this dual role of the influence function.

Let $T_n(X_1, \ldots, X_n)$ be a statistic based on a random sample $X_1, \ldots, X_n$ drawn from the distribution G. Let

$$G_n(x) = \frac{1}{n} \sum_{i=1}^{n} \chi(X_i \leq x),$$

be the empirical distribution function based on the data where $\chi(A)$ is the indicator of the set A. Formally, if T_n can be written as $T(G_n)$ where $T(\cdot)$ is independent of n, $T(\cdot)$ is called a statistical functional. This functional $T(\cdot)$ is a function defined on an appropriate space of distribution functions $\mathcal{D}^*$, which, for our purpose, will include the true distribution G, all model distributions as appropriate, all empirical distributions G_n, $n \geq 1$, and will be closed under convex combinations. In the general case we will consider the range of these functions to be $\mathbb{R}^p$; however, for ease of presentation we will concentrate on the case $p = 1$ unless otherwise mentioned. Statistical functionals were introduced by von Mises (1936, 1937, 1947). In practice, our parameter of interest is $T(G)$, and we want to study the properties of $T(G_n)$ as an estimator for $T(G)$.

A functional T is called linear, if for all $G, F \in \mathcal{D}^*$,

$$T(\alpha F + (1 - \alpha)G) = \alpha T(F) + (1 - \alpha)T(G), \ 0 \leq \alpha \leq 1.$$

It may be easily verified that a functional T having the form

$$T(G) = \int \tau(x) dG(x) \tag{1.10}$$

is a linear functional. In fact, it may be shown that any linear functional must be of the form (1.10). This functional is explicitly defined. Note that for a linear functional of the form (1.10),

$$n^{1/2}(T(G_n) - T(G)) \xrightarrow{\mathcal{D}} Z^* \sim N(0, \sigma^2),$$

provided $0 < \int \tau^2(x) dG(x) - (\int \tau(x) dG(x))^2 = \sigma^2 < \infty$. Thus, the estimator $T(G_n)$ is consistent and asymptotically normal in this case.

On the other hand, suppose that ψ is a real valued function of two variables and let $T_n = T(G_n)$ be defined implicitly as the solution of the equation

$$\sum_{i=1}^{n} \psi(X_i, T_n) = 0.$$

The corresponding functional $T(G) = \theta$ is the solution of

$$\int \psi(x, \theta) dG(x) = 0.$$

Estimators of this type are called M-estimators. See, for example, Huber (1981) and Hampel et al. (1986).

Suppose that G and F are two distributions in $\mathcal{D}^*$. Notice that

$$T(\alpha F + (1 - \alpha)G) = T(G + \alpha(F - G)).$$

The von-Mises derivative T'_G of T at G is defined by

$$T'_G(F - G) = \frac{d}{d\alpha} T(G + \alpha(F - G)) \Big|_{\alpha=0}$$

if there exists a real valued function $\phi_G(x)$, independent of F, such that

$$T'_G(F - G) = \int \phi_G(x) d(F - G)(x).$$

The function $\phi_G(x)$ is uniquely defined only up to an additive constant, since $d(F - G)$ has total measure zero. This non-uniqueness can be removed by adding the restriction

$$\int \phi_G(x) dG(x) = 0. \tag{1.11}$$

Our aim is to do an approximation of the form

$$T(G_n) - T(G) \approx T'_G(G_n - G)$$

in large samples.

The function $\phi_G(\cdot)$ is called the influence function (IF) of the functional $T(\cdot)$ at G, and is generally denoted by $\phi_G(y) = \mathrm{IF}(y, T, G)$. Condition (1.11) implies that the influence function has mean zero. Under appropriate conditions, the influence function can also be defined directly by

$$\phi_G(y) = \frac{d}{d\alpha} T((1 - \alpha)G + \alpha \wedge_y) \Big|_{\alpha=0} = \frac{d}{d\alpha} T(G + \alpha(\wedge_y - G)) \Big|_{\alpha=0}, \tag{1.12}$$

where $\wedge_y(x)$ is as defined in Section 1.1. Equation (1.12) shows that the influence function can be considered to be an approximation of the relative influence on T of small departures from G.

It is also easy to see, using Equation (1.12), that the mean functional defined by

$$T_{\text{mean}}(G) = \int x \, dG$$

has influence function

$$\text{IF}(y, T_{\text{mean}}, G) = y - T_{\text{mean}}(G). \tag{1.13}$$

Thus, its influence function is unbounded as a function of y.

Let $A(\alpha) = T(G + \alpha(F - G))$, $\alpha \in [0, 1]$. Consider its Taylor series expansion around $\alpha = 0$,

$$A(\alpha) = A(0) + \alpha A'(0) + \text{Higher Order Terms.}$$

Evaluating the above at $\alpha = 1$, one gets

$$T(F) = T(G) + T'_G(F - G) + \text{Higher Order Terms.}$$

Letting R_n represent the Higher Order Terms, and replacing F with G_n, we get

$$
\begin{aligned}
T(G_n) &= T(G) + T'_G(G_n - G) + R_n \\
&= T(G) + \int \phi_G(x) \, dG_n(x) + R_n
\end{aligned}
$$

so that

$$n^{1/2}(T(G_n) - T(G)) = \frac{1}{n^{1/2}} \sum_{i=1}^{n} \phi_G(X_i) + n^{1/2} R_n. \tag{1.14}$$

Thus, if $n^{1/2} R_n$ goes to zero in probability as $n \to \infty$, the asymptotic distribution of $n^{1/2}(T(G_n) - T(G))$ is normal with mean zero and variance equal to the variance of the influence function $\phi_G(X)$, i.e.

$$n^{1/2}(T(G_n) - T(G)) \xrightarrow{\mathcal{D}} W \sim N(0, \text{Var}(\phi_G(X))). \tag{1.15}$$

In fact, one often uses $\hat{\sigma}/\sqrt{n}$ as the standard error of the estimator $T(G_n)$ – the so called "sandwich" estimator – where

$$\hat{\sigma}^2 = (n-1)^{-1} \sum_{i=1}^{n} \phi_{G_n}^2(X_i).$$

The condition $n^{1/2} R_n \to 0$ is generally not easy to establish, but is often true.

When there is a parametric model $\{F_\theta : \theta \in \Theta \subseteq \mathbb{R}\}$, we may be interested in estimating the parameter θ using the functional $T(G_n)$. The functional $T(\cdot)$ is called Fisher consistent if $T(F_\theta) = \theta$.

In a parameteric model, the maximum likelihood estimate T_n solves the score equation so that

$$\sum_{i=1}^{n} u_{T_n}(X_i) = 0,$$

where $u_\theta(x)$ is the score function. Notice that the maximum likelihood estimator is also an M-estimator. The maximum likelihood functional $T = T_{\mathrm{ML}}(G)$ is defined implicitly by

$$\int u_T(x)dG(x) = 0. \tag{1.16}$$

From Equation (1.1) it is easy to see that $T_{\mathrm{ML}}(F_\theta) = \theta$, so that the maximum likelihood estimator is Fisher consistent. A straightforward differentiation of Equation (1.16) shows that the influence function of the maximum likelihood functional at the model equals $\mathrm{IF}(y, T_{\mathrm{ML}}, F_\theta) = u_\theta(y)/I(\theta)$ (this will equal $I^{-1}(\theta)u_\theta(y)$ in multiparameter situations where u is a vector and $I(\theta)$ is a matrix). From (1.15) it follows that the asymptotic variance of $n^{1/2}(T_{\mathrm{ML}}(G_n) - \theta)$ equals $I^{-1}(\theta)$.

Recall that the influence function is an approximation of the influence of small departures from the assumed distribution on the functional T. From the robustness point of view, an estimator is expected to be more resistant to outlying observations when its influence function is bounded. As the score function $u_\theta(y)$ is usually unbounded, the influence function of the maximum likelihood estimator is also generally unbounded; it is also well known that the maximum likelihood estimator suffers from severe robustness problems. Many of the other M-estimators are, however, specially constructed to have bounded influence functions and so they have better robustness properties than the maximum likelihood estimator. The M-estimator based on Huber's ψ function is a well known example for the location case (Huber, 1964).

The sample median $T_n = T(G_n)$, which can be described as the solution of the estimating equation

$$\int \psi(x, T_n)dG(x) = 0,$$

where $\psi(x, T) = [\chi(x > T) - \chi(x < T)]$, $\chi(\cdot)$ being the indicator function, is also an M-estimator. It can be shown that when T is the median functional, defined as the solution of the equation

$$\int \psi(x, T)dG = 0, \tag{1.17}$$

the influence function of T is given by

$$\phi_G(y) = \begin{cases} \dfrac{1}{2g(t)} & y > t \\ -\dfrac{1}{2g(t)} & y < t \end{cases}$$

where g is the density of G and $t = T(G)$ is the true population median. Notice that this influence function is bounded, indicating that the sample median may have better robustness properties than the sample mean as an estimator of location.

One of the major issues highlighted in this book is that although the influence function is one of the most important indicators of robustness in the classical robustness literature, its usefulness in describing the robustness of the minimum distance estimators is limited, and alternative measures have to be developed to study the robustness of these estimators.

See Serfling (1980) and Fernholz (1983) for further discussions on statistical functionals.

1.6   Outline of the Book

The major objective of the book is to present a general, comprehensive treatment of density-based "statistical distances" of the chi-square type and their uses in parametric inference, multivariate goodness-of-fit tests, and other applications.

In Chapter 2, we introduce the notion of statistical distances, describe the class of disparities and discuss their properties under the discrete model. This is the pivotal chapter in setting up the theme of this book. The asymptotic distribution of the minimum Hellinger distance estimator, and more generally the minimum disparity estimators, are also derived in this chapter.

In Chapter 3, the context described in Chapter 2 is extended to continuous models. The additional difficulties due to the incompatibility of measures, and the approaches for providing remedies for this difficult problem, are considered in this chapter. The differences between the approaches due to Beran (1977) and Basu and Lindsay (1994) are highlighted, and the asymptotic distributions are derived.

Chapter 4 describes the robustness properties of the minimum distance estimators considered in this book. The mainstream robustness concept of bounded influence functions is not useful in this context, since the minimum distance estimators studied here are all equivalent to the maximum likelihood estimator in terms of the influence function at the model. This chapter discusses the development and application of the other methods which provide useful discrimination between the robust minimum distance estimators and the maximum likelihood estimator.

Chapter 5 discusses the other fundamental paradigm of statistical inference – hypothesis testing – using minimum distance methods. Asymptotic properties of the disparity difference tests and their robustness indicators are studied for both discrete and continuous models.

In spite of their asymptotic efficiency, many of the minimum distance methods studied in this book suffer from a serious deficiency problem in small samples which severely limits their practical applicability. In Chapter 6 we present several modifications applicable in small samples which enhance the performance of these methods and make them competitive with the methods based on the likelihood function without compromising their robustness properties. The performance of the resulting estimators and tests are discussed and illustrated in this chapter.

A useful variant of minimum distance estimation is the method of weighted likelihood estimation. Maximum likelihood estimation is the cornerstone of classical statistical inference, and is often well understood and appreciated by scientists outside the core statistical community. A weighted likelihood estimation method described in Chapter 7 originates from the minimum distance idea, but branches out to an estimating equation technique to stand on its own; the basis for comparison is the maximum likelihood score equation. The asymptotic properties and the robustness aspect of these methods are also described in this chapter.

In Chapter 8 we consider the multivariate goodness-of-fit testing problem based on the class of disparities. The text by Read and Cressie (1988) already sets a basic standard in this area. Our aim here is not to simply repeat the issues already highlighted in Read and Cressie (1988) and Pardo (2006), but to provide additional information on these topics and provide new results of interest, thus supplementing the existing state of knowledge in this important area.

In Chapter 9 we discuss the density power divergence and its use in statistical inference. These divergences do not belong to the class of disparities except for very special cases. However, this technique avoids the use of any nonparametric smoothing as part of the estimation procedure, which is a major advantage in continuous models. This technique also provides a smooth bridge between the efficient maximum likelihood estimator and the highly robust (but somewhat inefficient) minimum L_2 distance estimator.

In Chapter 10 we briefly discuss several related problems. These problems extend the context of minimum distance estimation based on disparities beyond the scenario involving independently and identically distributed data. These include, but are not limited to, the analysis of mixture models, survival data and grouped data problems.

In Chapters 11 and 12 we shift gears and try to speak the language of information science and engineering in the context of the same density-based distances and explore their applications in the above disciplines. In Chapter 11 we study different types of information measures between two probability distributions and discuss their effective usage. We briefly describe the historical perspective, consider different distance like measures and derive their mathematical properties with a general outline of the uses of information theoretic measure from a practical point of view. In Chapter 12 we consider the minimum distance method in diverse applications of estimating target objectives using given observations and constraints. Some examples such as training neural networks, Fuzzy-theoretical divergences, and optical phase problems are described as typical applications of the minimum distance method.

2

Statistical Distances

2.1 Introduction

An important component of statistical modeling is the quantification of the amount of discrepancy between data and the model through an appropriate divergence. Based on a sample of n independent and identically distributed observations, such divergences may be constructed, for example, between the empirical distribution function and its population version, or a nonparametric density estimate obtained from the data (constructed, if necessary, using an appropriate density estimation method such as the kernel density estimation) and the probability density function at the model. The first one represents a divergence between two distribution functions, while the second one is the divergence between two probability density functions. In this book, our primary attention will be on the density-based approach. A prominent example of the early use of the density-based idea is the chi-square distance of Pearson (1900).

It is important to make it clear at this stage that many of the density-based divergences (and some of the other divergences as well) that are utilized for different purposes in the statistical literature are not mathematical distances in the sense of being metrics. Most of them are not symmetric in their arguments. This is not very important for statistical purposes. In fact, in many cases it is the asymmetry in the structure of these divergences which has a major role in imparting some of the desirable properties to the estimators generated by them. What is statistically important is that these measures should be nonnegative, and should be equal to zero if and only if the data match the model exactly. Any divergence which satisfies the above two properties will be referred to here as a "statistical distance." This entire book is devoted to the use of statistical distances in statistical inference, with emphasis on the density-based divergences. In a loose sense, we will often drop the term "statistical," and refer to these divergence measures as "distances." In effect, the word "distance" will be used interchangeably with the word "divergence" or, as will be defined later, with the word "disparity."

Unless specifically mentioned otherwise, the scenario under consideration will be the one where n independent and identically distributed observations have been obtained from a distribution G modeled by the parametric family $\mathcal{F}$ as defined in Section 1.1. Our major interest here is in inference problems involving the parameter θ. We will briefly discuss more specialized problems in later chapters.

Since we are interested in density-based distances, we will assume that the density function f_θ of each model element F_θ exists with respect to an appropriate dominating measure, as does the density function of the true distribution. We will, in fact, often represent the model family in terms of the family of densities $\{f_\theta : \theta \in \Theta \subseteq \mathbb{R}^p\}$.

2.2   Distances Based on Distribution Functions

Suppose that $X_1, \ldots, X_n$ represent a random sample of independent and identically distributed observations drawn from an unknown continuous distribution, and let

$$G_n(x) = \frac{1}{n} \sum_{i=1}^{n} \chi(X_i \leq x)$$

be the empirical distribution function where $\chi(A)$ represents the indicator function for the event A. Let $\{F_\theta : \theta \in \Theta \subseteq \mathbb{R}^p\}$ be a parametric family of model distributions used to describe the true distribution. A general measure of distance between G_n and F_θ will be denoted by $\rho(G_n, F_\theta)$. Then the weighted Kolmogorov–Smirnov distance is defined by

$$\rho_{\mathrm{KS}}(G_n, F_\theta) = \sup_{-\infty < z < \infty} |G_n(z) - F_\theta(z)| \sqrt{\psi(F_\theta(z))},$$

where $\psi(u) = 1$ gives the usual Kolmogorov–Smirnov distance measure. A minimum distance estimator corresponding to the Kolmogorov–Smirnov distance (or the weighted Kolmogorov–Smirnov distance if appropriate) can be obtained by minimizing the above distance over the parameter space Θ.

The Kolmogorov–Smirnov distance has been used for many different purposes. For example, Kolmogorov–Smirnov statistics have long been used in testing for one dimensional probability distributions by comparing the data with a known and fixed reference distribution. Consider the ordinary Kolmogorov–Smirnov distance

$$\sup_z |G_n(z) - G(z)| \tag{2.1}$$

for testing the null hypothesis that the known distribution $G(\cdot)$ represents the true data generating distribution. When G is continuous, and the null hypothesis is true, the statistic

$$D_n = n^{1/2} \sup_z |G_n(z) - G(z)|$$

has an asymptotic distribution linked to a standard Brownian bridge. The asymptotic critical values of the statistic D_n can be obtained (Kolmogorov, 1933) by noting the convergence

$$P(D_n \leq t) \to H(t) = 1 - 2\sum_{i=1}^{\infty}(-1)^{i-1}e^{-2i^2t^2}.$$

The test has wide applicability because the asymptotic distribution of the statistic does not depend on the null distribution G; also see Smirnov (1939). Historically, the Kolmogorov–Smirnov metric in (2.1) generated the first goodness-of-fit test which is (pointwise) consistent against any alternative (e.g., Lehmann and Romano, 2008). Massey (1951) appears to be the first to refer to it as the Kolmogorov–Smirnov test. The test is also useful for testing the equality of two distribution functions.

A useful adaptation of the Kolmogorov–Smirnov test is the Lilliefors test (Lilliefors, 1967), which provides a test for normality, and corrects for the bias in the ordinary Kolmogorov–Smirnov test for normality. However, it is generally outperformed by the Shapiro-Wilk statistic (Shapiro and Wilk, 1965) or the Anderson–Darling statistic (Anderson and Darling, 1952) as tests for normality, in terms of the attained power.

The minimum Kolmogorov–Smirnov distance estimator enjoys several nice properties. For example, Drossos and Philippou (1980) show that among many other minimum distance estimators, the minimum Kolmogorov–Smirnov distance estimator enjoys the invariance of the maximum likelihood estimator. Parr and Schucany (1980) have shown that the minimum distance estimator based on the Kolmogorov–Smirnov distance is competitive with several of its rivals. In other applications, Durbin (1975) considered the distribution of the Kolmogorov–Smirnov statistic in terms of a Fourier transform and produced explicit results for the exponential case. In Margolin and Maurer (1976), the results for the exponential case are derived from general results for order statistics, and computationally efficient approximations to these distribution functions are obtained.

The Kolmogorov–Smirnov distance is an extremely popular distance in statistics and many other scientific disciplines. It has been used, for example, in pattern recognition, image comparison and image segmentation, signature verification, credit scoring and library design, just to name a few application domains. Some examples of recent applications of the distance include Benson and Nikitin (1995), Rassokhin and Agrafiotis (2000) and Bockstaele et al. (2006). Weber et al. (2006) present and implement an algorithm for computing the parameter estimates in a univariate probability model for a continuous random variable that minimizes the Kolmogorov–Smirnov test statistic.

However, Donoho and Liu (1988b) have demonstrated certain "pathologies" of some minimum distance estimators, including the minimum Kolmogorov–Smirnov distance estimator; for example, the asymptotic variance of the minimum Kolmogorov–Smirnov distance estimator is unbounded over small Kolmogorov–Smirnov neighborhoods of the model, a consequence of the Kolmogorov–Smirnov distance being "non-Hilbertian." Also see Millar (1981) for some other properties of this method of estimation.

The ordinary Kolmogorov–Smirnov distance and the associated inference relate to continuous underlying distributions, and the overwhelming majority of the available literature in this area operates under the continuity assumption. Appropriate modifications are needed for the applicability of these methods in discrete and discontinuous distributions. This has been addressed by, among others, Schmid (1958), Noether (1963), Coberly and Lewis (1972), Conover (1972), Petitt and Stevens (1977), Wood and Altavela (1978) and Gleser (1985). Bartels et al. (1978) provide some computational details in this connection.

Another prominent member of the class of distances based on distribution functions is the Cramér–von Mises distance. The weighted Cramér–von Mises distance between the empirical distribution function G_n and the model distribution function F_θ is given by

$$\rho_{\text{CM}}(G_n, F_\theta) = \int_{-\infty}^{\infty} (G_n(z) - F_\theta(z))^2 \psi(F_\theta(z)) dF_\theta(z), \qquad (2.2)$$

where $\psi(u) = 1$ gives the usual Cramér–von Mises distance, and $\psi(u) = [u(1-u)]^{-1}$ generates the Anderson–Darling measure (Anderson and Darling, 1952). The ordinary Cramér–von Mises distance is often denoted by the symbol ω^2 in the literature.

The Cramér–von Mises criterion is widely used for testing goodness-of-fit, as well as for parametric estimation. To test the goodness-of-fit of a given, fixed distribution G, Cramér (1928) suggested an integral measure obtained by integrating a weighted version of the squared residual $(G_n(x) - G(x))^2$. It appears that von Mises (1931) suggested an equivalent test independently. The form of the Cramér–von Mises distance that is currently being used by practitioners includes the modifications of Smirnov (1936, 1937), and the test has also been called the Cramér-Smirnov test in the literature. Darling (1955, 1957) gives an excellent account of these tests. Also see Anderson (1962).

For the one sample case, given ordered observations $X_1, \ldots, X_n$ obtained through an independently and identically distributed sample, the Cramér–von Mises statistic may be described as

$$n\rho_{\text{CM}}(G_n, G) = n\omega^2 = \frac{1}{12n} + \sum_{i=1}^{n} \left[\frac{2i-1}{2n} - G(X_i) \right]. \qquad (2.3)$$

We reject that hypothesis that the data come from the distribution G for large observed values of the statistic.

In the two-sample case, let $X_1, \ldots, X_n$ and $Y_1, \ldots, Y_m$ be the observed values, arranged in increasing order, in independent and identically distributed samples from the two populations. Suppose we are interested in testing for the equality of the two populations. In this case, the analog of the test statistic in Equation (2.3) is given by

$$\frac{U}{nm(n+m)} - \frac{4mn-1}{6(m+n)},$$

where $U = n \sum_{i=1}^{n}(r_i - i)^2 + m \sum_{j=1}^{m}(s_j - j)^2$, and r_i and s_j are the ranks of the observations X_i and Y_j respectively in the pooled sample (Anderson, 1962). Available tables of the statistic U facilitate the comparison and conclusion based on this statistic.

Millar (1981) considered weighted Cramér–von Mises distance estimation from a decision theoretic standpoint. Boos (1981) used a weighted Cramér–von Mises distance to estimate the parameter in a location family. Parr and De Wet (1981) extended the results of Boos to more general settings outside the location model. Wiens (1987) discussed the weighted Cramér–von Mises distance estimation of a location parameter and developed robust estimators which have optimal minimax variance properties in gross error neighborhoods of the target model. Öztürk and Hettmansperger (1997) considered generalized weighted Cramér–von Mises distance estimators which have high efficiency at the true model and stable behavior at the neighborhood of the target model. Heathcote and Silvapulle (1981) and Hettmansperger et al. (1994) used the unweighted Cramér–von Mises distance for the simultaneous estimation of location and scale parameters.

In the k-sample one way analysis of variance problem, Brown (1982) developed a multiple response permutation procedure based on within-group sums of absolute rank differences. The relevant distribution turns out to be a convolution of $k-1$ copies of the usual Cramér–von Mises distribution.

The Anderson–Darling distance (Anderson and Darling, 1952), a special case of (2.2) for $\psi(u) = [u(1-u)]^{-1}$ provides one of the most powerful tests for normality against most alternatives (e.g., Stephens, 1974). Also see Boos (1982) for a general discussion. For details of the application of the Anderson–Darling test in normal, exponential, Gumbel and Weibull distributions, see Shorack and Wellner (1986). Stephens (1979) discusses the case of the logistic distribution. Tables of critical values for the statistic for some specific distributions are given in Pearson and Hartley (1972).

See Kiefer (1959) for another application of the Anderson–Darling statistic to the k-sample problem. For some other aspects of statistical analysis based on the Anderson–Darling statistic, the reader is referred to Scholz and Stephens (1987), Thas and Ottoy (2003) and Mansuy (2005).

2.3 Density-Based Distances

Within the class of density-based distances, our focus in this book will be on the family of chi-square type distances, generally called ϕ-divergences, f-

divergences, g-divergences, or disparities (see Csiszár, 1963, 1967a,b; Ali and Silvey, 1966; Lindsay, 1994; Pardo, 2006). The primary reason for focusing our attention on this family of distances is that all the minimum distance procedures based on these distances generate estimators which are asymptotically fully efficient and many of them have remarkably strong robustness properties. In addition, the likelihood based methods (maximum likelihood estimation, likelihood ratio test) can often be considered as special cases of this approach, so that the entire framework can be viewed as a generalized approach containing the likelihood based methods as well as other robust minimum distance procedures as special cases.

For ease of presentation, we will first introduce these distances for discrete models. The distances for the continuous models will be introduced subsequently. Some other distances will be discussed in later chapters.

2.3.1   The Distances in Discrete Models

Parametric estimation based on minimum chi-square type methods has been studied by many authors. Pardo (2006) provides a nice description of the minimum distance methods in connection with these distances; however, the coverage of the latter work is primarily limited to the case of discrete models with finite support based on the multinomial distribution. In addition, the robustness angle is not emphasized in Pardo (2006). In contrast, we will consider the more general countable support case, extend this later to the case of continuous models, and present the issue of robustness as the distinguishing theme of this book. To be specific, we will follow the approach of Lindsay (1994) to describe our minimum distance methodology; this is primarily because it allows a neat illustration of the adjustment between robustness and efficiency through the residual adjustment function and the Pearson residual to be described later in this chapter.

Let $X_1, \ldots, X_n$ represent a sequence of independent and identically distributed observations from a distribution G having a probability density function g with respect to the counting measure. Without loss of generality, we will assume that the support of the distribution G is $\mathcal{X} = \{0, 1, 2, \ldots, \}$. Let $d_n(x)$ represent the relative frequency of the value x in the random sample described in the previous paragraph. Let the parametric model family $\mathcal{F}$, which models the true data generating distrubtion G, be as defined in Section 1.1. We will denote by $\mathcal{G}$ the class of all distributions having densities with respect to the counting measure (or the appropriate dominating measure in other cases), and we will assume this class to be convex. We will also assume that both G and $\mathcal{F}$ belong to $\mathcal{G}$.

One of our main aims in this context is to estimate the parameter θ efficiently and robustly. In minimum distance estimation, as mentioned before, we estimate the parameter θ by determining the element of the model family which provides the closest match to the data in terms of the distance under consideration. In the discrete setup described here, this can be achieved by quantifying the separation between the vectors $\underset{\sim}{d} = (d_n(0), d_n(1), \ldots)^T$ and $\underset{\sim}{f_\theta} = (f_\theta(0), f_\theta(1), \ldots)^T$, which are probability vectors satisfying

$$\sum_{x=0}^{\infty} d_n(x) = \sum_{x=0}^{\infty} f_\theta(x) = 1. \tag{2.4}$$

One way to quantify the separation between the vectors $\underset{\sim}{d}$ and $\underset{\sim}{f_\theta}$ is through the class of disparities (Lindsay, 1994); see also Csiszár (1963, 1967a,b) and Ali and Silvey (1966). In the following, we formally define a disparity.

Definition 2.1. Let C be a thrice differentiable, strictly convex function on $[-1, \infty)$, satisfying

$$C(0) = 0. \tag{2.5}$$

Let the *Pearson residual* at the value x be defined by

$$\delta(x) = \frac{d_n(x)}{f_\theta(x)} - 1. \tag{2.6}$$

(We will denote it by $\delta_n(x)$ whenever the dependence on n has to be made explicit). Then the disparity between $\underset{\sim}{d}$ and $\underset{\sim}{f_\theta}$ generated by C is given by

$$\rho_C(d_n, f_\theta) = \sum_{x=0}^{\infty} C(\delta(x)) f_\theta(x). \tag{2.7}$$

For simplicity of notation we will refer to the disparity by the expression on the left-hand side of Equation (2.7), rather than as $\rho_C(\underset{\sim}{d}, \underset{\sim}{f_\theta})$.

The conditions imposed on the function $C(\cdot)$ by Definition 2.1 will be called the disparity conditions. In the spirit of the statistical distance notation, we will often refer to the quantity in (2.7) based on a function C as in Definition 2.1 as a distance satisfying the disparity conditions. We will refer to the function C as the disparity generating function.

Notice that the right-hand side of Equation (2.7) is the expectation of $C(\delta(X))$ with respect to the density $f_\theta(x)$. Since C is a strictly convex function, it follows from Jensen's inequality and Equations (2.4) and (2.5) that

$$\sum_{x=0}^{\infty} C(\delta(x)) f_\theta(x) \geq C(E_{f_\theta}(\delta(X))) = C(0) = 0,$$

establishing the result that the disparity defined in Equation (2.7) is always nonnegative. Notice that since $E_{f_\theta}(C(\delta(x))) = E_{f_\theta}(C(\delta(x)) - k\delta(x))$ for any scalar k, there may be different versions of the disparity generating function C generating the same disparity.

By the strict convexity of the disparity generating function C, the disparity in (2.7) is zero only when $d_n \equiv f_\theta$, identically. Thus, the disparity satisfies the basic requirements of a statistical distance. For simplicity of presentation, in the following we will write the expression on the right-hand side of Equation (2.7) as $\sum C(\delta) f_\theta$.

Specific forms of the function C generate many well known disparities. For example, $C(\delta) = (\delta + 1) \log(\delta + 1) - \delta$ generates the well known likelihood disparity (LD) given by

$$\text{LD}(d_n, f_\theta) = \sum [d_n \log(d_n/f_\theta) + (f_\theta - d_n)] = \sum d_n \log(d_n/f_\theta), \quad (2.8)$$

which is a form of the Kullback–Leibler divergence (Kullback and Leibler, 1951). However, it is the symmetric opposite of likelihood disparity that we will refer to as the Kullback–Leibler divergence (KLD), which has the form

$$\text{KLD}(d_n, f_\theta) = \sum [f_\theta \log(f_\theta/d_n) + (d_n - f_\theta)] = \sum f_\theta \log(f_\theta/d_n). \quad (2.9)$$

This distance corresponds to $C(\delta) = \delta - \log(\delta + 1)$. The (twice, squared) Hellinger distance (HD) has the form

$$\text{HD}(d_n, f_\theta) = 2 \sum [d_n^{1/2} - f_\theta^{1/2}]^2 \quad (2.10)$$

which corresponds to $C(\delta) = 2((\delta + 1)^{1/2} - 1)^2$. The Pearson's chi-square (divided by 2) is defined as

$$\text{PCS}(d_n, f_\theta) = \sum \frac{(d_n - f_\theta)^2}{2 f_\theta}, \quad (2.11)$$

where $C(\delta) = \delta^2/2$, and the Neyman's chi-square (divided by 2) is defined as

$$\text{NCS}(d_n, f_\theta) = \sum \frac{(d_n - f_\theta)^2}{2 d_n}, \quad (2.12)$$

where $C(\delta) = \frac{\delta^2}{2(\delta + 1)}$. For the rest of the chapter the terms HD, PCS and NCS will indicate the quantities in Equations (2.10), (2.11) and (2.12) respectively.

There are several important subfamilies of the class of disparities, each generating several common divergences. These include the Cressie–Read family (Cressie and Read, 1984) of power divergences (PD), indexed by a real parameter $\lambda \in (-\infty, \infty)$, and having the form

$$\text{PD}_\lambda(d_n, f_\theta) = \frac{1}{\lambda(\lambda + 1)} \sum d_n \left[\left(\frac{d_n}{f_\theta} \right)^\lambda - 1 \right]. \quad (2.13)$$

Cressie and Read (1984) denoted the power divergence with tuning parameter λ as I^λ. For the sake of a uniform notation, we will refer to it as PD_λ. Notice that for values of $\lambda = 1, 0, -1/2, -1$ and -2 the Cressie–Read form in Equation (2.13) generates the Pearson's chi-square (PCS), the likelihood disparity (LD), the Hellinger distance (HD), the Kullback–Leibler divergence (KLD) and the Neyman's chi-square (NCS) respectively. The LD and KLD are not directly obtainable from Equation (2.13) by replacing $\lambda = 0$ and -1 in its expression; however, they are the continuous limits of the expression on the right hand side of Equation (2.13) as $\lambda \to 0$ and $\lambda \to -1$ respectively.

It is also easy to see that the Hellinger distance ($\lambda = -1/2$) is the only distance metric within the Cressie–Read family, and the divergences that are equally spaced on either side of $\lambda = -1/2$ in the λ scale are symmetric opposites of each other obtained by interchanging d_n and f_θ. Thus, one gets the KLD ($\lambda = -1 = -1/2 - 1/2$) by interchanging d_n and f_θ in the expression for LD ($\lambda = 0 = -1/2 + 1/2$), and the NCS ($\lambda = -2 = -1/2 - 3/2$) by interchanging d_n and f_θ in the expression for the PCS ($\lambda = 1 = -1/2 + 3/2$). In fact, one can write $\mathrm{PD}_\lambda(d_n, f_\theta)$ alternatively as

$$\mathrm{PD}_\alpha^*(d_n, f_\theta) = \frac{4}{1 - \alpha^2} \sum_x d_n \left[1 - \left(\frac{f_\theta}{d_n} \right)^{(1+\alpha)/2} \right]. \qquad (2.14)$$

Notice that under this formulation $\mathrm{PD}_\lambda = \mathrm{PD}_\alpha^*$, with $\alpha = -(1 + 2\lambda)$. The PD_α^* distance is symmetric for $\alpha = 0$ (which is the HD), and the choices α and $-\alpha$ generate distances which are symmetric opposites of each other when $\alpha \neq 0$. Jimenez and Shao (2001) have called distances corresponding to α and $-\alpha$ adjoints of each other. The Hellinger distance is self adjoint. There are some advantages of the form in (2.14). However, in the rest of this book we will continue to use the form for the power divergence family given in (2.13) or the modified form in (2.15) to conform to the huge volume of research that has followed up on the inspired work of Cressie and Read (1984).

While we know that any distance generated by a function $C(\cdot)$ satisfying the disparity conditions is nonnegative, it is sometimes of substantial benefit to be able to express the disparity in such a way that each individual term in the summation of the right-hand side of Equation (2.7) is nonnegative (this is not an automatic property for an arbitrary version of the disparity generating function C). Often we can achieve this with a little extra effort. For example, with the Cressie–Read family, one can write the disparity PD_λ as

$$\mathrm{PD}_\lambda(d_n, f_\theta) = \sum \left\{ \frac{1}{\lambda(\lambda + 1)} d_n \left[\left(\frac{d_n}{f_\theta} \right)^\lambda - 1 \right] + \frac{f_\theta - d_n}{\lambda + 1} \right\}, \qquad (2.15)$$

and it is a simple matter to check that all the terms in the right-hand side of Equation (2.15) are nonnegative. In terms of the function $C(\cdot)$, this amounts to redefining the function, without changing the value of the disparity, so that $C(\delta)$ is a nonnegative convex function with $C(0) = 0$ as its minimum value. Since $C(\cdot)$ is convex, this essentially means introducing the additional restriction

$$C'(0) = 0, \qquad (2.16)$$

where C' represents the first derivative of C with respect to its argument (similarly C'' will represent the second derivative of C). This is the reason for considering, for example, the alternative expression $\sum[d_n \log(d_n/f_\theta) + (f_\theta - d_n)]$ when defining the likelihood disparity in Equation (2.8) (rather than just the usual $\sum d_n \log(d_n/f_\theta)$). Defining $C(\cdot)$ to be nonnegative aids the interpretation, and is also helpful in deriving some of the asymptotic properties that we will consider later. However, this is not a basic disparity condition, and depending on our mathematical convenience, we will also make use of disparity generating functions which are not necessarily of this type.

The C function for the Cressie–Read family of distances under the formulation in Equation (2.15) is given by

$$C_\lambda(\delta) = \frac{(\delta+1)^{\lambda+1} - (\delta+1)}{\lambda(\lambda+1)} - \frac{\delta}{\lambda+1}. \tag{2.17}$$

Other subfamilies within the class of disparities include the blended weight Hellinger distance (Lindsay, 1994; Basu and Lindsay, 1994; Shin, Basu and Sarkar, 1995), as a function of a tuning parameter $\alpha \in [0,1]$, as

$$\text{BWHD}_\alpha(d_n, f_\theta) = \frac{1}{2} \sum \frac{(d_n - f_\theta)^2}{(\alpha d_n^{1/2} + \bar\alpha f_\theta^{1/2})^2}, \tag{2.18}$$

where $\bar\alpha = 1 - \alpha$. The C function for this is given by

$$C_\alpha(\delta) = \frac{1}{2} \frac{\delta^2}{[\alpha(\delta+1)^{1/2} + \bar\alpha]^2}. \tag{2.19}$$

For $\alpha = 0, 1/2$, and 1, this family generates the Pearson's chi-square, the Hellinger distance and the Neyman's chi-square respectively. Although the above authors have used the BWHD_α only for $\alpha \in [0,1]$, in practice there is no conceptual difficulty in allowing the range of the tuning parameter to be $(-\infty, \infty)$. The disparities on the right-hand side of (2.18) satisfy the conditions for statistical distances for all $\alpha \in (-\infty, \infty)$, which is a consequence of the squared term in the denominator.

Another such family is the blended weight chi-square divergence (Lindsay, 1994; Shin, Basu and Sarkar, 1996); this is given, as a function of a tuning parameter α, as

$$\text{BWCS}_\alpha(d_n, f_\theta) = \frac{1}{2} \sum \frac{(d_n - f_\theta)^2}{\alpha d_n + \bar\alpha f_\theta}, \tag{2.20}$$

where $\bar\alpha = 1 - \alpha$. The C function for this is given by

$$C_\alpha(\delta) = \frac{1}{2} \frac{\delta^2}{[\alpha(\delta+1) + \bar\alpha]}. \tag{2.21}$$

This family generates the Pearson's chi-square and the Neyman's chi-square for $\alpha = 0$ and 1 respectively. Note that the particular choice $\alpha = 1/2$ generates the symmetric chi-square (SCS) measure, given by

$$\text{SCS}(d_n, f_\theta) = \sum \frac{(d_n - f_\theta)^2}{d_n + f_\theta}, \qquad (2.22)$$

which leads to another genuine distance within the class of disparities.

The family of generalized negative exponential disparities (Bhandari, Basu and Sarkar, 2006), as a function of the tuning parameter $\lambda \geq 0$, corresponds to the C function

$$C_\lambda(\delta) = \begin{cases} (e^{-\lambda \delta} + \lambda \delta - 1)/\lambda^2, & \text{if } \lambda > 0 \\ \delta^2/2 & \text{if } \lambda = 0. \end{cases} \qquad (2.23)$$

The $\lambda = 0$ case, which is obtained by taking the limit of the expression for $\lambda > 0$ as $\lambda \to 0$, corresponds to the Pearson's chi-square. The ordinary negative exponential disparity, which corresponds to $\lambda = 1$, has the form $\text{NED}(d_n, f_\theta) = \sum (e^{-\delta} + \delta - 1) f_\theta$.

The disparities for the generalized Kullback–Leibler (GKL) divergence family have been defined by Park and Basu (2003) as

$$\text{GKL}_\tau(d_n, f_\theta) = \sum \left[\frac{d_n}{\bar{\tau}} \log(d_n/f_\theta) - (\frac{d_n}{\bar{\tau}} + \frac{f_\theta}{\tau}) \log(\tau \frac{d_n}{f_\theta} + \bar{\tau}) \right], \qquad (2.24)$$

where $\bar{\tau} = 1 - \tau$. In this case, the C function is given by

$$C_\tau(\delta) = \frac{\delta + 1}{1 - \tau} \log(\delta + 1) - \frac{\tau \delta + 1}{\tau(1 - \tau)} \log(\tau \delta + 1). \qquad (2.25)$$

The range of the tuning parameter is $\tau \in [0, 1]$. The disparities for the cases $\tau = 0$ and $\tau = 1$ are the limiting cases as $\tau \to 0$ and $\tau \to 1$, and in these cases the disparities equal the LD and KLD measures respectively.

A motivation for the construction of the GKL family, which produces a smooth bridge between LD and KLD is as follows. It is easy to see that for any two probability density functions g and f,

$$\text{GKL}_\tau(g, f) = \min_p \{\tau \text{LD}(g, p) + (1 - \tau) \text{KLD}(p, f)\}, \qquad (2.26)$$

where the minimization is over the density p. The right-hand side of the above equation is actually the solution of the likelihood ratio testing problem which minimizes the likelihood disparity subject to $p \in B_f = \{t : \text{KLD}(t, f) \leq c\}$, and τ is an appropriate constant depending on c. A reversal of the roles of LD and KLD generates the celebrated power divergence family of Cressie and Read (1984).

In addition to the restriction imposed on the disparities in Equation (2.16), sometimes it is convenient, particularly in the goodness-of-fit testing scenario, to impose the restriction

$$C''(0) = 1 \qquad (2.27)$$

on the defining function $C(\cdot)$. Given any convex function C having the property $C(0) = 0$, the disparity (2.7) can be centered and rescaled to the form

$$\rho_{C^*}(d_n, f_\theta) = \sum C^*(\delta) f_\theta = \sum \left(\frac{C(\delta) - C'(0)\delta}{C''(0)} \right) f_\theta, \qquad (2.28)$$

provided $C''(0) \neq 0$. Notice that $C^*(0) = 0$, and if C is convex, so is C^*. In addition, C^* satisfies the conditions (2.16) and (2.27). This does not change the estimating properties of the disparity in the sense that if $\hat{\theta}$ is the minimizer of ρ_C, then it is also the minimizer of ρ_{C^*}. All the C functions presented in Equations (2.17), (2.19), (2.21), (2.23), and (2.25) satisfy conditions (2.16) and (2.27). In the rest of this book, unless otherwise mentioned, we will assume that the C functions are standardized to satisfy $C'(0) = 0$ and $C''(0) = 1$.

Some other less known disparities have been discussed in Park and Basu (2004).

A disparity function takes the value zero when the two arguments are identical. The following two lemmas establish the upper bound of the disparity function for two arbitrary densities g and f and some ancillary results; also see Vajda (1972) and our discussion in Chapter 11. Since these results below are general and encompass both discrete and continuous models, we express the disparities with integrals in the two lemmas below.

Lemma 2.1. *Consider two probability density functions g and f, and let $\rho_C(g, f) = \int C(g/f - 1)f$ represent a disparity between the densities g and f, where the function C satisfies the disparity conditions. Let $D(g, f) = C(g/f - 1)f$, and define $C'(\infty) = \lim_{\delta \to \infty}[C(\delta)/\delta]$. Then we have the following results.*

(i) $D(g, f) \leq D(0, f)\chi(g \leq f) + D(g, 0)\chi(f < g) \leq D(0, f) + D(g, 0),$

(ii) $D(g, f) \leq C(-1)f + C'(\infty)g,$

where $\chi(A)$ is the indicator for the set A.

Proof. First, for $g \in [0, f]$ with fixed f, look at $D(g, f)$ as a function of g.

$$\frac{\partial}{\partial g} D(g, f) = C'\left(\frac{g}{f} - 1\right) < 0, \quad \forall g \in (0, f)$$

since $C(\cdot)$ is decreasing for $\delta < 0$. Hence $D(g, f) \leq D(0, f)$ for $g \in [0, f]$. Note that $D(g, f) \leq C(-1)f$ for $\forall g \in [0, f]$.

Next, for $f \in (0, g)$ with fixed g, look at $D(g, f)$ as a function of f.

$$\frac{\partial}{\partial f} D(g, f) = -C'\left(\frac{g}{f} - 1\right)\frac{g}{f} + C\left(\frac{g}{f} - 1\right) = -A\left(\frac{g}{f} - 1\right) \leq 0, \quad \forall f \in (0, g)$$

since $A(\delta)$ is an increasing function with $A(0) = 0$. Hence $D(g, f) \leq D(g, 0)$ for $f \in [0, g)$. This proves part (i). Note that

$$D(g, 0) = \lim_{t \to 0} C(g/t - 1)t = C'(\infty)g,$$

so that part (ii) holds. $\qquad \square$

Lemma 2.2. *Suppose that $C(-1)$ and $C'(\infty)$ are finite. Then the disparity $\rho_C(g, f)$ is bounded above by $C(-1) + C'(\infty)$.*

Proof. This follows easily from Lemma 2.1 (ii). □

2.3.2 More on the Hellinger Distance

In many ways, the Hellinger distance is the focal point of the theme of this book. Notice that the actual Hellinger distance is equal to

$$\left\{ \sum \left(d_n^{1/2} - f_\theta^{1/2} \right)^2 \right\}^{1/2} = \left\{ \frac{1}{2} \mathrm{HD}(d_n, f_\theta) \right\}^{1/2} \tag{2.29}$$

where $\mathrm{HD}(d_n, f_\theta)$ is the measure defined in Equation (2.10). The distance in (2.29) is a genuine metric satisfying the triangle inequality, and the disparity measure $\mathrm{HD}(d_n, f_\theta)$ based on it is by far the most popular disparity in robust minimum distance literature. Notice that the measure in (2.10) may be written as

$$\mathrm{HD}(d_n, f_\theta) = 2 \left(2 - 2 \sum_x d_n^{1/2} f_\theta^{1/2} \right) = 4 \left(1 - \sum_x d_n^{1/2} f_\theta^{1/2} \right),$$

making it a one to one function of the term $\sum_x d_n^{1/2} f_\theta^{1/2}$. Observe that this term is also linked to the distance measure

$$\mathrm{B}(d_n, f_\theta) = - \log \left(\sum_x d_n^{1/2} f_\theta^{1/2} \right) \tag{2.30}$$

given by Bhattacharyya (1943), and the equivalence of $\mathrm{HD}(d_n, f_\theta)$ in (2.10) and the Bhattacharyya distance in (2.30) may be expressed as

$$\mathrm{HD}(d_n, f_\theta) = 4(1 - e^{-\mathrm{B}(d_n, f_\theta)}).$$

The term $\sum_x d_n^{1/2} f_\theta^{1/2}$ is sometimes referred to as the Bhattacharyya coefficient, and can be thought of as an approximate measure of the amount of overlap between two probability densities. Although it does not satisfy the triangle inequality, the Bhattacharyya distance is nonnegative, and equals zero if and only if $d_n \equiv f_\theta$, identically. See Kailath (1967), Djouadi et al. (1990), and Aherne et al. (1997), among others, for some interesting applications of the Bhattacharyya distance in solving different problems in various disciplines.

The Hellinger distance is named after Ernst David Hellinger, a German mathematician who was active in the first half of the twentieth century. He introduced a type of integral called the Hellinger integral (Hellinger, 1909) which led to the construction of the Hellinger distance. The Hellinger distance is also referred to as the Matusita distance (Matusita, 1954; Kirmani, 1971) or the Jeffreys–Matusita distance in the literature. Both the Bhattacharyya distance and the Matusita (Hellinger) distance are extensively used as measures of separation between probability densities in many practical problems such as remote sensing (for example Landgrebe, 2003; Canty, 2007). The disparity goodness-of-fit statistic (see Chapter 8) in multinomial models based on the Hellinger distance is also referred to as the Freeman-Tukey statistic in the literature (for example Freeman and Tukey, 1950; Read, 1993).

Bhattacharyya's distance may be looked upon as a special case of the Rényi divergence (Rényi, 1961; Leise and Vajda, 1987) given by

$$\mathrm{RD}_r(d_n, f_\theta) = \frac{1}{r(r-1)} \log \left(\sum_x d_n^r(x) f_\theta^{1-r}(x) \right), \quad r \neq 0, 1. \qquad (2.31)$$

The distances at $r = 0, 1$ are obtained as the limiting cases for those values, which gives
$$\mathrm{RD}_1(d_n, f_\theta) = \lim_{r \to 1} \mathrm{RD}_r(d_n, f_\theta) = \mathrm{LD}(d_n, f_\theta),$$
while
$$\mathrm{RD}_0(d_n, f_\theta) = \lim_{r \to 0} \mathrm{RD}_r(d_n, f_\theta) = \mathrm{KLD}(d_n, f_\theta),$$

where LD and KLD are as defined in Equations (2.8) and (2.9). The case $r = 1/2$ generates (four times) the Bhattacharyya distance.

The connection of the family of Rényi divergences for $r \in (0, 1)$ with that of the members of the generalized Hellinger distance family

$$\mathrm{GHD}(d_n, f_\theta) = \frac{1}{r(1-r)} \left(1 - \sum_x d_n^r(x) f_\theta^{1-r}(x) \right), \quad r \in (0, 1), \qquad (2.32)$$

(Simpson 1989a, and Basu, Basu and Chaudhuri, 1997) can be easily observed by comparing (2.31) and (2.32).

For the rest of the book, unless otherwise qualified, the term Hellinger distance will refer to the measure on the right-hand side of Equation (2.10).

2.3.3 The Minimum Distance Estimator and the Estimating Equations

We begin this section by demonstrating an important fact. The class of minimum distance estimators based on disparities contains the maximum likelihood estimator as a special case.

The minimum distance estimator $\hat{\theta}$ of θ, based on the disparity ρ_C, is defined by the relation

$$\rho_C(d_n, f_{\hat{\theta}}) = \min_{\theta \in \Theta} \rho_C(d_n, f_\theta) \qquad (2.33)$$

provided such a minimum exists.

Consider the estimation setup described earlier in this section. We have a discrete parametric model $\mathcal{F} = \{F_\theta : \theta \in \Theta \subseteq \mathbb{R}^p\}$, and suppose that a random sample $X_1, \ldots, X_n$ is available from the true distribution G with which we wish to estimate θ; the support of the random variables is assumed, without loss of generality, to be $\mathcal{X} = \{0, 1, 2, \ldots, \}$. Determination of the maximum likelihood estimator of θ corresponds to the maximization of

$$\log \prod_{i=1}^{n} f_\theta(X_i) = \sum_{i=1}^{n} \log f_\theta(X_i). \tag{2.34}$$

However, if we write the sum on the right-hand of the above equation in terms of the elements of the sample space, rather than the index i for the observation of the random sample, then this sum equals

$$n \sum_{x=0}^{\infty} d_n(x) \log f_\theta(x). \tag{2.35}$$

Dropping the dummy variable x, maximizing the above with respect to θ is equivalent to minimizing

$$-\sum d_n \log f_\theta,$$

and hence equivalent to minimizing

$$\sum d_n \log(d_n/f_\theta) \tag{2.36}$$

with respect to θ. Notice that this is the likelihood disparity between d_n and f_θ. Thus, the likelihood disparity is minimized by the maximum likelihood estimator of θ, which shows that the class of minimum distance estimators based on disparities includes the maximum likelihood estimator under discrete models.

As already described, the minimum distance estimator $\hat{\theta}$ based on the disparity ρ_C is obtained through the minimization described in Equation (2.33). In the present subsection we will describe the geometric structure of the disparities and their gradients to have a better insight on the robustness and the efficiency properties of these estimators. Under differentiability of the model, the minimum distance estimator of θ based on the disparity ρ_C is obtained by solving the estimating equation

$$-\nabla \rho_C(d_n, f_\theta) = \sum (C'(\delta)(\delta + 1) - C(\delta)) \nabla f_\theta = 0, \tag{2.37}$$

where ∇ represents the gradient with respect to θ. Letting

$$A(\delta) = C'(\delta)(\delta + 1) - C(\delta), \tag{2.38}$$

the estimating equation for θ has the form

$$-\nabla \rho_C(d_n, f_\theta) = \sum A(\delta) \nabla f_\theta = 0. \tag{2.39}$$

The function $A(\delta)$ can be suitably standardized, without changing the estimating properties of the minimum disparity estimators, so that it satisfies

$$A(0) = 0 \text{ and } A'(0) = 1. \tag{2.40}$$

These properties are automatic when the corresponding C function satisfies the conditions (2.16) and (2.27). The function $A(\delta)$, when thus standardized, is called the residual adjustment function (RAF) of the disparity. The residual adjustment function will be one of our key components in studying the properties of our minimum distance estimators.

The estimating equations of the different minimum distance estimators represented by Equation (2.39) differ only in the form of the residual adjustment function $A(\delta)$. Thus, the different properties of the minimum disparity estimators must be governed by the form of the function $A(\delta)$. Notice that $A'(\delta) = (\delta + 1)C''(\delta)$, and as $C(\cdot)$ is a strictly convex function, $A'(\delta) > 0$ for $\delta > -1$; hence $A(\cdot)$ is a strictly increasing function on $[-1, \infty)$.

As our primary motivation in proposing these minimum distance estimators is in developing a class of estimators, which has full asymptotic efficiency coupled with strong robustness properties, we will now proceed to describe the role of the RAF in determining the robustness properties of the estimators. For this purpose, we will consider a probabilistic – rather than a geometric – characterization of the outliers in a data set; however, these concepts often coincide. An element x of the sample space with a large positive value of the Pearson residual $\delta(x)$ represents a large outlier in relation to the parametric model in the sense that the actual observed proportion is much larger here than what is predicted by the model. We note that Davies and Gather (1993) also defined outliers in terms of their position relative to the model that most of the observations follow. In robust estimation, our aim is to downweight observations having large positive values of δ. This is achieved by such disparities for which the RAF $A(\delta)$ exhibits a severely dampened response to increasing δ. For a qualitative description, we will take the RAF of the likelihood disparity as the basis for comparison. Notice that the likelihood disparity is minimized by the maximum likelihood estimator. The estimating equation of the maximum likelihood estimator (the likelihood equation) can be expressed as

$$-\nabla \text{LD}(d_n, f_\theta) = \sum \delta \nabla f_\theta = 0. \tag{2.41}$$

Thus, one gets $A_{\text{LD}}(\delta) = \delta$, and the RAF is linear for the likelihood disparity, passing through the origin at 45 degrees to the X-axis or the δ-axis (as well as to the Y-axis). Hence the comparison of other minimum distance estimators with the maximum likelihood estimator must focus on how the other RAFs depart from linearity. The set of conditions given by (2.40) guarantees that all RAFs are tangential to the line $A_{\text{LD}}(\delta) = \delta$ at the origin ($\delta = 0$).

In Figure 2.1, we have presented the graphs of the residual adjustment functions of the five common distances within the class of disparities. Notice that the RAFs for the HD, the KLD and the NCS all provide strong

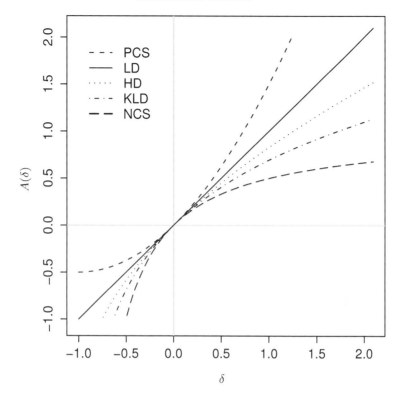

FIGURE 2.1
Residual Adjustment Functions for five common distances.

downweighting for large δ (relative to the likelihood disparity), with the NCS providing the strongest downweighting. On the other hand, the PCS actually magnifies the effect of large δ outliers rather than shrinking them. As a result, the minimum distance estimators based on the Pearson's chi-square distance are expected to be even worse than the maximum likelihood estimator in terms of robustness.

Expanding the estimating equation given by (2.39) in a Taylor series around $\delta = 0$, we get

$$-\nabla \rho_C(d_n, f_\theta) = \sum A(\delta) \nabla f_\theta = \sum \left\{ \delta + \frac{A_2}{2} \delta^2 + \dots \right\} \nabla f_\theta = 0. \quad (2.42)$$

Comparing with Equation (2.41) we see that the leading term in the estimating function of any disparity is the same as that of the likelihood disparity which gives some intuitive justification of the asymptotic equivalence of all minimum distance estimators based on disparities and the maximum likelihood estimator. As we will see later, the quantity $A_2 = A''(0)$ in the above equation turns out to be a key player in describing the properties of the minimum distance estimator.

2.3.4 The Estimation Curvature

A dampened response to increasing positive δ will imply that the RAF shrinks the effect of large outliers as δ increases. Under the standardizations given in Equation (2.28), all residual adjustment functions satisfy $A(0) = 0$ and $A'(0) = 1$. For the likelihood disparity, which has a linear residual adjustment function, the second and all successive derivatives of the RAF evaluated at $\delta = 0$ (or anywhere else) are equal to zero. In comparison, RAFs for which $A_2 = A''(0)$ is positive, curve locally upwards (in comparison to $A_{\mathrm{LD}}(\delta)$) at $\delta = 0$, while the reverse is observed when A_2 is negative. Thus, A_2 can be used as a measure of local robustness, with negative values of A_2 being preferred. A negative value for the A_2 parameter is achieved, for example, for the HD, the KLD, the NCS (see Figure 2.1), and all other members of the Cressie–Read family with $\lambda < 0$. Similarly this is achieved for all members of the BWHD and BWCS families with $\alpha > 1/3$. This is also true for all members of the GNED family with $\lambda > 1$, and all members of the GKL family with $\tau > 0$.

The estimation curvature A_2 also has some relevance in the context of the second-order efficiency of these minimum distance estimators. In case of multinomial models, the concept of second-order efficiency of Rao (1961) can be directly applied. The second-order efficiency E_2 of an estimator (the smaller E_2 is, the better the efficiency) T is measured by finding the minimum asymptotic variance of $U_n(\theta) - \alpha(\theta)[T - \theta] - \lambda(\theta)[T - \theta]^2$ over α and λ, where $U_n(\theta)$ is the score function in an independent and identically distributed sample. The minimum value is determined by the model and the parametrization. In the multinomial model this is minimized by the MLE (Rao, 1961, 1962). The deficiency of a minimum distance estimator within the class of disparities is a simple function of the estimation curvature A_2 and a nonnegative quantity D which depends on the model, but not on $A(\delta)$.

We present the following theorem in the context of models having a finite support linking the maximum likelihood estimator to the concept of second-order efficiency. See Rao et al. (1983), Read and Cressie (1988) and Lindsay (1994) for further discussion including a proof of the theorem.

Theorem 2.3. *Suppose the sample space $\mathcal{X} = \{0, 1, \ldots, K\}$ is finite ($K < \infty$). The second-order efficiency E_2 of a minimum distance estimator (MDE) based on a disparity having residual adjustment function $A(\delta)$ and estimation curvature A_2 is given by*

$$E_2(\mathrm{MDE}) = E_2(\mathrm{MLE}) + A_2^2 D,$$

where D is a nonnegative quantity depending on the model but not on the residual adjustment function $A(\delta)$.

For the PD family, the result was established by Read and Cressie (1988).

However, it turns out that the calculations presented therein depend only on the first and second derivatives of the estimating functions with respect to δ (at $\delta = 0$). If two RAFs have the same value of the estimation curvature, then they have the same first and second derivatives of the RAF at $\delta = 0$. Thus the above theorem holds generally for the class of minimum distance estimators based on disparities.

Thus, for models such as the multinomial, disparities satisfying $A_2 = 0$ generate second-order efficient estimators; apart from the likelihood disparity, this is achieved by the BWHD and BWCS families for $\alpha = 1/3$, and for the GNED for $\lambda = 1$. The latter disparity is called the negative exponential disparity (NED).

Notice that $A_2 = 0$ implies that the residual adjustment function of the corresponding distance has a second-order contact with that of the likelihood disparity at the origin. In this case, the right-hand side of Equation (2.42) is equivalent to the expression for the likelihood disparity up to the second order. In this sense, our working definition of second-order efficiency of a minimum distance estimator based on disparities is that the corresponding estimation curvature A_2 equals zero.

As mentioned in Section 1.3.3, the concept of second-order efficiency is not as simple as that of first-order efficiency, and is also subject to some controversy. Berkson (1980) has questioned the significance of this measure. Read and Cressie (1988) show that the MLE is no longer the optimal within the class of minimum distance estimators when considering the Hodges–Lehmann deficiency of the estimators.

2.3.5 Controlling of Inliers

Unlike the outliers and the large positive values of the Pearson residual generated by them, observations with fewer data than expected will generate negative values of δ, and such observations will be denoted as *inliers*. If one requires the RAF to shrink both positive and negative residuals (outliers and inliers) relative to maximum likelihood, the RAF should have the property

$$|A(\delta)| \leq |\delta| \tag{2.43}$$

for all δ. Such RAFs must cross the $A_{\mathrm{LD}}(\delta) = \delta$ line at $\delta = 0$, and hence should satisfy $A''(0) = 0$. The corresponding estimators must therefore be automatically second-order efficient. However, to satisfy the condition (2.43) the third derivative $A'''(0)$ must be negative (unless it is itself zero in which case one has to consider the derivative of the next higher order). The negative exponential disparity is an example of disparities satisfying (2.43), and it has remarkable robustness properties in spite of its second-order efficient behavior (Bhandari, Basu and Sarkar, 2006).

However, the negative exponential disparity is a rather special distance, and practically all the other robust distances within the class of disparities (including the Hellinger distance) magnify the effect of inliers while shrinking

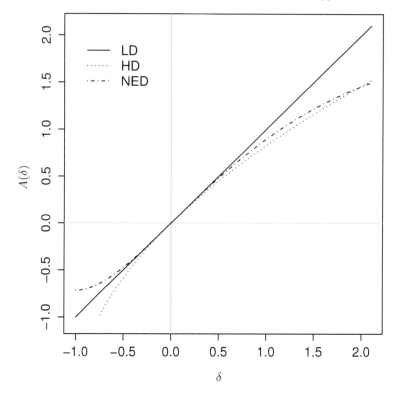

FIGURE 2.2
Residual Adjustment Functions for LD, HD and NED.

the effect of outliers. But it appears that the proper controlling of inliers is one of the keys for good small sample efficiency; as a result the estimators and tests based on these robust distances can be substantially less efficient compared to the likelihood based methods in small samples. In Chapter 6 we will consider some strategies to tackle the inlier problem, without compromising the robustness properties of these distances.

2.3.6   The Robustified Likelihood Disparity

The theory and interpretation of the minimum distance techniques used in this book are based on the two key functions; the first is the disparity generating function $C(\cdot)$, and the second is the residual adjustment function $A(\cdot)$. In the direct approach to estimation, we normally start with a distance function, and obtain the residual adjustment function by taking a derivative of the distance measure with respect to the parameter and suitably standardizing it. Sometimes we also take the reverse approach. In this case, we first define a residual adjustment function with the right properties, and then reconstruct the distance. This is done by solving the differential Equation (2.38) which recovers the disparity generating function $C(\cdot)$. Given any differentiable and increasing function $A(\delta)$, the corresponding disparity generating function obtained through the above procedure has the form

$$C(\delta) = \int_0^\delta \int_0^t A'(s)(1+s)^{-1} ds \, dt. \qquad (2.44)$$

The *strictly increasing* condition may be replaced by *nondecreasing* over parts of the domain bounded away from zero without any technical difficulty. Lack of differentiability at one or two points at such regions of the domain are also entirely fixable if the function is continuous.

In the previous sections of this chapter we have seen that most of the features of the minimum distance estimators are governed by the smoothness properties of the residual adjustment function $A(\delta)$ and the magnitude of its derivatives at $\delta = 0$. In particular when the estimation curvature parameter A_2 equals zero, the residual adjustment function $A(\delta)$ of the disparity has a second-order contact with the residual adjustment function $A_{\mathrm{LD}}(\delta)$ of the likelihood disparity. If successive derivatives of the residual adjustment function $A(\delta)$ at $\delta = 0$ continue to be zero up to the k-th order, $A(\delta)$ has a k-th order contact with $A_{\mathrm{LD}}(\delta)$. A comparison with Equation (2.42) reveals that the estimating equation for the disparity in question is equivalent to that of the likelihood disparity up to the k-th order in that case.

In an attempt to develop a structure where the estimating equations are completely equivalent to the likelihood equations in a neighborhood of $\delta = 0$, but have powerful outlier downweighting properties for large values of δ, Chakraborty, Basu and Sarkar (2001) proposed the residual adjustment function

$$A_{\alpha,\alpha^*}(\delta) = \begin{cases} \alpha & \text{for } -1 \le \delta \le \alpha \\ \delta & \text{for } \alpha < \delta < \alpha^* \\ \alpha^* & \text{for } \delta \ge \alpha^*. \end{cases} \qquad (2.45)$$

with numbers α^* and α satisfying $-1 \le \alpha < 0 < \alpha^* < \infty$. That the function will limit the impact of large δ outliers is obvious. The additional tuning parameter α provides further flexibility to the estimation procedure. The function A_{α,α^*} has the shape of the Huber's ψ function, and is constant over parts of its domain.

The corresponding disparity generating function obtained by solving the differential Equation (2.38) has the form

$$C_{\alpha,\alpha^*}(\delta) = \begin{cases} (\delta+1)\log(\alpha+1) - \alpha & \text{for } -1 \le \delta \le \alpha \\ (\delta+1)\log(\delta+1) - \delta & \text{for } \alpha < \delta < \alpha^* \\ (\delta+1)\log(\alpha^*+1) - \alpha^* & \text{for } \delta \ge \alpha^*. \end{cases} \qquad (2.46)$$

Clearly, $C_{\alpha,\alpha^*}(\delta)$ is strictly convex on (α, α^*) and linear on $(-\infty, \alpha]$ and $[\alpha^*, \infty)$ with the slopes at $\delta = \alpha$ and $\delta = \alpha^*$ well defined. Hence $C_{\alpha,\alpha^*}(\delta)$

TABLE 2.1

Forms of C functions and RAFs for common disparities. For BWHD and BWCS, $\bar{\alpha} = 1 - \alpha$; for GKL, $\bar{\tau} = 1 - \tau$. For GNED, the forms are given for $\lambda > 0$ ($\lambda = 0$ case is the PCS).

Disparity	C function	RAF
LD	$(\delta + 1)\log(\delta + 1) - \delta$	δ
HD	$2((\delta + 1)^{1/2} - 1)^2$	$2((\delta + 1)^{1/2} - 1)$
PCS	$\dfrac{\delta^2}{2}$	$\delta + \dfrac{\delta^2}{2}$
NCS	$\dfrac{\delta^2}{2(\delta + 1)}$	$1 - \dfrac{1}{\delta + 1}$
KLD	$\delta - \log(\delta + 1)$	$\log(\delta + 1)$
PD	$\dfrac{(\delta + 1)^{\lambda + 1} - (\delta + 1)}{\lambda(\lambda + 1)} - \dfrac{\delta}{\lambda + 1}$	$\dfrac{(\delta + 1)^{\lambda + 1} - 1}{\lambda + 1}$
BWHD	$\dfrac{1}{2} \dfrac{\delta^2}{[\alpha(\delta + 1)^{1/2} + \bar{\alpha}]^2}$	$\dfrac{\delta}{[\alpha(\delta + 1)^{1/2} + \bar{\alpha}]^2} + \dfrac{\bar{\alpha}\delta^2}{2[\alpha(\delta + 1)^{1/2} + \bar{\alpha}]^3}$
BWCS	$\dfrac{1}{2} \dfrac{\delta^2}{[\alpha(\delta + 1) + \bar{\alpha}]}$	$\dfrac{\delta}{1 + \alpha\delta} + \dfrac{\bar{\alpha}}{2}\left[\dfrac{\delta}{1 + \alpha\delta}\right]^2$
SCS	$\dfrac{\delta^2}{\delta + 2}$	$\dfrac{\delta(3\delta + 4)}{(\delta + 2)^2}$
NED	$e^{-\delta} - 1 + \delta$	$2 - (2 + \delta)e^{-\delta}$
GNED	$\dfrac{e^{-\lambda\delta} - 1 + \lambda\delta}{\lambda^2}$	$\dfrac{(\lambda + 1) - ((\lambda + 1) + \lambda\delta)e^{-\lambda\delta}}{\lambda^2}$
GKL	$\dfrac{\delta + 1}{\bar{\tau}}\log(\delta + 1) - \dfrac{\tau\delta + 1}{\tau\bar{\tau}}\log(\tau\delta + 1)$	$\dfrac{1}{\tau}\log(\tau\delta + 1)$
RLD	$\begin{cases} (\delta + 1)\log\bar{\alpha} + \alpha \\ (\delta + 1)\log(\delta + 1) - \delta \\ (\delta + 1)\log(1/\bar{\alpha}) - \alpha/\bar{\alpha} \end{cases}$	$\begin{cases} -\alpha &: \quad \delta < -\alpha \\ \delta &: \quad -\alpha \le \delta < \alpha/\bar{\alpha} \\ \alpha/\bar{\alpha} &: \quad \delta \ge \alpha/\bar{\alpha} \end{cases}$

is convex on the entire interval $[-1, \infty)$. We will refer to the disparity generated by C_{α,α^*} as the robustified likelihood disparity (with tuning parameters α and α^*). The minimizer of the robustified likelihood disparity $\mathrm{RLD}_{\alpha,\alpha^*}(d_n, f_\theta)$ over $\theta \in \Theta$ will be called the robustified likelihood estimator. The robustified likelihood estimators will have full asymptotic efficiency, and good robustness properties depending on the tuning parameters. The RLD has particular relevance in the context of weighted likelihood estimators, as we will see in Chapter 7.

A convenient one parameter formulation of RLD may by obtained by choosing α to be any number between $(0, 1)$, and letting $\alpha^* = \alpha/\bar{\alpha}$, where

$\bar{\alpha} = 1 - \alpha$. It is easy to see that one would recover the residual adjustment function of the likelihood disparity from (2.45) for the limiting case $\alpha \to 1$. Also, smaller values of α will expand the range over which the proposed down-weighting will be applied. It is this one parameter version of RLD that is presented in Table 2.1, where the disparity generating function $C(\cdot)$ and the residual adjustment function $A(\cdot)$ of several common disparities and families of disparities are provided.

2.3.7 The Influence Function of the Minimum Distance Estimators

As indicated in Chapter 1, the influence function of an estimator is a useful indicator of its asymptotic efficiency, as well as of its classical first-order robustness. Consider a generic disparity ρ_C, and let A represent its residual adjustment function. To find the influence function of our minimum distance estimators, we consider the ϵ contaminated version of the true density g given by

$$g_\epsilon(x) = (1 - \epsilon)g(x) + \epsilon\chi_y(x).$$

Similarly $G_\epsilon(x) = (1 - \epsilon)G(x) + \epsilon \wedge_y (x)$. Here $\chi_y(x)$ and $\wedge_y(x)$ are as defined in Section 1.1. Consider the minimum distance functional $T(G)$ representing the minimizer of $\rho_C(g, f_\theta)$. Let $\theta_\epsilon = T(G_\epsilon)$ be the functional obtained via the minimization of $\rho_C(g_\epsilon, f_\theta)$, which satisfies

$$\sum_x A(\delta_\epsilon(x))\nabla f_{\theta_\epsilon}(x) = 0, \tag{2.47}$$

where $\delta_\epsilon(x) = g_\epsilon(x)/f_{\theta_\epsilon}(x) - 1$. Then the influence function $\phi_G(y) = T'(y)$ of the functional T at the distribution G is the first derivative of θ_ϵ evaluated at $\epsilon = 0$. The form of the influence function of our minimum distance estimators is derived in the following theorem.

Theorem 2.4. [Lindsay (1994, Proposition 1)]. *For a disparity $\rho(\cdot, \cdot)$ associated with a corresponding estimating equation*

$$\sum_x A(\delta(x))\nabla f_\theta(x) = 0, \tag{2.48}$$

the influence function of the minimum distance functional T at G has the form

$$T'(y) = D^{-1}N,$$

where

$$N = A'(\delta(y))u_{\theta^g}(y) - E_g\left[A'(\delta(X))u_{\theta^g}(X)\right]$$

and

$$D = E_g\left[u_{\theta^g}(X)u_{\theta^g}^T(X)A'(\delta(X))\right] - \sum_x A(\delta(x))\nabla_2 f_{\theta^g}(x)$$

for $\theta^g = T(G)$ and $\delta(x) = g(x)/f_{\theta^g}(x) - 1$.

Proof. Direct differentiation of Equation (2.47) gives

$$\frac{\partial}{\partial \epsilon}\theta_\epsilon = D_\epsilon^{-1}N_\epsilon, \tag{2.49}$$

where

$$N_\epsilon = A'(\delta_\epsilon(y))u_{\theta_\epsilon}(y) - \sum A'(\delta_\epsilon(x))u_{\theta_\epsilon}(x)g(x)$$

and

$$D_\epsilon = \sum u_{\theta_\epsilon}(x)u_{\theta_\epsilon}^T(x)A'(\delta_\epsilon(x))g_\epsilon(x) - \sum A(\delta_\epsilon(x))\nabla_2 f_{\theta_\epsilon}(x).$$

When evaluated at $\epsilon = 0$ we get $\theta_\epsilon = \theta^g$ and $\delta_\epsilon(\cdot) = \delta(\cdot)$. Replacing these in Equation (2.49) we get the required result. □

The following is a direct corollary of the above theorem, which indicates the asymptotic efficiency of all the minimum distance estimators based on disparities at the model.

Corollary 2.5. *Consider the conditions of Theorem 2.4. When the true distribution G belongs to the parametric model, so that the density $g(x) = f_\theta(x)$ for some $\theta \in \Theta$, we get $\theta^g = \theta$, $\delta(x) = 0$ for all x, and the minimum distance estimator corresponding to the estimating equation $\sum A(\delta(x))\nabla f_\theta(x) = 0$ has influence function $T'(y) = I^{-1}(\theta)u_\theta(y)$, where $I(\theta)$ is the Fisher information matrix at θ.*

Observe that the above theorem and corollary could have been phrased simply in terms of the estimating equation, rather than linking them directly to a disparity. The influence function described in Theorem 2.4 describes all estimators obtained as the solution of estimating equations of the type (2.48) without any reference to a minimum distance problem. However, given any differentiable and increasing function $A(\delta)$, one can construct a corresponding disparity measure ρ_C having residual adjustment function $A(\delta)$ using relation (2.44). Thus, we prefer to present the above as properties of minimum distance estimators based on disparities, rather than as a class of estimators obtained as solutions of appropriate estimating equations. However, we will also consider the influence function of weighted likelihood estimators later on (Chapter 7), where the functionals are obtained as solutions of appropriate estimating equations, and may not directly correspond to the optimization of an objective function.

The most striking revelation of the above corollary is that all the minimum distance estimators based on disparities have the same influence function at the model as the maximum likelihood estimator, as is necessary if these estimators are to be asymptotically fully efficient. Thus, the influence functions of these minimum distance estimators are not useful indicators for describing their robustness. We will demonstrate in Chapter 4 that a distinction can be made in respect to the higher order influence terms which give a better description of the stability of the minimum distance estimators.

2.3.8 ϕ-Divergences

We have mentioned in Section 2.3.1 that in this book we will follow the approach of Lindsay (1994) and develop the minimum distance estimation procedure in terms of disparities; this is done primarily to exploit the geometry of the method, and to describe the tradeoff between the robustness and the efficiency of the procedure in terms of the residual adjustment function $A(\cdot)$ of the disparity. For the sake of completeness, here we briefly describe the structure in terms of the ϕ-divergences, as presented by Csiszár (1963, 1967a,b), Ali and Silvey (1966), Pardo (2006) and others.

Consider the densities d_n and f_θ under the notation and setup of Section 2.3.1. The ϕ-divergence measure between these densities is given by

$$D_\phi(d_n, f_\theta) = \sum_{x=0}^{\infty} \phi\left(\frac{d_n(x)}{f_\theta(x)}\right) f_\theta(x), \qquad (2.50)$$

where the function ϕ is a convex function defined on all nonnegative real values such that $\phi(1) = 0$. The function ϕ is also required to satisfy the conditions $0\phi(0/0) = 0$ and $0\phi(p/0) = p\lim_{u\to\infty} \phi(u)/u$. Comparing Equation (2.50) with Equation (2.7) we see that the definitions produce equivalent distances for $C(u - 1) = \phi(u)$. As in the case with disparities, establishing the asymptotic properties of the minimum ϕ-divergence estimator will require additional smoothness assumptions on the function; in particular, the function ϕ will be required to be thrice continuously differentiable.

Just as the representation of the disparity in terms of the disparity generating function C is not unique, the function ϕ can also be appropriately modified to guarantee that it satisfies additional useful properties without changing either the value of the divergence or the essential properties of the function. Thus, given any such function ϕ, one can define the function

$$\phi^*(u) = \frac{\phi(u) - \phi'(1)(u - 1)}{\phi''(1)}$$

in analogy with Equation (2.28). The function ϕ^* satisfies $\phi^{*'}(1) = 0$ and $\phi^{*''}(1) = 1$. When such a ϕ^* is used in the representation of the ϕ-divergence, each term in its summand is nonnegative.

A list of some specific cases of ϕ-divergences is provided in Pardo (2006, Section 1.2). There is some overlap of this list with the list of disparities presented in Table 2.1 of this book. As the other divergences in Pardo's (2006) list are of peripheral interest to us, we refer the reader to Pardo for a more expanded discussion of these divergences rather than repeating them here.

However, particular mention may be made here of the Bhattacharyya distance and the class of Rényi divergences; these distances have already been introduced in Section 2.3.2. Except for some special cases, these distances do not belong to the class of ϕ-divergences according to the definition presented here. However, in some cases the distance measure can be written in the form

$$D_\phi^h(d_n, f_\theta) = h(D_\phi(d_n, f))$$

where h is a real, increasing, differentiable function on the range of ϕ, where ϕ satisfies the usual properties. Such divergences have been called (h, ϕ) divergences by Menéndez et al. (1995), who have studied the properties of the corresponding estimators. The Bhattacharyya distance, the family of Rényi divergences, and the family of Sharma and Mittal divergences (Sharma and Mittal, 1977), belong to the class of (h, ϕ) divergences. See Pardo (2006, Section 1.2) for a definition of the Sharma and Mittal divergence.

Consider the problem of minimum distance estimation based on the disparity ρ_C and let G be the true, data generating distribution. Throughout the rest of the book, we will denote θ^g to be the *best fitting* value of the parameter if θ^g minimizes $\rho_C(g, f_\theta)$ over $\theta \in \Theta$. When $G = F_{\theta_0}$, so that the true distribution belongs to the model, θ_0 will be referred to as the *true* value of the parameter.

2.4   Minimum Hellinger Distance Estimation: Discrete Models

Beran (1977) considered the problem of parametric estimation based on the minimum Hellinger distance method. He established the asymptotic distribution of the minimum Hellinger distance estimator under the assumption that the distributions have densities with respect to the Lebesgue measure. This was a path-breaking paper which significantly influenced future research in this area. The existence and consistency results of the minimum Hellinger distance estimator will be discussed in this section following Beran's approach. However, Beran's asymptotic normality results for the minimum Hellinger distance estimator will be discussed in Chapter 3, where the development of the method in continuous models is described. Some of the robustness indicators considered by Beran will be discussed in Chapter 4.

Considerable simplifications over Beran's approach may be possible in the derivation of the asymptotic properties of the minimum Hellinger distance estimator in discrete models, and subsequent authors have further extended Beran's result in many ways. In this section we will briefly review the preliminary results of Beran, which establishes the consistency of the minimum Hellinger distance functional under appropriate conditions for general models. We will follow this up by presenting the additional results of Simpson (1987) and Tamura and Boos (1986) which enhance Beran's consistency results. Finally, we present Simpson's proof of the asymptotic normality of the estimator under discrete models.

2.4.1 Consistency of the Minimum Hellinger Distance Estimator

We consider the parametric setup and notation of Section 2.3.1. Let $\mathcal{G}$ be the class of all distributions having densities with respect to the dominating measure. The minimum Hellinger distance functional $T(G)$ is defined on $\mathcal{G}$ by the requirement that for every G in $\mathcal{G}$,

$$\mathrm{HD}(g, f_{T(G)}) = \inf_{\theta \in \Theta} \mathrm{HD}(g, f_\theta), \qquad (2.51)$$

where g is the density function corresponding to G, provided such a minimum exists.

Definition 2.2. We define a parametric model $\mathcal{F}$ as described in Section 1.1 to be identifiable, if for any $\theta_1, \theta_2 \in \Theta$, $\theta_1 \neq \theta_2$ implies $f_{\theta_1}(x) \neq f_{\theta_2}(x)$ on a set of positive dominating measure, where f_θ represents the density function of F_θ.

We now present Beran's proof of the consistency of the minimum Hellinger distance functional. Since the consistency result of Beran applies to a general model, here we will represent the Hellinger distance between the densities g and f as

$$\mathrm{HD}(g, f) = 2 \int (g^{1/2}(x) - f^{1/2}(x))^2 dx, \qquad (2.52)$$

with the integrals being replaced by sums when we are specifically dealing with the discrete model.

Lemma 2.6. [Beran (1977, Theorem 1)]. *Suppose that the model family is identifiable, and Θ is a compact subset of $\mathbb{R}^p$. Also suppose that $f_\theta(x)$ is continuous in θ for almost all x. Then,*

   (i) *for all $G \in \mathcal{G}$, there exists a $T(G)$ satisfying (2.51).*

   (ii) *if $T(G)$ is unique, the functional T is continuous at G in the Hellinger topology (i.e., $T(G_n) \to T(G)$ whenever $g_n \to g$ in the Hellinger metric, where g_n is the density of G_n).*

(iii) *$T(F_\theta) = \theta$, uniquely, for every $\theta \in \Theta$.*

Proof. (i) Consider a sequence of parameter values t_n such that $t_n \to t$. From the definition of the distance and Cauchy-Schwarz inequality, it follows that

$$|\mathrm{HD}(g, f_{t_n}) - \mathrm{HD}(g, f_t)| = 4 \left| \int [f_{t_n}^{1/2}(x) - f_t^{1/2}(x)] g^{1/2}(x) dx \right|$$

$$\leq 4 \left(\int \left[f_{t_n}^{1/2}(x) - f_t^{1/2}(x) \right]^2 dx \right)^{1/2}.$$

The last term converges to zero from the pointwise continuity assumption and a generalized version of the dominated convergence theorem. The above shows that $h(t) = \mathrm{HD}(g, f_t)$ is a continuous function of its argument. By the compactness of Θ, $h(t)$ achieves a minimum over $t \in \Theta$.

(ii) Suppose that the sequence $\{G_n\}$ converges to G in the Hellinger topology, i.e., $\mathrm{HD}(g_n, g) \to 0$ as $n \to \infty$. Let $h_n(t) = \mathrm{HD}(g_n, f_t)$. By writing $\theta = T(G)$ and $\theta_n = T(G_n)$, will show that $h(\theta_n) \to h(\theta)$.

Note that

$$|h_n(t) - h(t)| = 4 \left| \int [g_n^{1/2}(x) - g^{1/2}(x)] f_t^{1/2}(x) dx \right|$$

$$\leq 4 \left(\int \left[g_n^{1/2}(x) - g^{1/2}(x) \right]^2 dx \right)^{1/2}.$$

Since g_n converges to g in the Hellinger metric, the right-hand side in the above equation converges to zero. Thus

$$\lim_{n \to \infty} \sup_t |h_n(t) - h(t)| = 0. \tag{2.53}$$

Now if $h(\theta) \geq h_n(\theta_n)$, then

$$h(\theta) - h_n(\theta_n) \leq h(\theta_n) - h_n(\theta_n),$$

and if $h_n(\theta_n) \geq h(\theta)$, then

$$h_n(\theta_n) - h(\theta) \leq h_n(\theta) - h(\theta).$$

Thus, we have

$$|h_n(\theta_n) - h(\theta)| \leq |h_n(\theta_n) - h(\theta_n)| + |h_n(\theta) - h(\theta)|$$
$$\leq 2 \sup_t |h_n(t) - h(t)|. \tag{2.54}$$

Combining (2.53) and (2.54), we get

$$\lim_{n \to \infty} h(\theta_n) = h(\theta). \tag{2.55}$$

We will show that $\theta_n \to \theta$ is necessarily implied by the above. If not, by compactness of Θ there exists a subsequence $\{\theta_m\} \subset \{\theta_n\}$ such that $\theta_m \to \theta_1 \neq \theta$. By continuity of h, this implies $h(\theta_m) \to h(\theta_1)$, and by (2.55), $h(\theta_1) = h(\theta)$, which contradicts the uniqueness of the functional $T(G)$. Thus, $\theta_n \to \theta$, and the functional T is continuous in the Hellinger topology.

(iii) As the parametric family is identifiable in the sense of Definition 2.2, $\mathrm{HD}(f_\theta, f_t)$ assumes the value zero at $t = \theta$, which uniquely minimizes $\mathrm{HD}(f_\theta, f_t)$ over Θ. Thus, $T(F_\theta) = \theta$, uniquely. Thus, under the assumption that the identifiability condition holds, the existence of the minimum Hellinger distance functional $T(G)$, where G belongs to the model family, is automatic. $\qquad\square$

Remark 2.1. There is nothing special about the Hellinger distance in the proof of Lemma 2.6 (iii). At a model element, the existence of any minimum distance estimator based on a disparity satisfying the disparity conditions is automatic under the identifiability condition, and $T(F_\theta) = \theta$ uniquely for the corresponding functional $T(\cdot)$.

Remark 2.2. The compactness assumption for the parameter space in Lemma 2.6 is somewhat restrictive. Beran (1977) argued that the result also applies if the parameter space Θ can be embedded in a compact space $\bar{\Theta}$, provided the distance $HD(g, f_\theta)$, viewed as a function of θ, can be extended to a continuous function on $\bar{\Theta}$. We illustrate this point with the location-scale family

$$\left\{ \frac{1}{\sigma} f\left(\frac{x - \mu}{\sigma}\right) \right\}.$$

The parameter space $(-\infty, \infty) \times [0, \infty)$ is not compact by itself. However, consider the transformation $\mu = \tan(\beta_1)$, $\sigma = \tan(\beta_2)$, and the parameter space for (β_1, β_2) equals $(-\pi/2, \pi/2) \times (0, \pi/2)$. Therefore h can be extended to a continuous function on

$$\bar{\Theta} = [\pi/2, \pi/2] \times [0, \pi/2],$$

which is compact and the extended function attains a minimum in $\bar{\Theta}$. However, the minimum must occur in the interior of θ, since otherwise one must have $h(t) = 4$ for all $t \in \theta$, which is clearly impossible. Therefore the conclusions of the theorem remain valid for this location-scale model.

Although the above efficiently validates the use of Beran's technique for a much wider class of models than those where the parameter space is restricted to a compact set, there are examples where this structure fails to hold, particularly in multiparameter situations.

In this connection we present the following lemma under the additional conditions of Simpson (1987), which extends Beran's existence and continuity result. Under the existing setup and notation of this section, let $\mathcal{G}$ now denote the class of distributions G having densities with respect to the dominating measure, and for which

$$\inf_{\theta \in \Theta - H} HD(g, f_\theta) > HD(g, f_{\theta^*}) \tag{2.56}$$

for some compact $H \subset \Theta$ and some $\theta^* \in H$. If Θ is compact, then $H = \Theta$, and Lemma 2.6 applies with $\mathcal{G}$ containing all distributions having densities that are not singular to the model distribution.

Lemma 2.7. *Suppose that $f_\theta(x)$ is continuous in θ for each x. Then we have*

(i) *For all $G \in \mathcal{G}$, $T(G)$ exists.*

(ii) *If $T(G)$ is unique, then $T(G_n) \to T(G)$ as $n \to \infty$ when the sequence of corresponding densities g_n converge to the density g in the Hellinger metric.*

(iii) *Let the parametric family $\mathcal{F}$ be identifiable. If $G = F_\theta$ for some $\theta \in \Theta$,*
   $T(G_n) \to \theta$ as $n \to \infty$ for any sequence g_n converging to $g = f_\theta$ in the
   Hellinger metric.

Proof. Part (i) of this Lemma is obvious from Lemma 2.6 (i) and condition
(2.56). For part (ii), note that if $HD(g_n, g)$ converges to zero, $G_n \in \mathcal{G}$ eventually; consequently $T(G_n)$ exists and eventually belongs to H. Then an application of Lemma 2.6 (ii) with the parameter space restricted to H establishes
the result.

For part (iii) note that by the identifiability of the parametric family,
$T(G) = T(F_\theta) = \theta$, uniquely, and

$$\inf_{t \in \Theta - H} HD(g, f_t) = \inf_{t \in \Theta - H} HD(f_\theta, f_t) > 0$$

for any compact subset H of Θ containing θ as an interior point. Then the
result follows from part (ii).   □

Let the parametric family $\mathcal{F}$ be identifiable and be supported on $\mathcal{X} =$
$\{0, 1, \ldots, \}$. Let the true data generating distribution G be a count distribution, and suppose that the parameter space is either compact, or condi-
tion (2.56) is satisfied. It then follows that $\sum_x |d_n(x) - g(x)| \to 0$ almost
surely (see Devroye and Gyorfi, 1985, p. 10) where d_n is the density estimate
defined in Section 2.3.1. By Lemmas 2.6 and 2.7, a minimizer of $HD(g, f_\theta)$
exists. Suppose that this minimizer $T(G)$ is unique. Since

$$HD(d_n, g) = 2 \sum_x (d_n^{1/2}(x) - g^{1/2}(x))^2 \leq 2 \sum_x |d_n(x) - g(x)| \to 0, \quad (2.57)$$

the minimum Hellinger distance estimator (the minimizer of $HD(d_n, f_\theta)$ over
Θ) converges to $T(G)$ in probability.

The following example, presented by Simpson (1987), provides a case
in question where the extension of Beran's continuity condition fails and
Lemma 2.7 has to be appealed to for establishing the consistency of the minimum Hellinger distance estimator.

Example 2.1. Consider the two parameter negative binomial case. Here the
model is given by

$$f_\theta(x) = \frac{\Gamma(x + c^{-1})}{x! \Gamma(c^{-1})} \left(\frac{cm}{1 + cm} \right)^x \left(\frac{1}{1 + cm} \right)^{c^{-1}}, \quad x = 0, 1, \ldots,$$

where $\theta = (m, c), 0 < m < \infty$, and $0 \leq c < \infty$. The above model generates the
Poisson density with mean m for $c = 0$; as $m \to \infty$ with $c = 0$, the model eventually becomes singular with any fixed G, and $HD(g, f_\theta) \to 4$. But it can be
shown that $f_\theta(0) \to 1$ as $m \to \infty$ with m/c fixed, so $HD(g, f_\theta) \to 4 - 4g^{1/2}(0)$
in this case. Hence as a function of θ, $HD(g, f_\theta)$ does not extend continuously
to the limit points of Θ, and a compaction of Θ via its conformal mapping

onto a sphere would fail to satisfy the conditions under which Lemma 2.6 can be applied.

On the other hand, it can be shown that

$$\lim_{n \to \infty} \inf_{\Theta - H_n} \mathrm{HD}(g, f_\theta) = 4 - 4g^{1/2}(0),$$

with $H_n = \{\theta = (m, c) : n^{-1} \leq m \leq n, 0 \leq c \leq n\}$. Thus, in this case, all one has to show is $\mathrm{HD}(g, f_{\theta^*}) < 4 - 4g^{1/2}(0)$ for some θ^*. If g is the model family, then the condition reduces to $f_\theta(0) < 1$, which is satisfied for all θ, and one can apply Lemma 2.7 to establish the consistency of the minimum Hellinger distance estimator under the appropriate convergence and uniqueness results. In particular, if G is a count distribution, $T(G)$ is unique, and d_n is the density estimate defined in Section 2.3.1, the functional T is consistent. ∥

Let $\| \cdot \|_2$ represent the L_2 norm, and suppose that the true distribution belongs to the model family. In this case, Tamura and Boos provide a simple proof of the consistency of the minimum Hellinger distance estimator under a slightly stronger version of the identifiability condition in Definition 2.2. The result is presented below.

Lemma 2.8. *Suppose that the true distribution belongs to the model and let θ represent the true value of the parameter. Consider a sequence of densities $\{g_n\}$ such that*

$$\|g_n^{1/2} - f_\theta^{1/2}\|_2 \to 0 \tag{2.58}$$

almost surely. Let $\{G_n\}$ be the corresponding sequence of distributions. Also, for any sequence $\{\theta_n : \theta_n \in \Theta\}$, suppose that

$$\|f_{\theta_n}^{1/2} - f_\theta^{1/2}\|_2 \to 0$$

implies $\theta_n \to \theta$. In addition, suppose that the minimum Hellinger distance estimator $T(G_n)$ exists for n sufficiently large. Then $T(G_n) \to \theta$ as $n \to \infty$.

Proof. Let $\theta_n = T(G_n)$. Using the triangle inequality we get

$$\|f_{\theta_n}^{1/2} - f_\theta^{1/2}\|_2 \leq \|f_{\theta_n}^{1/2} - g_n^{1/2}\|_2 + \|g_n^{1/2} - f_\theta^{1/2}\|_2. \tag{2.59}$$

But from the definition of the minimum Hellinger distance

$$\|f_{\theta_n}^{1/2} - g_n^{1/2}\|_2 \leq \|g_n^{1/2} - f_\theta^{1/2}\|_2,$$

so that Equation (2.59) reduces to

$$\|f_{\theta_n}^{1/2} - f_\theta^{1/2}\|_2 \leq 2\|g_n^{1/2} - f_\theta^{1/2}\|_2, \tag{2.60}$$

so that the result follows from the given conditions. □

2.4.2 Asymptotic Normality of the Minimum Hellinger Distance Estimator

To derive the asymptotic normality of the minimum Hellinger distance estimator (MHDE), we impose smoothness conditions on the model. For notational simplicity, let $s_\theta = f_\theta^{1/2}$. Suppose that for θ in the interior of Θ, s_θ is twice differentiable in L_2. These conditions may be expressed as

$$\|s_t - s_\theta - \dot{s}_\theta^T(t-\theta)\|_2 = o(|t-\theta|) \tag{2.61}$$

and

$$\frac{\dot{s}_t - \dot{s}_\theta - \ddot{s}_\theta(t-\theta)}{|t-\theta|} \to 0 \tag{2.62}$$

componentwise in L_2 as $|t-\theta| \to 0$. Here $\dot{s}_\theta$ $(p \times 1)$ and $\ddot{s}_\theta$ $(p \times p)$ are the indicated first and second derivatives which are in L_2, and $|a| = \max(|a_1|, |a_2|, \ldots, |a_p|)$.

We now present Simpson's proof of the asymptotic normality of the minimum Hellinger distance estimator. First we provide a preliminary result which will be a useful tool in our future calculations.

Lemma 2.9. *For any $x \in \mathcal{X}$ with $0 < g(x) < 1$, $n^{1/4}(d_n^{1/2}(x) - g^{1/2}(x)) \to 0$ with probability 1.*

Proof. Since $0 < g(x) < 1$, we also have $0 < g(x)(1-g(x)) < 1$. Thus, by the strong law of larger numbers, $d_n(x) - g(x) \to 0$ with probability 1.

From Theorem 3 of Feller (1971, page 239), we have

$$n^{\frac{1}{2}-\epsilon}(d_n(x) - g(x)) \to 0$$

for $\epsilon > 0$. In particular for $\epsilon = 1/4$ one gets

$$n^{1/4}(d_n(x) - g(x)) \to 0 \tag{2.63}$$

with probability 1. By a Taylor series expansion we then get

$$n^{1/4}(d_n^{1/2}(x) - g^{1/2}(x)) = n^{1/4}(d_n(x) - g(x))\frac{1}{2g^{1/2}(x)} + o(n^{1/4}|d_n(x) - g(x)|),$$

so that the result follows by using Equation (2.63). $\qquad\square$

The next theorem, presented by Simpson (1987), is the primary component of the asymptotic normality proof of the minimum Hellinger distance estimator.

Theorem 2.10. *Assume the setup, conditions and definitions of Section 2.3.1, and let $X_1, \ldots, X_n$ be n independent and identically distributed observations from the true distribution G. Let the model $\mathcal{F}$ and the true distribution G be supported on $\mathcal{X} = \{0, 1, \ldots\}$. Let θ be a zero of $\nabla HD(g, f_t)$ (i.e.,*

θ solves the equation $\nabla \mathrm{HD}(g, f_t) = 0$), and suppose that condition (2.61) holds at θ. If $\dot{s}_\theta \in L_1$, then

$$-\nabla \mathrm{HD}(d_n, f_\theta) \;=\; n^{-1} \left[2 \sum_{i=1}^{n} \dot{s}_\theta(X_i) g^{-1/2}(X_i) \right] + o_p(n^{-1/2}). \quad (2.64)$$

Proof. Let

$$R_n = -\nabla \mathrm{HD}(d_n, f_\theta) - n^{-1} \left[2 \sum_{i=1}^{n} \dot{s}_\theta(X_i) g^{-1/2}(X_i) \right]$$

$$= -\nabla \mathrm{HD}(d_n, f_\theta) - 2 \sum_{x=0}^{\infty} \dot{s}_\theta(x) g^{-1/2}(x) d_n(x).$$

We will show that $n^{1/2} R_n = o_p(1)$, which will establish the result. Since θ is a zero of $\nabla \mathrm{HD}(g, f_t)$, we have

$$-4 \sum_{x=0}^{\infty} g^{1/2}(x) \dot{s}_\theta(x) = 0.$$

Then some simple algebra shows that

$$R_n = -2 \sum_{x=0}^{\infty} \dot{s}_\theta(x) g^{-1/2}(x) \left[d_n^{1/2}(x) - g^{1/2}(x) \right]^2.$$

If R_{ni} denotes the ith component of R_n, and $\dot{s}_{i\theta}(x)$ denotes the ith component of $\dot{s}_\theta(x)$, we get

$$E\left\{ n^{1/2} |R_{ni}| \right\} \le 2 \sum_{x=0}^{\infty} |\dot{s}_{i\theta}(x)| g^{-1/2}(x) \times n^{1/2} E\{d_n^{1/2}(x) - g^{1/2}(x)\}^2. \quad (2.65)$$

Let $H_n(x) = n^{1/4} \left(d_n^{1/2}(x) - g^{1/2}(x) \right)$. From Lemma 2.9, $H_n(x) \to 0$ as $n \to \infty$, and by continuity the same convergence holds for

$$H_n^2(x) = n^{1/2} (d_n^{1/2}(x) - g^{1/2}(x))^2.$$

Since $(a^{1/2} - b^{1/2})^2 \le |a - b|$ for $a, b \ge 0$,

$$
\begin{aligned}
E\{d_n^{1/2}(x) - g^{1/2}(x)\}^2 &\le E|d_n(x) - g(x)| \\
&\le [E\{d_n(x) - g(x)\}^2]^{1/2} \\
&= n^{-1/2} [g(x)(1 - g(x))]^{1/2}, \quad (2.66)
\end{aligned}
$$

and hence the summand in (2.65) is dominated by $|\dot{s}_{i\theta}(x)|$. Since $\dot{s}_\theta \in L_1$, it will follow that $E\left\{ n^{1/2} |R_{ni}| \right\} \to 0$ as $n \to 0$ for each i if we can show that $EH_n^2(x) \to 0$ as $n \to 0$. Now

$$E|H_n(x)|^{2(1+\epsilon)} \le [g(x)(1 - g(x))]^{(1+\epsilon)/2} < \infty$$

for $0 < \epsilon < 1$. Thus, $\{H_n^2(x)\}$ is uniformly integrable (Serfling, 1980, p. 14), and $EH_n^2(x) \to 0$ as $n \to \infty$.

The convergence of $n^{1/2}R_n$ to 0 follows from the convergence of $E[n^{1/2}|R_n|]$ by Markov's inequality.   □

We need one more small result before establishing the normality proof. This is given in the lemma below.

Lemma 2.11. *Suppose conditions (2.61) and (2.62) hold. Then*
$$\nabla_2 \mathrm{HD}(d_n, f_\theta) = \nabla_2 \mathrm{HD}(g, f_\theta) + o_p(1),$$
where θ is a zero of $\nabla \mathrm{HD}(g, f_t)$.

Proof. We have $\nabla_2 \mathrm{HD}(d_n, f_\theta) = -4 \sum_x d_n^{1/2}(x) \ddot{s}_\theta(x)$. Thus,
$$|\nabla_2 \mathrm{HD}(d_n, f_\theta) - \nabla_2 \mathrm{HD}(g, f_\theta)| = \left| \sum_x (d_n^{1/2}(x) - g^{1/2}(x)) \ddot{s}_\theta(x) \right|. \qquad (2.67)$$

Our result will be proved if we can show that the right-hand side the above equation tends to zero in probability. By Cauchy Schwarz inequality, the right-hand side of (2.67) is bounded by $M\{\sum_x (d_n^{1/2}(x) - g^{1/2}(x))^2\}^{1/2}$, where M is the maximum of componentwise L_2 norms of $\ddot{s}_\theta$. But by Equation (2.57)
$$\sum_x (d_n^{1/2}(x) - g^{1/2}(x))^2 \leq \sum_x |d_n(x) - g(x)| \to 0$$

as $n \to \infty$, so that the right-hand side of (2.67) converges to zero in probability.   □

With this background we are now ready to state and prove the asymptotic normality of the minimum Hellinger distance estimator.

Theorem 2.12. [Simpson (1987, Theorem 2)]. *Let the true distribution G and the model $\mathcal{F}$ be supported on $\mathcal{X} = \{0, 1, \ldots, \}$. Let $X_1, \ldots X_n$ be independent and identically distributed observations from G. Suppose that (2.61) and (2.62) hold, and that $\nabla \mathrm{HD}(g, f_t)$ has a zero θ in the interior of Θ, $\nabla_2 \mathrm{HD}(g, f_\theta)$ is nonsingular, and $\dot{s}_\theta \in L_1$. Then the weak convergence of the minimum Hellinger distance estimator $\hat{\theta}_n$ to θ implies that $n^{1/2}(\hat{\theta}_n - \theta)$ has an asymptotic p-dimensional multivariate normal distribution with mean vector zero and variance $V_\theta = [\nabla_2 \mathrm{HD}(g, f_\theta)]^{-1} I(\theta)[\nabla_2 \mathrm{HD}(g, f_\theta)]^{-1}$. In particular, if $G = F_\theta$ for some $\theta \in \Theta$, $\nabla_2 \mathrm{HD}(g, f_\theta) = I(\theta)$, so that $V_\theta = I^{-1}(\theta)$.*

Proof. Under the given conditions,
$$\nabla \mathrm{HD}(d_n, f_{\hat{\theta}_n}) = 0 = \nabla \mathrm{HD}(d_n, f_\theta) + \nabla_2 \mathrm{HD}(d_n, f_\theta)(\hat{\theta}_n - \theta) + o(|\hat{\theta}_n - \theta|).$$

Together with Lemma 2.11, this immediately gives
$$(\hat{\theta}_n - \theta) = -[\{\nabla_2 \mathrm{HD}(g, f_\theta)\}^{-1} + o_p(1)] \nabla \mathrm{HD}(d_n, f_\theta)$$

Then Theorem 2.10 and a multivariate central limit theorem yields the result. When the true distribution $G = F_\theta$ belongs to the model, direct straightforward calculations show that $V_\theta = I^{-1}(\theta)$.   □

2.5 Minimum Distance Estimation Based on Disparities: Discrete Models

Consider the setup of Section 2.3.1. Let ∇_j represent the gradient with respect to θ_j, the j-th component of the parameter vector θ. Similarly let ∇_{jk} and ∇_{jkl} represent the joint partial derivatives with respect to the corresponding parameters.

In an obvious extension of the notation of Equation (1.2), we define

$$u_{j\theta}(x) = \nabla_j(\log f_\theta(x))$$
$$u_{jk\theta}(x) = \nabla_{jk}(\log f_\theta(x))$$
$$u_{jkl\theta}(x) = \nabla_{jkl}(\log f_\theta(x)).$$

Let $\delta_n(x)$ be as defined in Equation (2.6). Let g represent the density of the true, data generating distribution G. The Pearson residual $\delta_g(x)$ corresponding to $g(x)$ will be defined by

$$\delta_g(x) = \frac{g(x) - f_\theta(x)}{f_\theta(x)}. \tag{2.68}$$

Also suppose that given the true density $g(x)$ and a disparity ρ_C, there exists a unique θ^g which minimizes the disparity $\rho_C(g, f_\theta)$, and that it solves the minimum disparity estimating equation

$$\sum_x A(\delta_g(x))\nabla f_\theta(x) = 0.$$

Assumption 2.1. Assume that the parametric family $f_\theta(x), \theta \in \Theta \subseteq \mathbb{R}^p$ has support $\mathcal{X}$ which is independent of θ, and $f_\theta(x) > 0$ for all $x \in \mathcal{X}$ and all $\theta \in \Theta$. Let the true density $g(x)$ also have the same support $\mathcal{X}$, where $g(x) > 0$ for all $x \in \mathcal{X}$.

Definition 2.3. The residual adjustment function $A(\delta)$ will be called regular, if it is twice differentiable and $A'(\delta)$ and $A''(\delta)(1+\delta)$ are bounded on $[-1, \infty)$, where $A'(\cdot)$ and $A''(\cdot)$ represent the first and second derivatives of $A(\cdot)$ with respect to its argument.

For parts of the argument in proving the consistency and the multivariate normality of the minimum distance estimator based on disparities, it helps to use the Hellinger residuals rather than the Pearson residuals. We define the Hellinger residuals as

$$\Delta_n(x) = \frac{d_n^{1/2}(x)}{f_\theta^{1/2}(x)} - 1,$$

$$\Delta_g(x) = \frac{g^{1/2}(x)}{f_\theta^{1/2}(x)} - 1.$$

Let $Y_n(x) = n^{1/2}(\Delta_n(x) - \Delta_g(x))^2$. Then the following lemma provides some bounds which are useful later in the main theorem of this chapter.

Lemma 2.13. *For any $k \in [0, 2]$ we have*

(i) $E[Y_n^k(x)] \leq n^{k/2} E[|\delta_n(x) - \delta_g(x)|]^k \leq \left[\dfrac{\{g(x)(1 - g(x))\}^{1/2}}{f_\theta(x)} \right]^k$.

(ii) $E[|\delta_n(x) - \delta_g(x)|] \leq \dfrac{2g(x)(1 - g(x))}{f_\theta(x)}$.

Proof. (i) For $a, b \geq 0$, we get $(a^{1/2} - b^{1/2})^2 \leq |a - b|$. Therefore,

$$
\begin{aligned}
E[Y_n^k(x)] &= n^{k/2} E\left[\left(\frac{d_n^{1/2}(x)}{f_\theta^{1/2}(x)} - \frac{g^{1/2}(x)}{f_\theta^{1/2}(x)} \right)^2 \right]^k \\
&\leq n^{k/2} E\left[\left| \frac{d_n(x)}{f_\theta(x)} - \frac{g(x)}{f_\theta(x)} \right| \right]^k \\
&= n^{k/2} E\left[|\delta_n(x) - \delta_g(x)| \right]^k.
\end{aligned}
$$

To prove the second inequality of (i), we use the Lyapounov's inequality

$$E[|X|^\alpha]^{1/\alpha} \leq E[|X|^\beta]^{1/\beta}$$

for $\alpha < \beta$. Thus

$$
\begin{aligned}
E\left[|\delta_n(x) - \delta_g(x)| \right]^k &\leq [E(\delta_n(x) - \delta_g(x))^2]^{k/2} \\
&= \frac{1}{f_\theta^k(x)} [E(d_n(x) - g(x))^2]^{k/2} \\
&= \frac{1}{f_\theta^k(x)} \left[\frac{g(x)(1 - g(x))}{n} \right]^{k/2},
\end{aligned}
$$

and the result follows.

(ii) By definition,

$$
\begin{aligned}
E[|\delta_n(x) - \delta_g(x)|] &= \frac{1}{f_\theta(x)} E\left[\left| \frac{1}{n} \sum_{i=1}^n \chi(X_i = x) - g(x) \right| \right] \\
&\leq \frac{1}{f_\theta(x)} \frac{1}{n} \sum_{i=1}^n E\left[|\chi(X_i = x) - g(x)| \right] \\
&= \frac{2g(x)(1 - g(x))}{f_\theta(x)}.
\end{aligned}
$$

The last equality follows from the fact that for a Bernoulli random variable X with parameter p, $E|X - p| = 2p(1 - p)$. $\qquad \square$

Next we prove the limiting result for the expectation of Y_n^k for values of k in $[0, 2)$.

Lemma 2.14. $\lim_n E[Y_n^k(x)] = 0$, *for* $k \in [0, 2)$.

Proof. By Lemma 2.9, $n^{1/4}(d_n^{1/2}(x) - g^{1/2}(x)) \to 0$ with probability 1 for each $x \in \mathcal{X}$. As $f(x) = x^2$ is a continuous function $n^{1/2}(d_n^{1/2}(x) - g^{1/2}(x))^2 \to 0$ with probability 1. Dividing by $f_\theta(x)$ we see that $Y_n(x)$ goes to zero with probability 1 for each x. Now by Lyapounov's inequality $E[Y_n^k(x)] \le E[Y_n^2(x)]^{k/2}$ for $k \in [0, 2)$. But

$$\sup_n E[Y_n^2(x)] \le \frac{g(x)(1 - g(x))}{f_\theta^2(x)}$$

from Lemma 2.13 (i), so that $\sup_n E[Y_n^k(x)]$ is bounded. The result then follows from Theorem 4.5.2 of Chung (1974). □

Lemma 2.15. [Lindsay (1994, Lemma 25)]. *Suppose that* $A(\delta)$ *is a regular RAF as described in Definition 2.3. Then there exists a finite* $B > 0$ *such that for all positive c and d, we have*

$$|A(c^2 - 1) - A(d^2 - 1) - (c^2 - d^2)A'(d^2 - 1)| \le B(c - d)^2.$$

Proof. Consider a second-order Taylor series – in c, around d – of the function within the absolute values. Note that the function is zero at $c = d$, and its first derivative with respect to c is $2cA'(c^2 - 1) - 2cA'(d^2 - 1)$, which is also zero at $c = d$. Thus, all that is needed is to show that the second derivative of the function within absolute values is bounded. This second derivative equals $4c^2 A''(c^2 - 1) + 2A'(c^2 - 1) - 2A'(d^2 - 1)$, which is bounded since $A(\delta)$ is a regular residual adjustment function. □

Let

$$a_n(x) = A(\delta_n(x) - A(\delta_g(x)) \quad \text{and} \quad b_n(x) = (\delta_n(x) - \delta_g(x))A'(\delta_g(x)).$$

Later we will need the limiting distribution of $S_{1n} = n^{1/2} \sum_x a_n(x) \nabla f_\theta(x)$, and Lemma 2.16 below shows that it is the same as the limiting distribution of $S_{2n} = n^{1/2} \sum_x b_n(x) \nabla f_\theta(x)$.

Assumption 2.2. We assume that $\sum_x g^{1/2}(x)|u_{j\theta}(x)|$ is finite for all $j = 1, 2, \ldots, p$.

Lemma 2.16. $E|S_{1n} - S_{2n}| \to 0$ *as* $n \to \infty$.

Proof. Let $\tau_n(x) = n^{1/2}|a_n(x) - b_n(x)|$. Then

$$E|S_{1n} - S_{2n}| = En^{1/2} \left| \sum_x (a_n(x) - b_n(x)) \nabla f_\theta(x) \right|$$

$$\le \sum_x E(\tau_n(x))|\nabla f_\theta(x)|. \tag{2.69}$$

But by Lemma 2.15, $\tau_n(x) \leq Bn^{1/2}(\Delta_n(x) - \Delta_g(x))^2 = BY_n(x)$. By Lemma 2.14, $E(\tau_n(x)) \to 0$. Also by Lemma 2.13 (i),

$$E(\tau_n(x)) \leq BE(Y_n(x)) \leq B\frac{g^{1/2}(x)}{f_\theta(x)},$$

so that $Bg^{1/2}(x)|u_\theta(x)|$ bounds the summand on the right-hand of Equation (2.69). The required result then follows from Assumption 2.2. An application of Markov's inequality shows that $S_{1n} - S_{2n} \to 0$ in probability. □

Corollary 2.17. *Assume that the RAF $A(\delta)$ is regular, and Assumption 2.2 holds. Then, if*

$$V = \mathrm{Var}_g[A'(\delta_g(X))u_\theta(X)] \tag{2.70}$$

is finite, we get

$$S_{1n} \xrightarrow{D} Z^* \sim N(0, V).$$

Proof. The quantity in question is $n^{1/2}\sum_x a_n(x)\nabla f_\theta(x)$. By Lemma 2.16, the asymptotic distribution of this is the same as that of $n^{1/2}\sum_x b_n(x)\nabla f_\theta(x)$, which can be written as

$$n^{1/2}\sum_x (\delta_n(x) - \delta_g(x))A'(\delta_g(x))\nabla f_\theta(x)$$

$$= n^{1/2}\sum_x (d_n(x) - g(x))A'(\delta_g(x))u_\theta(x)$$

$$= n^{1/2}\frac{1}{n}\sum_{i=1}^n \sum_x [\chi(X_i = x) - g(x)]A'(\delta_g(x))u_\theta(x)$$

$$= n^{1/2}\frac{1}{n}\sum_{i=1}^n [A'(\delta_g(X_i))u_\theta(X_i) - E_g(A'(\delta_g(X))u_\theta(X))].$$

The required result then follows from a simple application of the central limit theorem. □

When all the relevant expressions are evaluated at $\theta = \theta^g$, one gets

$$S_{1n} = n^{1/2}\sum (A(\delta_n^g(x)) - A(\delta_g^g(x)))\nabla f_{\theta^g}(x)$$

$$= n^{1/2}\sum A(\delta_n^g(x))\nabla f_{\theta^g}(x)$$

$$= -n^{1/2}\nabla\rho_C(d_n, f_\theta)|_{\theta=\theta^g}, \tag{2.71}$$

where $\delta_n^g(x) = d_n(x)/f_{\theta^g}(x) - 1$ and $\delta_g^g(x) = g(x)/f_{\theta^g}(x) - 1$. But from

Lemma 2.16, $S_{1n} = S_{2n} + o_p(1)$, so that

$$
\begin{aligned}
&- n^{1/2} \nabla \rho_C(d_n, f_\theta)|_{\theta=\theta^g} \\
&= S_{2n} + o_p(1) \\
&= n^{1/2} \left\{ \frac{1}{n} \sum_{i=1}^{n} \left[A'(\delta_g^g(X_i)) u_{\theta^g}(X_i) - E_g(A'(\delta_g^g(X)) u_{\theta^g}(X)) \right] \right\} + o_p(1).
\end{aligned}
$$
(2.72)

On the other hand, when the true distribution belongs to the model, $G = F_\theta$ for some $\theta \in \Theta$; in this case, $\theta^g = \theta$, $A'(\delta_g^g(x)) = A'(0) = 1$, so that under f_θ, Equation (2.72) reduces to

$$
\begin{aligned}
-n^{1/2} \nabla \rho_C(d_n, f_\theta) &= n^{1/2} \left[\frac{1}{n} \sum_{i=1}^{n} u_\theta(X_i) \right] + o_p(1) \\
&= Z_n(\theta) + o_p(1)
\end{aligned}
$$
(2.73)

where

$$
Z_n(\theta) = n^{-1/2} \sum_{i=1}^{n} u_\theta(X_i).
$$
(2.74)

In particular for the likelihood disparity – leading to the likelihood equation – the above relation is exact (the additional $o_p(1)$ term is absent). The relation (2.73) gives partial indication of the asymptotic efficiency of the minimum distance estimator based on the disparity ρ_C.

Going back to Lemma 2.16, the following corollary obtains the representation of the minimum Hellinger distance estimator in Theorem 2.10 of Section 2.4 as the special case of Lemma 2.16. When evaluated at the true parameter θ^g, we get, from the above lemma,

$$
\begin{aligned}
-\nabla \rho_C(d_n, f_\theta)|_{\theta=\theta^g} &= n^{-1/2} S_{1n}|_{\theta=\theta^g} \\
&= n^{-1/2} S_{2n}|_{\theta=\theta^g} + o_p(n^{-1/2}) \\
&= \sum_x (\delta_n^g(x) - \delta_g^g(x)) A'(\delta_g^g(x)) \nabla f_{\theta^g}(x) + o_p(n^{-1/2}).
\end{aligned}
$$
(2.75)

We then have the following corollary.

Corollary 2.18. *Suppose $\rho_C \equiv$ HD is the Hellinger distance in (2.10), and let θ^g be the best fitting parameter which solves $\nabla \mathrm{HD}(g, f_\theta) = 0$. Then*

$$
\begin{aligned}
-\nabla \rho_C(d_n, f_\theta)|_{\theta=\theta^g} &= -\nabla \mathrm{HD}(d_n, f_\theta)|_{\theta=\theta^g} \\
&= \frac{1}{n} \sum_{i=1}^{n} 2 \dot{s}_{\theta^g}(X_i) g^{-1/2}(X_i) + o_p(n^{-1/2}),
\end{aligned}
$$

where $s_t = f_t^{1/2}$, and $\dot{s}_t$ represents the first derivative of s_t with respect to t.

Proof. By (2.75) we have

$$-\nabla HD(d_n, f_\theta)|_{\theta=\theta^g} = \sum_x (\delta_n^g(x) - \delta_g^g(x)) A'_{HD}(\delta_g^g(x)) \nabla f_{\theta^g}(x) + o_p(n^{-1/2}).$$
(2.76)

By replacing the values of δ_n^g, δ_g^g, $A'_{HD}(\cdot)$, and by observing that $\sum_{x=0}^{\infty} \dot{s}_{\theta^g}(x) g^{1/2}(x) = 0$, we get

$$\sum_x (\delta_n^g(x) - \delta_g^g(x)) A'_{HD}(\delta_g^g(x)) \nabla f_{\theta^g}(x) = \sum_x \dot{s}_{\theta^g}(x) g^{-1/2}(x) d_n(x).$$

Writing the term on the right-hand side of the above equation as a sum of the observation index i, and replacing the same in Equation (2.76), the required result follows. Thus, the representation presented by Simpson in (2.64) is a special case of the representation in (2.75) for the Hellinger distance. □

Definition 2.4. Suppose that the densities in the parametric family $\{f_\theta : \theta \in \Theta\}$ have common support K^*. Then the true density g is called compatible with the family $\{f_\theta\}$ if K^* is also the support of g.

Suppose $X_1, \ldots, X_n$ are n independent and identically distributed observations from a discrete distribution G modeled by $\mathcal{F} = \{F_\theta : \theta \in \Theta \subseteq \mathbb{R}^p\}$ and let $\mathcal{X} = \{0, 1, \ldots, \}$. Let g and $\{f_\theta\}$ represent the corresponding densities. Consider a disparity $\rho_C(d_n, f_\theta)$, where C is the disparity generating function, and let $A(\cdot)$ be the associated residual adjustment function. Let θ^g be the best fitting value of the parameter. We make the following assumptions for the proof of our main theorem.

(A1) The model family $\mathcal{F}$ is identifiable in the sense of Definition 2.2.

(A2) The probability density functions f_θ of the model distributions have common support so that the set $\mathcal{X} = \{x : f_\theta(x) > 0\}$ is independent of θ. Also, the true distribution g is compatible with the model family $\{f_\theta\}$ of densities in the sense of Definition 2.4.

(A3) There exists an open subset ω of Θ for which the best fitting parameter θ^g is an interior point and for almost all x the density $f_\theta(x)$ admits all third derivatives of the type $\nabla_{jkl} f_\theta(x)$ for all $\theta \in \omega$.

(A4) The matrix

$$J_g = E_g[u_{\theta^g}(X) u_{\theta^g}^T(X) A'(\delta_g^g(X))] - \sum A(\delta_g^g(x)) \nabla_2 f_{\theta^g}(x)$$

is positive definite where $\delta_g^g(x) = g(x)/f_{\theta^g}(x) - 1$ and $\nabla_2 f_\theta(x)$ is the $p \times p$ matrix of second derivatives of $f_\theta(x)$ having $\nabla_{ij} f_\theta(x)$ as its (i, j)th element. Notice that J_g is the same as the matrix D defined in Theorem 2.4.

(A5) The quantities

$$\sum_x g^{1/2}(x)|u_{j\theta}(x)| \,,\, \sum_x g^{1/2}(x)|u_{j\theta}(x)||u_{k\theta}(x)| \text{ and } \sum_x g^{1/2}(x)|u_{jk\theta}(x)|$$

are bounded for all j and k and all $\theta \in \omega$.

(A6) For almost all x there exist functions $M_{jkl}(x), M_{jk,l}(x), M_{j,k,l}(x)$ that dominate, in absolute value,

$$u_{jkl\theta}(x), \quad u_{jk\theta}(x)u_{l\theta}(x) \quad \text{and} \quad u_{j\theta}(x)u_{k\theta}(x)u_{l\theta}(x)$$

for all j, k, l, and that are uniformly bounded in expectation with respect to g and f_θ for all $\theta \in \omega$.

(A7) The RAF $A(\delta)$ is regular in the sense of Definition 2.3, and K_1 and K_2 represent the bounds of $A'(\delta)$ and $A''(\delta)(1 + \delta)$ respectively.

Theorem 2.19. [Lindsay (1994, Theorem 33)]. *Suppose that Assumptions (A1)–(A7) hold. Then there exists a consistent sequence θ_n of roots to the minimum disparity estimating equations in (2.39). Also the asymptotic distribution of $n^{1/2}(\theta_n - \theta^g)$ is p-dimensional multivariate normal with mean vector 0 and covariance matrix $J_g^{-1}V_g J_g^{-1}$, where V_g is the quantity defined in (2.70) evaluated at $\theta = \theta^g$.*

Proof. To prove consistency, we will use the arguments of Lehmann (1983, page 430). Consider the behavior of $\rho_C(d_n, f_\theta)$ on a sphere Q_a which has radius a and center at θ^g. We will show that for a sufficiently small a, the probability tends to 1 that $\rho_C(d_n, f_\theta) > \rho_C(d_n, f_{\theta^g})$ for θ on the surface of Q_a, so that the disparity has a local minimum with respect to θ in the interior of Q_a. At a local minimum the estimating equations must be satisfied. Therefore, for any $a > 0$ sufficiently small, the minimum disparity estimating equations have a solution θ_n within Q_a with probability tending to 1 as $n \to \infty$.

Taking a Taylor series expansion of $\rho_C(d_n, f_\theta)$ about $\theta = \theta^g$ we get

$$
\begin{aligned}
\rho_C&(d_n, f_{\theta^g}) - \rho_C(d_n, f_\theta) \\
&= -\Big[\sum_j (\theta_j - \theta_j^g)\nabla_j \rho_C(d_n, f_\theta)\Big|_{\theta=\theta^g} \\
&\quad + \frac{1}{2}\sum_{j,k}(\theta_j - \theta_j^g)(\theta_k - \theta_k^g)\nabla_{jk}\rho_C(d_n, f_\theta)\Big|_{\theta=\theta^g} \\
&\quad + \frac{1}{6}\sum_{j,k,l}(\theta_j - \theta_j^g)(\theta_k - \theta_k^g)(\theta_l - \theta_l^g)\nabla_{jkl}\rho_C(d_n, f_\theta)\Big|_{\theta=\theta^*}\Big] \\
&= S_1 + S_2 + S_3 \ (\text{say}), \hspace{3cm} (2.77)
\end{aligned}
$$

where θ^* lies on the line segment joining θ^g and θ; θ_j and θ_j^g represent the j-th component of θ and θ^g respectively. We will inspect the linear, quadratic, and cubic terms one by one and determine their proper limits.

For the linear term S_1 in (2.77), we have

$$\nabla_j \rho_C(d_n, f_\theta)|_{\theta=\theta^g} = -\sum_x A(\delta_n^g(x)) \nabla_j f_{\theta^g}(x) \tag{2.78}$$

where $\delta_n^g(x)$ is $\delta_n(x)$ evaluated at $\theta = \theta^g$, and we will show that the right-hand side converges to $-\sum_x A(\delta_g^g(x)) \nabla_j f_{\theta^g}(x)$ by showing that the difference converges to 0 in probability. Since $A'(\delta)$ is bounded by K_1, the absolute value of the difference is bounded by

$$K_1 \sum_x |\delta_n^g(x) - \delta_g^g(x)| |\nabla_j f_{\theta^g}(x)|. \tag{2.79}$$

We will show that the expected value of the above quantity goes to zero. By Lemma 2.13 (i), $E[|\delta_n^g(x) - \delta_g^g(x)|]$ is bounded above by $n^{-1/2}$ times a finite quantity. Thus, $E[|\delta_n^g(x) - \delta_g^g(x)|] \to 0$ as $n \to \infty$. Again from the bound in Lemma 2.13 (ii), we get

$$E[K_1 \sum_x |\delta_n^g(x) - \delta_g^g(x)| |\nabla f_{\theta^g}(x)|] \le 2K_1 \sum_x g^{1/2}(x) |u_{\theta^g}(x)|,$$

and the right-hand side is finite by assumption. Thus, by the dominated convergence theorem, the expectation of the quantity in (2.79) goes to zero. By Markov's inequality, the quantity in (2.79) itself goes to zero in probability. Thus,

$$\sum_x A(\delta_n^g(x)) \nabla f_{\theta^g}(x) \to \sum_x A(\delta_g^g(x)) \nabla f_{\theta^g}(x), \tag{2.80}$$

in probability. But the quantity on the right-hand side of (2.80) is zero by the definition of the minimum disparity estimating equations. Thus, with probability tending to 1, $|S_1| < pa^3$, where p is the dimension of θ and a is the radius of the sphere Q_a.

For the quadratic term S_2 in (2.77), we have

$$\nabla_{jk} \rho_C(d_n, f_\theta)_{\theta=\theta^g} = -[\nabla_k \sum_x A(\delta_n(x)) \nabla_j f_\theta(x)|_{\theta=\theta^g}].$$

We will show that $\nabla_k \sum_x A(\delta_n(x)) \nabla_j f_\theta(x)|_{\theta=\theta^g}$ converges to $-J_g^{jk}$, the negative of the (j, k)-th term of J_g, with probability tending to 1 so that $2S_2$ converges to a negative definite quadratic form. The first term of $\nabla_k \sum_x A(\delta_n(x)) \nabla_j f_\theta(x)|_{\theta=\theta^g}$ equals

$$-\sum_x A'(\delta_n^g(x))(1 + \delta_n^g(x)) u_{j\theta^g}(x) u_{k\theta^g}(x) f_{\theta^g}(x) \tag{2.81}$$

which will be shown to converge to the term

$$-\sum_x A'(\delta_g^g(x))(1 + \delta_g^g(x)) u_{j\theta^g}(x) u_{k\theta^g}(x) f_{\theta^g}(x). \tag{2.82}$$

Consider the absolute difference between the two terms. By doing a one term Taylor series expansion of the difference

$$\left| A'(\delta_n^g(x))(1 + \delta_n^g(x)) - A'(\delta_g^g(x))(1 + \delta_g^g(x)) \right|$$

in δ_n around δ_g we see that

$$(K_1 + K_2) \sum_x |\delta_n^g(x) - \delta_g^g(x)| |u_{j\theta^g}(x)| |u_{k\theta^g}(x)| f_{\theta^g}(x)$$

bounds the absolute difference between (2.81) and (2.82). Since by assumption

$$\sum_x g^{1/2}(x) |u_{j\theta^g}(x)| |u_{k\theta^g}(x)| < \infty,$$

the absolute difference between (2.81) and (2.82) goes to zero in probability by an argument similar to that of the linear term.

Next we will show that

$$\sum_x A(\delta_n^g(x)) \nabla_{jk} f_{\theta^g}(x) \text{ converges to } \sum_x A(\delta_g^g(x)) \nabla_{jk} f_{\theta^g}(x).$$

The absolute difference is bounded by $K_1 \sum_x |\delta_n^g(x) - \delta_g^g(x)| |\nabla_{jk} f_{\theta^g}(x)|$. But note that $\frac{\nabla_{jk} f_\theta(x)}{f_\theta(x)} = u_{jk\theta}(x) + u_{j\theta}(x) u_{k\theta}(x)$. Hence by assumption this difference goes to zero in probability by a similar argument. Thus, $\nabla_k \sum_x A(\delta_n(x)) \nabla_j f_\theta(x)|_{\theta=\theta^g}$ converges to $-J_g^{jk}$. Therefore

$$2S_2 = \sum_{j,k} \left\{ [\nabla_k \sum_x A(\delta_n(x)) \nabla_j f_{\theta^g}(x)] - [-J_g^{jk}] \right\} (\theta_j - \theta_j^g)(\theta_k - \theta_k^g)$$

$$+ \sum_{j,k} \left\{ -J_g^{jk}(\theta_j - \theta_j^g)(\theta_k - \theta_k^g) \right\}.$$

The absolute value of the first term is less than $p^2 a^3$ with probability tending to 1. The second term is a negative definite quadratic form in the variables $(\theta_j - \theta_j^g)$. Letting λ_1 be the largest eigenvalue of J_g, the quadratic form is less than $\lambda_1 a^2$. Combining the two terms, we see that there exists $c > 0$ and $a_0 > 0$, such that for $a < a_0$, $S_2 < -ca^2$ with probability tending to 1.

For the cubic term S_3 in (2.77), we have

$$\nabla_{jkl} \rho_C(d_n, f_\theta)|_{\theta=\theta^*} = -\nabla_{kl} \sum_x A(\delta_n(x)) \nabla_j f_\theta(x)|_{\theta=\theta^*}.$$

In this case, the terms are calculated at θ^* (not θ^g). We will show that the quantities are bounded in absolute value. Let us look at the terms of $\nabla_{kl} \sum_x A(\delta_n(x)) \nabla_j f_\theta(x)|_{\theta=\theta^*}$ one by one. We use the notation

$$\delta_n^*(x) = \frac{d_n(x) - f_{\theta^*}(x)}{f_{\theta^*}(x)}.$$

$\sum_x A''(\delta_n^*(x))(\delta_n^*(x) + 1)^2 u_{j\theta^*}(x) u_{k\theta^*}(x) u_{l\theta^*}(x) f_{\theta^*}(x)$: The sum will be bounded in absolute value by a constant times $|\delta_n^*(x) + 1| M_{j,k,l}(x) f_{\theta^*}(x) = d_n(x) M_{j,k,l}(x)$. By the central limit theorem, the sum converges to the expectation of $M_{j,k,l}(X)$ (with respect to the density g). Thus, by assumption this term is bounded.

$\sum_x A'(\delta_n^*(x))(\delta_n^*(x) + 1) u_{j\theta^*}(x) u_{k\theta^*}(x) u_{l\theta^*}(x) f_{\theta^*}(x)$: The sum will again be bounded in absolute value by a constant times $|\delta_n^*(x) + 1| M_{j,k,l}(x) f_{\theta^*}(x) = d_n(x) M_{j,k,l}(x)$. The boundedness of this term therefore follows as in the case of the previous term.

$\sum_x A'(\delta_n^*(x))(\delta_n^*(x) + 1) u_{jk\theta^*}(x) u_{l\theta^*}(x) f_{\theta^*}(x)$: The sum will be bounded in absolute value by a constant times $|\delta_n^*(x) + 1| M_{jk,l}(x) f_{\theta^*}(x) = d_n(x) M_{jk,l}(x)$. Its boundedness follows similarly.

$\sum_x A(\delta_n^*(x)) \nabla_{jkl} f_{\theta^*}(x)$: We can write

$$|A(\delta)| = | \int_0^\delta A'(x) dx | \le K_1 |\delta|,$$

so that

$$|A(\delta_n^*(x))| \le K_1 |d_n(x)/f_{\theta^*}(x) - 1|$$
$$\le K_1 |d_n(x)/f_{\theta^*}(x) + 1| = \frac{K_1}{f_{\theta^*}(x)} (d_n(x) + f_{\theta^*}(x)).$$

Also note that

$$\left| \frac{\nabla_{jkl} f_\theta(x)}{f_\theta(x)} \right| = u_{jkl\theta}(x) + u_{jk\theta}(x) u_{l\theta}(x) + u_{jl\theta}(x) u_{k\theta}(x)$$
$$+ u_{j\theta}(x) u_{kl\theta}(x) + u_{j\theta}(x) u_{k\theta}(x) u_{l\theta}(x).$$

Thus, the summand is bounded by $K_1 |d_n(x) + f_{\theta^*}(x)| M(x)$, where

$$M(x) = M_{jkl}(x) + M_{jk,l}(x) + M_{jl,k}(x) + M_{j,kl}(x) + M_{j,k,l}(x),$$

so that this sum is also bounded with probability tending to 1. Hence we have $|S_3| < ba^3$ on the sphere Q_a with probability tending to 1.

Combining the three inequalities, we see that

$$\max(S_1 + S_2 + S_3) < -ca^2 + (b + p)a^3,$$

which is less than zero for $a < c/(b + p)$.

Thus, for any sufficiently small a there exists a sequence of roots $\theta_n = \theta_n(a)$ to the minimum disparity estimating equations such that $P(||\theta_n - \theta^g|| < a)$ converges to 1, where $||\cdot||_2$ represents the L_2 norm. It remains to show that we can determine such a sequence independently of a. Let θ_n^* be the root which is closest to θ^g. This exists because the limit of a sequence of roots is again a root by the continuity of the disparity as a function of the parameter. This completes the proof of the consistency part.

For the multivariate normality, let us expand $\sum_x A(\delta_n(x))\nabla_j f_\theta(x)$ about $\theta = \theta^g$ to obtain

$$
\sum_x A(\delta_n(x))\nabla_j f_\theta(x)
$$
$$
= \sum_x A(\delta_n^g(x))\nabla_j f_{\theta^g}(x)
$$
$$
+ \sum_k (\theta_k - \theta_k^g)\nabla_k \sum_x A(\delta_n(x))\nabla_j f_\theta(x)|_{\theta=\theta^g}
$$
$$
+ \frac{1}{2}\sum_{k,l}(\theta_k - \theta_k^g)(\theta_l - \theta_l^g)\nabla_{kl}\sum_x A(\delta_n(x))\nabla_j f_\theta(x)|_{\theta=\theta'}
$$

where $\theta = \theta'$ is a point on the line segment connecting θ and θ^g. Next we will replace θ by θ_n where θ_n is a solution of the minimum disparity estimating equation, which can be assumed to be consistent by the previous part. The left-hand side of the above equation then becomes zero and the equation can be rewritten as

$$
-n^{1/2}\sum_x A(\delta_n^g(x))\nabla_j f_{\theta^g}(x)
$$
$$
= n^{1/2}\sum_k (\theta_{nk}-\theta_k^g)\Big[\nabla_k\sum_x A(\delta_n(x))\nabla_j f_\theta(x)|_{\theta=\theta^g}
$$
$$
+ \frac{1}{2}\sum_l(\theta_{nl}-\theta_l^g)\nabla_{kl}\sum_x A(\delta_n(x))\nabla_j f_\theta(x)|_{\theta=\theta'}\Big]
$$

$$(2.83)$$

But

$$
n^{1/2}\sum_x A(\delta_n^g(x))\nabla_j f_{\theta^g}(x) = n^{1/2}\sum_x \big\{A(\delta_n^g(x)) - A(\delta_g^g(x))\big\}\nabla_j f_{\theta^g}(x)
$$

has a multivariate normal distribution with mean zero and variance V_g, by Corollary 2.17. As argued in the consistency part, the first term within the bracketed quantity in the right-hand side of (2.83) converges to J_g with probability tending to 1, while the second term within the brackets is an $o_p(1)$ term. Then it follows from Lehmann (1983, Lemma 4.1) that the asymptotic distribution of $n^{1/2}(\theta_n - \theta^g)$ is multivariate normal with mean zero and covariance matrix $J_g^{-1}V_g J_G^{-1}$. □

Corollary 2.20. *Assume the conditions of Theorem 2.19. In addition, suppose that the true distribution belongs to the model ($G = F_\theta$ for some $\theta \in \Theta$). If θ_n represents the minimum distance estimator corresponding to a disparity satisfying the conditions in Definition 2.1, then $n^{1/2}(\theta_n - \theta)$ has an asymptotic normal distribution with mean vector 0 and covariance matrix $I^{-1}(\theta)$, where $I(\theta)$ is the Fisher information about θ in f_θ.*

Proof. When $G = F_\theta$, we get $\theta^g = T(G) = T(F_\theta) = \theta$. In this case, $\delta_g(x) = 0$, $A(\delta_g(x)) = 0$, $A'(\delta_g(x)) = 1$ and $J_g = I(\theta)$. Also $V_g = I(\theta)$, so that one gets

$$J_g^{-1} V_g J_g^{-1} = I^{-1}(\theta)$$

and the result holds. □

When the model is true, i.e., $G = F_\theta$ for some $\theta \in \Theta$, Equations (2.71), (2.72) and (2.73) show that the left-hand side of Equation (2.83) equals $Z_n(\theta) + o_p(1)$. In addition, since the bracketed quantity on the right-hand side of Equation (2.83) now converges in probability to $I(\theta)$, Equation (2.83) leads to the relation

$$\theta_n = \theta + n^{-1/2} I^{-1}(\theta) Z_n(\theta) + o_p(n^{-1/2}), \tag{2.84}$$

as one would get when the estimator θ_n is first-order efficient, where $Z_n(\theta)$ is as in Equation (2.74).

Corollary 2.21. *The influence function approximation (1.14) is valid for the minimum disparity functionals.*

Proof. From Equation (2.71) and Lemma 2.16, we get

$$-n^{1/2} \nabla \rho_C(d_n, f_\theta)|_{\theta=\theta^g} = n^{12} \sum (\delta_n^g - \delta_g^g) A'(\delta_g^g) \nabla f_{\theta^g} + o_p(1).$$

Using Equation (2.83), it then follows,

$$
\begin{aligned}
n^{1/2}(\hat{\theta}_n - \theta^g) &= -n^{1/2} J_g^{-1} \nabla \rho_C(d_n, f_\theta)|_{\theta=\theta^g} + o_p(1). \\
&= n^{1/2} J_g^{-1} \sum (\delta_n^g - \delta_g^g) A'(\delta_g^g) \nabla f_{\theta^g} + o_p(1). \\
&= n^{1/2} J_g^{-1} \sum (d_n - g) A' (g/f_{\theta^g} - 1) u_{\theta^g} + o_p(1). \\
&= n^{1/2} \left[\frac{1}{n} \sum_{i=1}^{n} T'(X_i) \right] + o_p(1) \tag{2.85}
\end{aligned}
$$

where

$$T'(y) = J_g^{-1} \left[A'(\delta_g^g(y)) u_{\theta^g}(y) - E_g(A'(\delta_g^g(X)) u_{\theta^g}(X)) \right],$$

and J_g and δ_g^g are as defined in Assumption (A4) of this section. Note that the form of $T'(y)$ above is exactly same as the one obtained in Theorem 2.4. Thus, the linearization of the minimum distance estimators based on the influence function approximation (1.14) holds. □

Remark 2.3. The most important component of the consistency and the asymptotic normality proofs of Theorem 2.19 are the convergences of the linear, quadratic, and cubic terms of the derivatives of the distance in its Taylor series expansion. In fact, once the three convergences have been established, the remaining steps in the proof of the theorem above are routine. In the subsequent chapters when we undertake consistency and normality proofs of different variants of our minimum distance estimator, we will simply work out the proofs of these three components. When the true distribution g belongs to the model, i.e., $g = f_{\theta_0}$ and $\theta^g = \theta_0$ for some $\theta_0 \in \Theta$, these convergences may be stated, under the assumptions and notation of Theorem 2.19, as:

1. In case of the linear term one has the convergence result

$$\nabla_j \rho_C(d_n, f_\theta)|_{\theta=\theta_0} = -\sum_x d_n(x) u_{j\theta_0}(x) + o_p(n^{-1/2}). \tag{2.86}$$

   This provides the key distributional result in the asymptotic normality of the minimum distance estimator. This result has been encountered several times in this chapter; see, for example, Equation (2.73). Theorem 2.19 proves the more general version of the result for an arbitrary density g not necessarily in the model.

2. For the quadratic term, one has the convergence

$$\nabla_{jk} \rho_C(d_n, f_\theta)|_{\theta=\theta_0} = \sum_x f_{\theta_0}(x) u_{j\theta_0}(x) u_{k\theta_0}(x) + o_p(1) = I_{jk}(\theta_0) + o_p(1),$$

$$\tag{2.87}$$

   so that the $p \times p$ matrix of second derivatives of the disparity converges to the Fisher information matrix $I(\theta_0)$. Theorem 2.19 proves the more general convergence to J_g, which reduces to $I(\theta_0)$ under model conditions.

3. In case of the cubic term there exists a finite positive constant γ such that, with probability tending to 1,

$$\left| \nabla_{jkl} \rho_C(d_n, f_\theta)|_{\theta=\theta^*} \right| < \gamma, \tag{2.88}$$

   where θ^* lies on the line segment joining θ_0 and $\hat{\theta}_n$, the minimum disparity estimator.

2.6 Some Examples

Example 2.2. Here we consider a chemical mutagenicity experiment. These data were analyzed previously by Simpson (1987). The details of the experimental protocol are available in Woodruff et al. (1984). In a sex linked recessive lethal test in Drosophila (fruit flies), the experimenter exposed groups of male flies to different doses of a chemical to be screened. Each male was then mated with unexposed females. Sampling 100 daughter flies from each male (roughly), the number of daughters carrying a recessive lethal mutation on

TABLE 2.2
Fits of the Poisson model to the Drosophila data using several estimation methods: First experimental run.

| | Recessive lethal count | | | | | | |
	0	1	2	3	4	≥ 5	$\hat{\theta}$
Observed	23	3	0	1	1	0	
LD	19.59	7.00	1.25	0.15	0.01	-	0.357
LD + D	24.95	2.88	0.17	0.01	-	-	0.115
HD	24.70	3.09	0.19	0.01	-	-	0.125
$PD_{-0.9}$	26.17	1.77	0.06	-	-	-	0.068
PCS	13.89	9.74	3.42	0.80	0.14	0.02	0.701
NED	24.79	3.02	0.18	0.01	-	-	0.122
$BWHD_{1/3}$	21.44	5.73	0.76	0.07	-	-	0.267
SCS	24.87	2.95	0.18	0.01	-	-	0.119
$BWCS_{0.2}$	24.30	3.45	0.24	0.01	-	-	0.142
$GKL_{1/3}$	24.73	3.07	0.19	0.01	-	-	0.124
$RLD_{1/3}$	24.92	2.90	0.17	0.01	-	-	0.117

TABLE 2.3
Fits of the Poisson model to the Drosophila data using several estimation methods: Second experimental run.

| | Recessive lethal count | | | | | | |
	0	1	2	3	4	≥ 5	$\hat{\theta}$
Observed	23	7	3	0	0	1 (91)	
LD	1.60	4.88	7.47	7.61	5.82	6.62	3.0588
LD + D	22.93	9.03	1.78	0.23	0.02	-	0.3939
HD	23.63	8.59	1.56	0.19	0.02	-	0.3637
$PD_{-0.9}$	25.79	7.13	0.98	0.09	0.01	-	0.2763
PCS	-	-	-	-	-	34	32.5649
NED	22.85	9.08	1.80	0.24	0.02	-	0.3973
$BWHD_{1/3}$	22.99	9.00	1.76	0.23	0.02	-	0.3913
SCS	23.24	8.84	1.68	0.21	0.02	-	0.3805
$BWCS_{0.2}$	22.58	9.24	1.89	0.26	0.03	-	0.4094
$GKL_{1/3}$	23.22	8.85	1.69	0.21	0.02	-	0.3813
$RLD_{1/3}$	23.75	8.52	1.53	0.18	0.02	-	0.3588

the X chromosome was noted. The data set consisted of the observed frequencies of males having $0, 1, 2, \ldots$ recessive lethal daughters. For our purpose, we consider two specific experimental runs, those on day 28 and the second run of day 177. In this example, we will refer to them as the first and the second experimental runs. The data are presented in Tables 2.2 and 2.3.

Poisson models are fitted to the data for both experimental runs using several different methods of parameter estimation within our minimum distance class. A quick look at the observed frequencies for the two experimental runs reveals that there are two mild outliers in the first experimental run. In comparison the second experimental run contains a huge outlier – an exceptionally large count – where one male is reported to have produced 91 daughters with the recessive lethal mutation. Thus, between the fitted models of these two experimental runs, all kinds of robust behavior (or lack thereof) of the different minimum distance techniques can be demonstrated.

In each of Tables 2.2 and 2.3, the expected frequencies corresponding to the different methods are provided in the body of the table, while the distances (together with the tuning parameters) and the parameter estimates are described in the first and the last column of the table respectively. Thus, the expected frequencies and the estimator for the LD row are based on full data maximum likelihood; the LD + D row represents the results of fitting the model by the method of maximum likelihood after a qualitative deletion of the outlier(s), i.e., removing the observations at 3 and 4 in case of the first experimental run, and the observation at 91 for the second experimental run. A '-' represents an expected frequency smaller than 0.01. All the abbreviations of the distance names are as described in this chapter.

Several things deserve mention. First we look at Table 2.2 and enumerate the striking observations.

1. The difference between the maximum likelihood estimate (minimum LD estimate) and the outlier deleted maximum likelihood estimate is substantial.

2. Other than the maximum likelihood estimate, the minimum PCS estimate, and, to lesser extent the minimum $BWHD_{1/3}$ estimate appear to be significantly influenced by the outlying values. In terms of robustness, the minimum PCS estimate is clearly the worst, by far, among all the minimum distance estimates presented in this example.

3. All the other estimates in our list successfully withstand the effect of the outliers. Each of these estimates provide an excellent fit to the first three cells of the observed data while effectively ignoring the large values.

4. The set of estimates which effectively discount the outlying observations include the minimum $BWCS_{0.2}$ estimate. Interestingly, the latter estimator has a positive estimation curvature ($A_2 = 0.4$) and by itself that would have predicted a nonrobust outcome in this case. That it does not happen demonstrates that while the estimation curvature is a useful local measure, it does not necessarily capture or represent the full global characteristics of a particular distance. In Chapter 4 we will see that the graphical interpretation based on combined residuals can give further insight on the robustness of some of the estimators not captured by the estimation curvature.

5. The minimum $PD_{-0.9}$ estimate appears to be the most conservative

in our list of robust estimates, and seems to downweight not just the outliers, but some of the more legitimate values as well. While there is no robustness issue here, this is indicative of another problem – that involving inliers – which we will discuss in Chapter 6.

The observations in Table 2.3 are generally similar. However, in this case, the large observation is a wildly discrepant value and not just a mild outlier. All our minimum distance estimators apart from the minimum LD estimator and the minimum PCS estimator are entirely successful in effectively ignoring this observation. These include both the $BWHD_{1/3}$ and the $BWCS_{0.2}$ estimators. This demonstrates that while sometimes a robust minimum distance method may provide a tentative treatment for a marginal outlier, an extreme outlier is solidly dealt with.   ‖

Example 2.3. The data set for this example involves the incidence of peritonitis for 390 kidney patients. The data, presented in Table 2.4, were provided by Professor Peter W. M. John (personal communication) of the Department of Mathematics, University of Texas at Austin, USA. The observed frequencies resulting from the number of cases of peritonitis are reported at the top of the table. A visual inspection suggests that a geometric distribution with parameter θ (success probability) around 0.5 may fit the data well. We fit a geometric model with parameter θ to this data using several of our minimum distance methods. In this case, the estimates do not show any dramatic outlier effect. The two observations at 10 and 12 are mild to moderate outliers. But unlike Table 2.2, the sample size in this case is substantially higher, so that the relative impact of these moderate outliers are expected to be less severe. Indeed a comparison of the $\hat{\theta}$ column (the estimates for the full data) with the $\hat{\theta}_D$ column (the estimates for the cleaned data after removing the two outliers) shows that the full data estimates and the outlier deleted estimates are fairly close for practically all the methods, and even for the LD and the $BWHD_{1/3}$ case the impacts are relatively minor. However, the outliers do appear to have a fair influence on the minimum PCS estimate, again outlining its robustness problems. The minimum $PD_{-0.9}$ estimate, on the other hand, is still highly conservative and appears to drag the estimate the other way, underscoring the inlier problem once again.   ‖

TABLE 2.4

Observed frequencies of the number of cases of peritonitis for each of 390 kidney patients, together with the expected frequencies under different estimation methods for the geometric model.

	Number of cases													$\hat{\theta}$	$\hat{\theta}_D$
	0	1	2	3	4	5	6	7	8	9	10	11	≥ 12		
Observed	199	94	46	23	17	4	4	1	0	0	1	0	1		
LD	193.5	97.5	49.1	24.7	12.5	6.3	3.2	1.6	0.8	0.4	0.2	0.1	0.1	0.496	0.509
PD$_{-0.9}$	212.4	96.7	44.1	20.1	9.1	4.2	1.9	0.9	0.4	0.2	0.1	-	-	0.544	0.551
PCS	179.8	96.9	52.2	28.2	15.2	8.2	4.4	2.4	1.3	0.7	0.4	0.2	0.2	0.461	0.501
NED	196.4	97.5	48.4	24.0	11.9	5.9	2.9	1.5	0.7	0.4	0.2	0.1	0.1	0.504	0.507
HD	199.1	97.5	47.7	23.4	11.4	5.6	2.7	1.3	0.7	0.3	0.2	0.1	0.1	0.510	0.518
BWHD$_{1/3}$	194.3	97.5	48.9	24.6	12.3	6.2	3.1	1.6	0.8	0.4	0.2	0.1	0.1	0.498	0.510
BWCS$_{0.2}$	193.0	97.5	49.2	24.9	12.6	6.3	3.2	1.6	0.8	0.4	0.2	0.1	0.1	0.495	0.505
GKL$_{1/3}$	197.1	97.5	48.2	23.8	11.8	5.8	2.9	1.4	0.7	0.3	0.2	0.1	0.1	0.506	0.512
RLD$_{1/3}$	197.8	97.5	48.0	23.7	11.7	5.7	2.8	1.4	0.7	0.3	0.2	0.1	0.1	0.507	0.509
SCS	197.8	97.5	48.0	23.7	11.7	5.8	2.8	1.4	0.7	0.3	0.2	0.1	0.1	0.507	0.511

3

Continuous Models

3.1   Introduction

In the previous chapter we have considered models having densities with respect to the counting measure. In this section we will consider the case of continuous random variables. Let $\mathcal{G}$ represent the class of all distributions having densities with respect to the Lebesgue measure. We will assume that the true, data generating distribution G and the model family $\mathcal{F} = \{F_\theta : \theta \in \Theta \subseteq \mathbb{R}^p\}$ belong to $\mathcal{G}$. Suppose that G and F_θ have densities g and f_θ with respect to the Lebesgue measure.

Let $X_1, \ldots, X_n$ be a random sample from the distribution G which is modeled by $\mathcal{F}$, and we wish to estimate the value of the model parameter θ. As in the case with discrete models, our aim here is to estimate the unknown parameter θ by choosing the model density which gives the closest fit to the data. Unlike the discrete case, however, this poses an immediate challenge; the data are discrete, but the model is continuous, so now there is an obvious incompatibility of measures in constructing a distance between the two. One cannot simply use relative frequencies to represent a nonparametric density estimate of the true data generating distribution in this case.

One strategy for constructing a distance in this case can be to consider a histogram of fixed bin width, say h. If the support of the random variable in question is the real line, one would need a countably infinite sequence of such bins to cover the entire support. One can then compute the empirical probabilities for the bins, and minimize their distance from the corresponding model based bin probabilities. This structure can be routinely extended to multidimensions. Usually, though, an artificial discretization of this type would entail a loss of information.

Instead of discretizing the model, another approach could be to construct a continuous density estimate using some appropriate nonparametric density estimation method such as the one based on kernels. In this case, let

$$g_n^*(x) = \frac{1}{n} \sum_{i=1}^{n} K(x, X_i, h_n) = \int K(x, y, h_n) dG_n(y) \qquad (3.1)$$

denote a nonparametric kernel density estimator where $K(x, y, h_n)$ is a smooth kernel function with bandwidth h_n and G_n is the empirical distribution function as obtained from the data. Very often the kernel is chosen as a symmetric density with scale h_n, i.e.,

$$K(x, X_i, h_n) = \frac{1}{h_n} w \left(\frac{x - X_i}{h_n} \right)$$

where $w(\cdot)$ is a symmetric nonnegative function satisfying

$$\int_{-\infty}^{\infty} w(x) dx = 1.$$

We can then estimate θ by minimizing a distance based on disparities defined as

$$\rho_C(g_n^*, f_\theta) = \int C(\delta(x)) f_\theta(x) dx,$$

where the Pearson residual $\delta(x)$ now equals

$$\delta(x) = \frac{g_n^*(x) - f_\theta(x)}{f_\theta(x)}.$$

Under differentiability of the model, the estimating equation now has the form

$$-\nabla \rho_C(g_n^*, f_\theta) = \int_x A(\delta(x)) \nabla f_\theta(x) dx = 0, \tag{3.2}$$

where the residual adjustment function $A(\delta)$ is as described in Equation (2.38). Under appropriate regularity conditions, the solution of this equation represents the minimum distance estimator (based on the disparity ρ_C). In spirit, the rest of the estimation procedure is similar to the discrete case. The residual adjustment function $A(\delta)$ and the disparity generating function $C(\delta)$ continue to provide the same interpretation as in the discrete case in terms of controlling probabilistic outliers.

In practice, however, the addition of the kernel density estimation process leads to substantial difficulties. The theoretical derivation of the asymptotic normality of the minimum distance estimators based on disparities and the description of their other asymptotic properties are far more complex in this case. However, under suitable conditions, the minimum distance estimators based on disparities continue to be first order efficient at the model. In this chapter we present the different approaches in which this problem has been addressed by different authors. Historically, the minimum Hellinger distance estimator remains the first estimator within the class of disparities to be developed for the dual purpose of robustness and asymptotic efficiency. Beran (1977) considered minimum Hellinger distance estimation in continuous models. As mentioned before, his contribution significantly influenced future research in this area. Many authors have considered other aspects of this work in continuous models; see Stather (1981), Tamura and Boos (1986), Simpson (1989b), Eslinger and Woodward (1991), Basu and Lindsay (1994), Cao et al. (1995) and Toma (2008). Wu and Karunamuni (2009) and Karunamuni and Wu (2009) have applied and extended the work of Beran (1977) to semiparametric models. Some researchers have considered the extension of the techniques based on the Hellinger distance to other disparities; see Basu, Sarkar, Vidyashankar (1997), Park and Basu (2003), Bhandari, Basu and Sarkar (2006), Broniatowski and Leorato (2006) and Broniatowski and Keziou (2009). Park and Basu (2004) derived general results for a subclass of disparities under certain conditions.

Donoho and Liu (1988a) argued that minimum distance estimators are "automatically robust" in the sense that the minimum distance functional based on a particular metric changes very little over small neighborhoods of the model based on the same metric subject to Fisher consistency; it also has good breakdown properties with respect to such contamination. In particular, the minimum Hellinger distance estimator has the best stability against Hellinger contamination among Fisher-consistent functionals. This is a very powerful result which provides strong justification of the use of the Hellinger distance in many practical problems. However, the approach of Donoho and Liu explicitly uses the properties of a mathematical metric such as the triangle inequality, and hence does not appear to have straightforward extensions to general classes of statistical distances.

Other authors have considered different applications of these methods in more specialized and extended models, some of which we will briefly describe in Chapter 10. Yang (1991) and Ying (1992) have applied the minimum Hellinger distance and related methods to the case of survival data and censored observations. Woodward, Whitney and Eslinger (1995) and Cutler and Cordero-Braña (1996) have applied these techniques to the case of mixture models. Pak (1996) and Pak and Basu (1998) have considered the linear regression problem through this minimum distance approach. Victoria-Feser and Ronchetti (1997) and Lin and He (2006) have extended these methods to the case of grouped data. Sriram and Vidyashankar (2000), among others, have applied the same to different stochastic process applications; Takada (2009) has applied the simulated minimum Hellinger distance estimator in case of stochastic volatility models. Cheng and Vidyashankar (2006) have approached the problem of adaptive estimation through minimum Hellinger distance techniques.

3.2   Minimum Hellinger Distance Estimation

3.2.1   The Minimum Hellinger Distance Functional

To present Beran's proof of the asymptotic normality of the minimum Hellinger distance estimator, we carry on from where we left off in Section 2.4.1 and first establish additional preliminary results about the minimum Hellinger distance functional. We denote by $T(G)$ the minimum Hellinger distance functional corresponding to the distribution $G \in \mathcal{G}$, which is defined by

$$\mathrm{HD}(g, f_{T(G)}) = \inf_{\theta \in \Theta} \mathrm{HD}(g, f_\theta)$$

provided such a minimum exists. Here the Hellinger distance measure between the continuous densities g and f is defined by

$$\mathrm{HD}(g, f) = 2 \int (g^{1/2} - f^{1/2})^2. \tag{3.3}$$

Suppose that the model family $\{\mathcal{F}\}$ is identifiable, and the conditions of Lemma 2.6 are satisfied. Then, for all $G \in \mathcal{G}$, $T(G)$ exists; also if $T(G)$ is unique, and if G_n be a sequence of distributions for which the corresponding sequence of densities g_n converge to g in the Hellinger metric, the minimum Hellinger distance functional $T(G_n)$ converges to $T(G)$ as $n \to \infty$. The continuity of the minimum Hellinger distance functional ensures the consistency of the minimum Hellinger distance estimator when the corresponding kernel density estimate has the appropriate convergence properties.

When the assumptions of Lemma 2.6 do not strictly hold, but those of Lemma 2.7 can be assumed for an appropriately defined subset of $\mathcal{G}$, the above conclusions remain valid for all G in the above subset of distributions.

We use the notation $s_t = f_t^{1/2}$, and as in Section 2.4, we make further assumptions to make the functional T differentiable. For specified $t \in \Theta$ we assume that there exists a $p \times 1$ vector $\dot{s}_t(x)$ with components in L_2 and a $p \times p$ matrix $\ddot{s}_t(x)$ with components in L_2 such that for every $p \times 1$ real vector e of unit euclidean length and for every scalar α in a neighborhood of zero,

$$s_{t+\alpha e}(x) = s_t(x) + \alpha e^T \dot{s}_t(x) + \alpha e^T \gamma_\alpha(x) \tag{3.4}$$

$$\dot{s}_{t+\alpha e}(x) = \dot{s}_t(x) + \alpha \ddot{s}_t(x)e + \alpha v_\alpha(x)e \tag{3.5}$$

where $\gamma_\alpha(x)$ is $p \times 1$ vector, $v_\alpha(x)$ is a $p \times p$ matrix, and each component of γ_α and v_α converges to zero in L_2 as $\alpha \to 0$. We work under these conditions to present the results derived by Beran (1977).

Theorem 3.1. [Beran (1977, Theorem 2)]. *Suppose that (3.4) and (3.5) hold for every $t \in \Theta$, $T(G)$ exists, is unique, and lies in the interior of Θ, $\int \ddot{s}_{T(G)}(x)g^{1/2}(x)dx$ is a nonsingular matrix, and the functional T is continuous at G in the Hellinger topology. Then, for every sequence of densities $\{g_n\}$ converging to g in the Hellinger metric, the corresponding sequence of functionals $T(G_n)$ has the expansion*

$$T(G_n) = T(G) + \int \vartheta_g(x)[g_n^{1/2}(x) - g^{1/2}(x)]dx$$

$$+ \, \xi_n \int \dot{s}_{T(G)}(x)[g_n^{1/2}(x) - g^{1/2}(x)]d(x) \tag{3.6}$$

where

$$\vartheta_g(x) = - \left[\int \ddot{s}_{T(G)}(x)g^{1/2}(x)dx \right]^{-1} \dot{s}_{T(G)}(x) \tag{3.7}$$

and ξ_n is a real $p \times p$ matrix which tends to zero as $n \to \infty$. When the true distribution belongs to the model, so that $g = f_\theta$ for some $\theta \in \Theta$, we get

$$\vartheta_g(x) = \vartheta_{f_\theta}(x) = -\left[\int \ddot{s}_\theta(x)s_\theta(x)dx\right]^{-1}\dot{s}_\theta(x)$$

$$= \left[\int \dot{s}_\theta(x)\dot{s}_\theta^T(x)dx\right]^{-1}\dot{s}_\theta(x)$$

$$= \left[\frac{1}{4}I(\theta)\right]^{-1}\dot{s}_\theta(x), \tag{3.8}$$

where $I(\theta)$ represents the Fisher information matrix.

Proof. Let $\theta = T(G)$, and $\theta_n = T(G_n)$. As the differentiability conditions in (3.4) and (3.5) hold, the functionals θ and θ_n must satisfy

$$\int \dot{s}_\theta(x)g^{1/2}(x)dx = 0, \quad \text{and} \quad \int \dot{s}_{\theta_n}(x)g_n^{1/2}(x)dx = 0$$

respectively. Using (3.5), we get

$$0 = \int \dot{s}_{\theta_n}(x)g_n^{1/2}(x)dx$$

$$= \int [\dot{s}_\theta(x) + \ddot{s}_\theta(x)(\theta_n - \theta) + v_n(x)(\theta_n - \theta)]g_n^{1/2}(x)dx$$

where the components of the $p \times p$ matrix $v_n(x)$ converge to zero in L_2 as $n \to \infty$. Thus, for n sufficiently large,

$$\theta_n - \theta = -\left[\int (\ddot{s}_\theta(x) + v_n(x))g_n^{1/2}(x)dx\right]^{-1}\int \dot{s}_\theta(x)g_n^{1/2}(x)dx$$

$$= -\left[\int \ddot{s}_\theta(x)g^{1/2}(x)]dx\right]^{-1}\int \dot{s}_\theta(x)[g_n^{1/2}(x) - g^{1/2}(x)]dx$$

$$+ \xi_n \int \dot{s}_\theta(x)[g_n^{1/2}(x) - g^{1/2}(x)]dx$$

as was to be proved.

Now suppose that $g = f_\theta$ belongs to the model. Clearly $\int \dot{s}_\theta(x)s_\theta(x)dx = 0$ for all θ in the interior of Θ. It then follows that for every sufficiently small α and every unit vector e,

$$0 = \int \alpha^{-1}[\dot{s}_{\theta+\alpha e}(x)s_{\theta+\alpha e}(x) - \dot{s}_\theta(x)s_\theta(x)]dx$$

$$= \int \alpha^{-1}\{[\dot{s}_{\theta+\alpha e}(x) - \dot{s}_\theta(x)]s_\theta(x) + [s_{\theta+\alpha e}(x) - s_\theta(x)]\dot{s}_{\theta+\alpha e}(x)\}dx$$

$$= \left[\int \ddot{s}_\theta(x)s_\theta(x)dx + \int \dot{s}_\theta(x)\dot{s}_\theta^T(x)dx\right]e + o(1)$$

which shows $\left[\int \dot{s}_\theta(x)\dot{s}_\theta^T(x)dx\right]^{-1}\dot{s}_\theta(x) = -\left[\int \ddot{s}_\theta(x)s_\theta(x)dx\right]^{-1}\dot{s}_\theta(x)$. The other equalities of (3.8) are obvious.   □

Remark 3.1. In actual computation, a kernel density estimator g_n^* with the right properties will be used to estimate the unknown parameter based on $\rho_C(g_n^*, f_\theta)$. In our functional notation, the estimator may be expressed as $T(G_n^*)$. However, G_n^* is simply the convolution of the empirical with the fixed, known, kernel and we will continue to refer to the estimator obtained by minimizing $\rho_C(g_n^*, f_\theta)$ as $T(G_n)$ in our functional notation.

3.2.2   The Asymptotic Distribution of the Minimum Hellinger Distance Estimator

In Theorem 3.1 we have provided the representation of the functional which we will exploit to find the minimum Hellinger distance estimator of the parameter. Suppose we have a random samples $X_1, \ldots, X_n$ from the distribution G, which is modeled by the parametric family $\mathcal{F}$. Beran suggested the use of the kernel density estimate g_n^* of g given by

$$g_n^*(x) = \frac{1}{n(h_n s_n)} \sum_{i=1}^n w\left(\frac{x - X_i}{h_n s_n}\right). \tag{3.9}$$

Here w is a smooth density on the real line, and the bandwidth is the product of h_n and s_n, where the quantity s_n is a robust estimator of scale, while the sequence h_n converges to zero at an appropriate rate. The following theorem lists the conditions under which the convergence of the kernel density estimate g_n^* to g in the Hellinger metric is guaranteed.

Theorem 3.2. [Beran (1977, Theorem 3)]. *Suppose that the kernel density estimate is given by (3.9), and the relevant quantities satisfy the following conditions:*

(i) w is absolutely continuous and has compact support; w' is bounded.

(ii) g is uniformly continuous.

(iii) $\lim_{n\to\infty} h_n = 0$; $\lim_{n\to\infty} n^{1/2}h_n = \infty$.

(iv) As $n \to \infty$, $s_n \to s$, a positive constant depending on g.

Then the kernel density estimate g_n^* converges to G in the Hellinger metric (i.e., $\mathrm{HD}(g_n^*, g) \to 0$ as $n \to \infty$). Thus, if T is a functional which is continuous in the Hellinger metric, then $T(G_n) \to T(G)$ in probability.

Proof. Let the empirical distribution based on the random sample $X_1, \ldots, X_n$ be denoted by G_n. Thus, the kernel density estimate in (3.9) may be expressed as

$$g_n^*(x) = \frac{1}{h_n s_n} \int w\left(\frac{x - y}{h_n s_n}\right) dG_n(y).$$

Also let the density $\tilde{g}_n = (h_n s_n)^{-1} \int w[(h_n s_n)^{-1}(x - y)]dG(y)$ represent the kernel smoothed version of the true model density. We will use $\tilde{g}_n$ as the intermediate tool in establishing the convergence of g_n^* to g. Integration by parts gives

$$|g_n^*(x) - \tilde{g}_n(x)| \leq n^{-1/2}(h_n s_n)^{-1} \sup_x |R_n(x)| \int |w'(x)|dx,$$

where $R_n(x) = n^{1/2}[G_n(x) - G(x)]$. On the other hand, suppose $a > 0$ is such that the interval $[-a, a]$ contains the support of w, then

$$|\tilde{g}_n(x) - g(x)| \leq \sup_{|t| < a} |g(x - h_n s_n t) - g(x)|.$$

From the above two results it is clear that an appropriately defined $g_n^*(x)$ satisfies $\sup_x |g_n^*(x) - g(x)| \to 0$ with probability 1. Since the functions g_n^* and g are densities, it then follows that

$$\int (g_n^{*1/2}(x) - g^{1/2}(x))^2 dx \to 0$$

with probability 1 for such g_n^* as was to be shown. $\qquad\square$

Finally, the asymptotic distribution of the minimum Hellinger distance estimator $T(G_n)$ is considered in the next theorem. This requires quite strong assumptions. Several authors have attempted to derive similar results under weaker conditions, some of which are presented in later sections. Here we present a sketch of Beran's original 1977 proof.

Theorem 3.3. [Beran (1977, Theorem 4)]. *Assume the following conditions.*

(i) *w is symmetric about 0 and has compact support.*

(ii) *w is twice absolutely continuous; w'' is bounded.*

(iii) *The minimum Hellinger distance functional T satisfies (3.6) and ϑ_g has compact support K^* on which it is continuous.*

(iv) *$g > 0$ on K^*; g is twice absolutely continuous and g'' is bounded.*

(v) *$\lim_{n\to\infty} n^{1/2}h_n = \infty$; $\lim n^{1/2}h_n^2 = 0$.*

(vi) *There exists a positive finite constant s depending on g such that $n^{1/2}(s_n - s)$ is bounded in probability.*

Then the limiting distribution of $n^{1/2}[T(G_n) - T(G)]$ under g as $n \to \infty$ is

$$N\left(0, \int \psi_g(x)\psi_g^T(x)g(x)dx\right),$$

where $\psi_g(x) = \vartheta_g(x)/(2g^{1/2}(x))$, and $\vartheta_g(x)$ is as defined in Equation (3.7). In particular, if $g = f_\theta$, the limiting distribution of $n^{1/2}[T(G_n) - T(G)]$ is $N(0, I^{-1}(\theta))$.

Sketch of the Proof. Let $T(G) = \theta$ and $T(G_n) = \theta_n$. Under the given conditions, g_n^* converges, in probability, to g in the Hellinger metric. Now from (3.6)

$$T(G_n) = T(G) + \int \vartheta_g(x)[g_n^{*1/2}(x) - g^{1/2}(x)]dx$$

$$+ \xi_n \int \dot{s}_{T(G)}[g_n^{*1/2}(x) - g^{1/2}(x)]dx$$

where $\xi_n \to 0$ in probability. Thus, the asymptotic normality of

$$n^{1/2}(\theta_n - \theta) = n^{1/2}[T(G_n) - T(G)]$$

is driven by the distribution of the term $n^{1/2} \int \vartheta_g(x)[g_n^{*1/2}(x) - g^{1/2}(x)]dx$. Notice that $\int \vartheta_g(x)g^{1/2}(x)dx = 0$, i.e., $\vartheta_g(x)$ is orthogonal to $g^{1/2}(x)$; also $\vartheta \in L_2$. Thus, our required result will be established if it can be shown that the limiting distribution of $n^{1/2} \int \sigma(x)[g_n^{*1/2}(x) - g^{1/2}(x)]dx$, with $\sigma \in L_2$, σ orthogonal to $g^{1/2}$, and σ supported on K^*, is $N\left(0, \int \varsigma(x)\varsigma^T(x)g(x)dx\right)$, where $\varsigma(x) = \sigma(x)/(2g^{1/2}(x))$.

By an application of the algebraic identity

$$b^{1/2} - a^{1/2} = (b-a)/(2a^{1/2}) - (b-a)^2/[2a^{1/2}(b^{1/2} + a^{1/2})^2] \qquad (3.10)$$

we get

$$n^{1/2} \int \sigma(x)[g_n^{*1/2}(x) - g^{1/2}(x)]^2 dx = n^{1/2} \int \sigma(x)[g_n^*(x) - g(x)]/(2g^{1/2}(x))dx + \zeta_n$$

where ζ_n is the remainder term. To complete the proof, one has to show the following two essential steps.

$$n^{1/2} \int \sigma(x)[g_n^{*1/2}(x) - g^{1/2}(x)]dx$$

$$- n^{1/2} \int \sigma(x)[g_n^*(x) - g(x)]/(2g^{1/2}(x))dx = o_p(1), \qquad (3.11)$$

so that the remainder term is $o_p(1)$, and

$$n^{1/2} \int \sigma(x)[g_n^*(x) - g(x)]/(2g^{1/2}(x))dx$$

$$- n^{1/2} \int \frac{\sigma(x)}{2g^{1/2}(x)}d(G_n - G)(x) = o_p(1). \qquad (3.12)$$

The results (3.11) and (3.12), taken together, immediately yield the required result given the conditions on σ.

The proofs of (3.11) and (3.12) are technical and involve complicated mathematics. The reader is referred to Beran (1977, p. 451–452) for complete details of the proof of (3.11). Here we provide the following outline of the proof of (3.12). Let $\varsigma(x) = \sigma(x)/(2g^{1/2}(x))$. Notice that

$$
n^{1/2} \int \frac{\sigma(x)}{(2g^{1/2}(x))}[g_n^*(x) - g(x)]dx = n^{1/2} \int \varsigma(x)[g_n^*(x) - g(x)]dx
$$

$$
= n^{1/2} \int \varsigma(x)T_n(x)dx + o_p(1),
$$

where $n^{1/2}T_n(x) = (h_n s)^{-1} \int w((h_n s)^{-1}(x - y))dR_n(y)$, and s is the limiting value of s_n as defined in the statement of the theorem. But

$$
n^{1/2} \int \varsigma(x)T_n(x)dx = \int dR_n(y) \int \varsigma(y + h_n sz)w(z)dz. \tag{3.13}
$$

and

$$
E\left[\int dR_n(y) \int \varsigma(y + h_n sz)w(z)dz - \int \varsigma(y)dR_n(y)\right]^2
$$

$$
\leq \int w(z)dz \int [\varsigma(y + h_n sz) - \varsigma(y)]^2 g(y)dy. \tag{3.14}
$$

The right-hand side of the last equation tends to zero as $n \to \infty$, which establishes (3.12). In Section 3.3 we will prove the asymptotic results for the minimum Hellinger distance estimators of multivariate location and covariance under more explicit assumptions which directly lead to the convergence of the term in (3.14); see condition (T6) of Theorem 3.4. □

Remark 3.2. The above derivation shows that

$$
n^{1/2}(\theta_n - \theta) = n^{1/2} \int \frac{\vartheta_g(x)}{2g^{1/2}(x)}d(G_n - G) + o_p(1).
$$

When $g = f_\theta$ belongs to the model family, we get

$$
n^{1/2}(\theta_n - \theta) = n^{1/2}\left[\frac{1}{4}I(\theta)\right]^{-1} \int \frac{\dot{s}_\theta(x)}{2f_\theta^{1/2}(x)}d(G_n - F_\theta) + o_p(1)
$$

so that the relation

$$
\theta_n = \theta + n^{-1/2}I^{-1}(\theta)Z_n(\theta) + o_p(n^{-1/2})
$$

continues to hold, where $Z_n(\theta)$ is as in Equation (2.74).

Since Beran's (1977) seminal paper, several other authors have made significant contributions to the problem of minimum Hellinger distance estimation, or more generally, to the problem of minimum distance estimation based on disparities for the continuous model (e.g., Stather, 1981; Tamura and Boos, 1986; Simpson, 1989b; Cao, Cuevas and Fraiman, 1995; Basu, Sarkar and Vidyashankar, 1997b; Park and Basu, 2004; Toma, 2008). Each set of authors makes a contribution in their own way making the literature richer, solving a particular component of the problem, in a particular setting, under their own particular conditions. However, a fully general framework for describing the problem of minimum distance estimation for continuous models does not exist yet, as it does for the discrete case. We will describe the existing techniques under their different settings.

Stather's work, presented in his Ph.D. thesis, represents an important extension of Beran's work. Unfortunately his results are unpublished, so one has to depend on a difficult-to-obtain Ph.D. thesis. Tamura and Boos (1986) studied minimum Hellinger distance estimation in the multivariate context where the location and covariance parameters are of interest. Beran's approach allowed the smoothing parameter to be random but restricted the true distribution to have compact support. Stather allowed infinite support for g, but chose nonrandom bandwidth h_n. Tamura and Boos (1986) also work with a nonrandom h_n and allow an infinite support for g, but their approach has similarities with Beran's proof as well. For actual implementation of the minimum distance method in continuous models based on kernel density estimates it certainly makes sense to use a bandwidth which is a multiple of a robust scale estimate and hence is random. However, the random component is somewhat of a distraction in establishing the asymptotic properties of the estimator. Beran also considered the limiting quantities of the random component to arrive at the final result. Most of the other authors also use nonrandom smoothing parameters. On the whole, the addition of the density estimation component makes the technique of minimum distance estimation significantly more complex compared to the case of discrete models.

The work of Cao et al. (1995) presents another example of the application of the density-based minimum distance method which uses a nonparametric density estimator obtained from the data. However, it is, to a certain degree, different from most of the other cases discussed here, in that these authors only consider the L_1, L_2 and L_∞ metrics as their choice of distances. It is not entirely clear why the authors excluded minimum Hellinger distance estimation from their discussion — a topic which would have fitted admirably with the theme of their paper. Beran (1977), Tamura and Boos (1986) and Simpson (1989b) are not mentioned in the paper. Be that as it may, their approach leads to several interesting results. However, neither of the three metrics considered by the authors belong to the class of disparities, and do not generate fully efficient estimators. Later (Chapter 9) we will discuss another useful method of density-based minimum distance estimation, based on Basu et al. (1998), of which at least the minimum L_2 metric method will be a part. The primary motivation of the method described by Basu et al. (1998) is that in certain cases including the L_2 metric case it is possible to perform the density-based minimum distance estimation routine for continuous models by avoiding direct density estimation. We will see that the asymptotic distribution of the estimator derived in Chapter 9 following Basu et al. (1998) – without data smoothing – matches the asymptotic distribution derived by Cao, Cuevas and

Fraiman (1995, Theorem 2, p. 616) – with data smoothing – for the minimum L_2 case. An estimation process which neither leads to full asymptotic efficiency, nor avoids the complications due to the density estimation, is not among the focused techniques of this book and we do not elaborate further on the approach of Cao, Cuevas and Fraiman (1995) in this book.

3.3 Estimation of Multivariate Location and Covariance

Tamura and Boos (1986) considered minimum Hellinger distance estimation for the multivariate case. The most commonly used kernel density estimate for k dimensional data is

$$g_n^*(x) = (nh_n^k)^{-1} \sum_{i=1}^{n} w((x - X_i)/h_n) \qquad (3.15)$$

where w is a density on $\mathbb{R}^k$ and $\{h_n\}$ is a bandwidth sequence with suitable properties. The focus of their work was on determining affine equivariant and affine covariant estimates of multivariate location and covariance respectively (see Tamura and Boos, 1986, Section 2, for relevant definitions). They considered elliptical models having densities of the form

$$f_{\mu,\Sigma}(x) = C_k |\Sigma|^{-1/2} \varrho[(x - \mu)^T \Sigma^{-1}(x - \mu)],$$

for suitable functions C_k and $\varrho(\cdot)$. If an initial affine covariant estimate $\hat{\Sigma}_0$ of covariance is available, a density estimate of the radial type can be constructed as

$$g_n^*(x) = (nh_n^k)^{-1} |\hat{\Sigma}_0|^{-1/2} \sum_{i=1}^{n} w\left(h_n^{-1} ||x - X_i||_{\hat{\Sigma}_0}\right) \qquad (3.16)$$

where $||x||_{\Sigma}^2 = x^T \Sigma^{-1} x$. Such a kernel satisfies the conditions under which the corresponding minimum Hellinger distance estimators of the location and covariance are affine equivariant and affine covariant respectively (Tamura and Boos, 1986, Lemma 2.1). In actual computation Tamura and Boos used the kernel function corresponding to a uniform random vector on the k dimensional unit sphere.

The convergence of the kernel density estimate g_n^* to the true density g in the Hellinger metric is a condition required by all minimum Hellinger distance approaches for consistency and other results. This convergence is implied by the L_1 convergence of g_n^* to g. The L_1 convergence of g_n^* to g is, in turn, implied by the condition $h_n + (nh_n^k)^{-1} \to 0$. Beran (1977), Tamura and Boos (1986) and Simpson (1989b) have all used different bandwidth conditions on he kernel, although all give the necessary convergence in the Hellinger metric.

There is an additional problem faced by the minimum distance method in the multivariate case. The order of the variance of the kernel density estimator is $O((nh_n^k)^{-1})$. However, the maximum absolute bias for the multivariate kernel density estimator, $\sup_x |Eg_n^*(x) - g(x)|$ goes to zero at a fixed rate h_n^2, independently of the dimension k, so that the asymptotic bias term in the minimum Hellinger distance estimator in the multivariate case is NOT $o(n^{-1/2})$. The asymptotic distribution presented by Tamura and Boos, therefore, relate to that of

$$n^{1/2}(T(G_n) - T(G) - B_n)$$

where B_n is the bias term, rather than just that of $n^{1/2}(T(G_n) - T(G))$.

Let $s_\theta = f_\theta^{1/2}$, and let $\dot{s}_\theta$ and $\ddot{s}_\theta$ be the corresponding first and second derivative. Let $\vartheta_g(x)$ be as defined in Theorems 3.1 and 3.3. Let

$$\psi_g(x) = \frac{\vartheta_g(x)}{2g^{1/2}(x)}. \tag{3.17}$$

Let $g_n^*(x)$ be an appropriate kernel density estimator, and let $\tilde{g}_n(x) = E[g_n^*(x)]$. Let $|x|$ and x^2 denote the $k \times 1$ vectors of elementwise absolute and squared values of x, respectively, and $||\cdot||$ represent a norm in $\mathbb{R}^k$. Let $\{a_n\}$ be a sequence of positive numbers tending to infinity with $\lambda_n(x) = \chi_{\{||x||\leq a_n\}}\psi_g(x)$ and $\eta_n(x) = \chi_{\{||x||>a_n\}}\psi_g(x)$.

Under the above notation, we list below the conditions required by Tamura and Boos for the asymptotic normality proof of the minimum Hellinger distance estimates of multivariate location and scatter.

(T1) Let g_n^* be as defined in Equation (3.15), where w is a symmetric square integrable density on $\mathbb{R}^k$ with compact support S. The bandwidth h_n satisfies $h_n + (nh_n^k)^{-1} \to 0$.

(T2) Either the true distribution belongs to the parametric model and the conditions of Lemma 2.8 are satisfied; or Θ is a compact subset of $\mathbb{R}^p$, the parametric model is identifiable in the sense of Definition 2.2, $f_\theta(x)$ is continuous in θ for almost all x, and $T(G)$ is unique.

(T3) $n \sup_{t\in S} P(||X_1 - h_n t|| > a_n) \to 0$ as $n \to \infty$.

(T4) $(n^{1/2}h_n^k)^{-1} \int |\lambda_n(x)|dx \to 0$ as $n \to \infty$.

(T5) $M_n = \sup_{||x||<a_n} \sup_{t\in S}\{g(x + h_n t)/g(x)\} = O(1)$ as $n \to \infty$.

(T6) The matrix $\int \psi_g(x)\psi_g^T(x)g(x)dx$ is finite (element-wise), $\sup_{||a||<b} \int \psi_g^2(x + a)g(x)dx$ is finite for some $b > 0$, and $\int [\psi_g(x + a) - \psi_g(x)]^2 g(x)dx \to 0$ as $||a|| \to 0$.

(T7) For each θ in the interior of Θ, $s_\theta = f_\theta^{1/2}$ satisfies the derivative conditions in (3.4) and (3.5), $T(G)$ lies in the interior of Θ, and the matrix $\int \ddot{s}_\theta(x)g^{1/2}(x)dx$ is nonsingular.

Theorem 3.4. [Tamura and Boos (1986, Theorem 4.1)]. *Suppose that conditions (T1) –(T7) hold. Let $X_1, \ldots, X_n$ represent a sample of independent and identically distributed k-vectors having probability density function g. Let $g_n^*(x)$ be a k-dimensional kernel density estimate of the form (3.15). Let G be the true distribution not necessarily in the parametric model. Then the minimum Hellinger distance estimator $T(G_n)$ is consistent, and allows the asymptotic distribution*

$$n^{1/2}[T(G_n) - T(G) - B_n] \to Z^* \sim N\left(0, \int \psi_g(x)\psi_g(x)^T g(x)dx\right),$$

where $B_n = 2C_n^* \int \psi_g(x)\tilde{g}_n^{1/2}(x)g^{1/2}(x)dx$, with $C_n^* \to I$, where I is the p dimensional identity matrix.

Sketch of Proof. Consistency follows from conditions (T1) and (T2) using Lemmas 2.6 and 2.8.

Condition (T7) gives the Taylor series expansion

$$T(G_n) - T(G) = \int \vartheta_g(x)[g_n^{*1/2}(x) - g^{1/2}(x)]dx$$

$$+ \xi_n \int \dot{s}_{T(G)}[g_n^{*1/2}(x) - g^{1/2}(x)]dx \qquad (3.18)$$

where $\xi_n \to 0$ in probability as $n \to \infty$. This expansion has already been encountered in Equation (3.6). Note that $\int \vartheta_g(x)g^{1/2}(x)dx = 0$, so that $\int \psi_g(x)g(x)dx$ is also zero where $\vartheta_g(x) = \psi_g(x)2g^{1/2}(x)$. Since $\dot{s}_{T(G)}(x)$ is proportional to $\vartheta_g(x)$, the terms on the right-hand side of Equation (3.18) may be combined, which gives

$$T(G_n) - T(G) = (I + \xi_n^*) \int \vartheta_g(x)[g_n^{*1/2}(x) - g^{1/2}(x)]dx, \qquad (3.19)$$

where $\xi_n^* \to 0$ in probability, and I is the p-dimensional identity matrix. This then leads to the result

$$T(G_n) - T(G) - B_n = (I + \xi_n^*) \int \vartheta_g(x)[g_n^{*1/2}(x) - \tilde{g}_n^{1/2}(x)]dx,$$

where

$$B_n = (I + \xi_n^*) \int \vartheta_g(x)[\tilde{g}_n^{1/2}(x) - g^{1/2}(x)]dx.$$

Using the relations $\int \vartheta_g(x)g^{1/2}(x)dx = 0$ and $\vartheta_g(x) = \psi_g(x)2g^{1/2}(x)$, and the fact that $\xi_n^* \to 0$ in probability, the bias can be written as $2C_n^* \int \psi_g(x)\tilde{g}_n^{1/2}(x)g^{1/2}(x)dx$, where $C_n^* \to I$, the p dimensional identity matrix, in probability.

The main task then is to show that $n^{1/2} \int \vartheta_g(x)[g_n^{*1/2} - \tilde{g}_n^{1/2}(x)]dx$ has the appropriate limiting normal distribution. For this purpose, the algebraic identity

$$b^{1/2} - a^{1/2} = (b-a)/(2a^{1/2}) - (b^{1/2} - a^{1/2})^2/(2a^{1/2})$$

is made use of. Note that this identity is the same as in (3.10) employed in Theorem 3.3, but written in a slightly different form. This leads to the relation

$$
\int \vartheta_g(x)[g_n^{*1/2} - \tilde{g}_n^{1/2}(x)]dx = \int \psi_g(x)[g_n^*(x) - \tilde{g}_n(x)]dx
$$
$$
- \int \psi_g(x)[g_n^{*1/2} - \tilde{g}_n^{1/2}(x)]^2 dx.
$$

The major steps in the proof involve showing:

(a) $\left| n^{1/2} \int \psi_g(x)[g_n^*(x) - \tilde{g}(x)]dx - n^{1/2}\dfrac{1}{n}\sum_{i=1}^{n}\psi_g(X_i) \right| \to 0$ as $n \to \infty$.

(b) $n^{1/2} \int \psi_g(x)[g_n^{*1/2}(x) - g^{1/2}(x)]^2 dx \to 0$ as $n \to \infty$.

To establish (a), let $\tau_{1n}(x) = \int \psi_g(x)[g_n^*(x) - \tilde{g}(x)]dx$. Then

$$
n^{1/2}\tau_{1n} = \int\int \psi_g(x)h_n^{-k}w(h_n^{-1}(x-y))dR_n(y)dx, \qquad (3.20)
$$

where $R_n(y) = n^{1/2}(G_n(y) - G(y))$. Using the Cauchy-Schwarz inequality and condition (T6), the above can be approximated in mean square by $\int \psi_g(y)dR_n(y)$. Notice that the term on the right-hand side of Equation (3.20) is basically the analog of the term $n^{1/2} \int \varsigma(x)T_n(x)dx$ in Theorem 3.3, and the manipulation in item (a) uses relations (3.13) and (3.14) as well.

Let $\tau_{2n} = \int \psi_g(x)[g_n^{*1/2}(x) - \tilde{g}_n^{1/2}(x)]^2 dx$. To establish part (b), notice that

$$
n^{1/2}\tau_{2n} = n^{1/2} \int \lambda_n(x)[g_n^{*1/2}(x) - \tilde{g}_n^{1/2}(x)]^2 dx
$$
$$
+ n^{1/2} \int \eta_n(x)[g_n^{*1/2}(x) - \tilde{g}_n^{1/2}(x)]^2 dx. \qquad (3.21)
$$

For the term $\int \eta_n(x)[g_n^{*1/2}(x) - \tilde{g}_n^{1/2}(x)]^2 dx$, one completes the square under the integral and looks at all the individual terms. Conditions (T3) and (T6), together with coordinate-wise applications of the Cauchy-Schwarz inequality to each component of the p-vectors show that each of the individual terms is of the order $o_p(n^{-1/2})$, so that the second term on the right-hand side of (3.21) is $o_p(1)$. To prove the convergence of the first term of the right-hand side of (3.21) to zero, it is enough to show that

$$
n^{1/2} \int |\lambda_n(x)|[g_n^*(x) - \tilde{g}_n(x)]^2 g^{-1}(x)dx \to 0
$$

in probability. However the expected value of the quantity on the left hand side of the above equation is less than

$$(n^{1/2}h_n^k)^{-1} \int |\lambda_n(x)| \int_S g(x + h_n z)g^{-1}(x)w^2(z)dzdx$$
$$\leq M_n(n^{1/2}h_n^k)^{-1} \int |\lambda_n(x)|dx \int_S w^2(z)dz,$$

and the result then follows from conditions (T4) and (T5). □

Tamura and Boos (1986) have briefly discussed the conditions under which Theorem 3.4 is developed. Conditions (T1), (T2) are easily seen to be the standard conditions necessary for the consistency of the estimator. Conditions (T3), (T4) and (T5) relate to the rate at which $\alpha_n \to 0$. For a sequence $\alpha_n = n^l$, $l > 0$, condition (T3) is approximately equivalent to the finiteness of $E||X_1||^{1/l}$. However, (T5) is easier to verify in case of the multivariate normal distribution. (T6) represents some routine model conditions, while (T7) lists the usual smoothness conditions on s_θ.

Condition (T4) also imposes a key restriction on the bandwidth. The conditions of Theorem 3.4 are satisfied for $c_n \sim n^{-(1/2k)+\epsilon}$. Notice that for $k = 1$ this will imply $n^{1/2}c_n \to \infty$. Tamura and Boos argue that the optimal rate of $c_n \sim n^{-1/(k+4)}$ can only be used for $k \leq 3$.

The presence of the bias term B_n is a matter of minor irritation, and for the theorem to be useful, it must be demonstrated that for reasonable sample sizes the bias remains small compared to the variance. Tamura and Boos (1986) have observed this to be true, in simulations from the bivariate normal, for sample sizes smaller than 400.

3.4 A General Structure

Unlike the discrete case, it seems to be difficult to achieve a completely general structure where the theory flows freely for the whole class of minimum distance estimators for the continuous case for all disparities under some general conditions on the model and the residual adjustment function. It is not easy to modify the proofs of Beran (1977), Tamura and Boos (1986) and others so that such a goal may be attained. For the present time we provide a proof by Park and Basu (2004) which does a partial generalization and works under a set of strong conditions on the residual adjustment function. Under the given setup, this produces a general approach and works for all disparities satisfying the assumed conditions. The conditions are strong, and are not satisfied by several common disparities, but the result is still useful and the list of disparities that do satisfy the conditions is also quite substantial. In Section 3.5 we

will present an approach based on model smoothing which will encompass a
larger class of disparities.

Let the random variable have a distribution G on the real line. Suppose
that $X_1, \ldots, X_n$ represent an independently and identically distributed sample
generated from G modeled by a parametric family $\mathcal{F}$ as defined in Section 1.1.
We assume that the true distribution G and the model distributions F_θ have
densities with respect to the Lebesgue measure, and let $\mathcal{G}$ be the class of all
distributions with respect to the Lebesgue measure. Let

$$g_n^*(x) = \frac{1}{nh_n} \sum_{i=1}^{n} w\left(\frac{x - X_i}{h_n}\right) \tag{3.22}$$

be the kernel density estimate based on the given data. Let $\rho_C(g_n^*, f_\theta)$ be a
distance based on a function C satisfying the disparity conditions. We prove
the existence of the corresponding minimum distance functional T and related
results in Lemma 3.5. Let $C'(\infty)$ be as defined in Lemma 2.1.

Lemma 3.5. [Park and Basu (2004, Theorem 3.1)]. *We assume that (a) the
parameter space Θ is compact, (b) the family of distributions is identifiable,
(c) $f_\theta(x)$ is continuous in θ for almost every x, and (d) $C(-1)$ and $C'(\infty)$
(as defined in Lemma 2.1) are finite. Then*

(i) for any $G \in \mathcal{G}$, there exists $\theta \in \Theta$ such that $T(G) = \theta$, and

(ii) for any $F_{\theta_0} \in \mathcal{F}$, $T(F_{\theta_0}) = \theta_0$, uniquely.

Proof. (i) Existence: Denote $D(g, f_\theta) = C(g/f_\theta - 1)f_\theta$. Let $\{\theta_n : \theta_n \in \Theta\}$
be a sequence such that $\theta_n \to \theta$ as $n \to \infty$. Since $D(g, f_{\theta_n}) \to D(g, f_\theta)$
by Assumption (c) and since $\int D(g, f_{\theta_n})$ is finite by Assumption (d) and
Lemma 2.2, we have

$$\rho_C(g, f_{\theta_n}) = \int D(g, f_{\theta_n}) \to \int D(g, f_\theta) = \rho_C(g, f_\theta),$$

by a generalized version of the dominated convergence theorem (Royden, 1988,
p. 92). Hence $\rho_C(g, f_t)$ is continuous in t and achieves an infimum for $t \in \Theta$
since Θ is compact.

(ii) Uniqueness of the functional at the model: This is immediate from
the identifiability assumption on the parametrization and the properties of a
statistical distance.   □

As in Remark 2.2, the compactness assumption above can be relaxed to
include such parameter spaces that may be embedded in a compact space Θ,
provided the distance $\rho_C(g, f_t)$, viewed as a function of t can be extended to
a continuous function on $\bar{\Theta}$.

Theorem 3.6. [Park and Basu (2004, Theorem 3.2)]. *Let $G \in \mathcal{G}$ be the true distribution with density g. Suppose that $|C'(\cdot)|$ is bounded on $[-1, \infty]$. Let $\{g_n\}$ be a sequence of densities, and let $\{G_n\}$ be the corresponding distribution functions. If $T(G)$ is unique, then under the assumptions of Lemma 3.5, the functional $T(\cdot$ is continuous at G in the sense that $T(G_n)$ converges to $T(G)$ whenever $g_n \to g$ in L_1.*

Proof. Suppose that $g_n \to g$ in L_1. Define $\varrho(t) = \rho_C(g, f_t)$ and $\varrho_n(t) = \rho_C(g_n, f_t)$. It follows that

$$|\varrho_n(t) - \varrho(t)| \leq \int |C(\delta_n) - C(\delta)| \, f_t,$$

where $\delta_n = g_n/f_t - 1$ and $\delta = g/f_t - 1$. By the mean value theorem and the boundedness of C' we get,

$$|\varrho_n(t) - \varrho(t)| \leq M \int |\delta_n - \delta| f_t = M \int |g_n - g| \quad \text{for all } t \in \Theta,$$

where $M = \max_\delta |C'(\delta)|$. Since $g_n \to g$ in L_1, we get

$$\sup_t |\varrho_n(t) - \varrho(t)| \to 0 \quad \text{as } n \to \infty. \tag{3.23}$$

Denote $\theta_n = \arg\inf_t \varrho_n(t)$ and $\theta = \arg\inf_t \varrho(t)$. If $\varrho(\theta) \geq \varrho_n(\theta_n)$, then $\varrho(\theta) - \varrho_n(\theta_n) \leq \varrho(\theta_n) - \varrho_n(\theta_n)$, and if $\varrho_n(\theta_n) \geq \varrho(\theta)$, then $\varrho_n(\theta_n) - \varrho(\theta) \leq \varrho_n(\theta) - \varrho(\theta)$. Therefore we have

$$|\varrho_n(\theta_n) - \varrho(\theta)| \leq |\varrho_n(\theta_n) - \varrho(\theta_n)| + |\varrho_n(\theta) - \varrho(\theta)| \leq 2 \sup_t |\varrho_n(t) - \varrho(t)|,$$

which implies $\varrho_n(\theta_n) \to \varrho(\theta)$ as $g_n \to g$ in L_1. Using this and (3.23), we obtain

$$\lim_{n \to \infty} \varrho(\theta_n) = \varrho(\theta). \tag{3.24}$$

Then the proof of the result $\theta_n \to \theta$ as $n \to \infty$ proceeds as in Lemma 2.6 (ii). $\qquad \square$

The following is a simple corollary of Theorem 3.6.

Corollary 3.7. [Park and Basu (2004, Corollary 3.3)]. *If the conditions of Theorem 3.6 hold and $g = f_\theta$, then $\theta_n = T(G_n) \to \theta$ and $\rho_C(f_{\theta_n}, f_\theta) \to 0$.*

The second part of the statement follows from the convergence of $\int |f_{\theta_n} - f_\theta|$ to zero from Glick's Theorem (Devroye and Györfi, 1985, p. 10). Also, if Θ cannot be embedded in a compact set, one can extend the results of Theorem 3.6 to the reduced set $\mathcal{G}$, the class of distributions having densities with respect to the dominating measure and satisfying

$$\inf_{t \in \Theta - H} \rho_C(g, f_t) > \rho_C(g, f_{\theta^*})$$

where θ^* belongs to H for some compact subset H of Θ, in the spirit of Lemma 2.7.

For deriving the asymptotic normality of the minimum distance estimator, we will assume that the model is identifiable, $f_\theta(x)$ is twice continuously differentiable with respect to θ, and for any $G \in \mathcal{G}$, $\rho_C(g, f_\theta)$ can be twice differentiated with respect to θ under the integral sign. Sufficient conditions for the above are given in, for example, Park and Basu (2004) as well as many standard texts. The main theorem, presented below, considers the case where the true distribution $G = F_{\theta_0}$ belongs to the model. As in Section 1.1, $\chi(A)$ denotes the indicator for the set A. The proof follows the approach of Tamura and Boos (1986) presented in Theorem 3.4.

Theorem 3.8. [Park and Basu (2004, Theorem 3.4)]. *Let θ_0 be the true value of the parameter, $\{\phi_n\}$ denote any sequence of estimators such that $\phi_n = \theta_0 + o_p(1)$, and $\{\alpha_n\}$ be any sequence of positive real numbers going to ∞. We make the following assumptions:*

(a) *$\int |\nabla_{ij} f_{\phi_n} - \nabla_{ij} f_{\theta_0}| = o_p(1)$ and $\int |u_{i\phi_n} u_{j\phi_n} f_{\phi_n} - u_{i\theta_0} u_{j\theta_0} f_{\theta_0}| = o_p(1)$, $i, j = 1, \ldots, p$, for all $\{\phi_n\}$ as defined above.*

(b) *The matrix $I(\theta_0)$ is finite (element wise), and $\int u_{i\theta_0}(x + a) u_{j\theta_0}(x + a) f_{\theta_0}(x) dx - \int u_{i\theta_0}(x) u_{j\theta_0}(x) f_{\theta_0}(x) dx \to 0$ as $|a| \to 0$, $i, j = 1, \ldots, p$.*

(c) *Let the kernel density estimate be as given in Equation (3.22), where $w(\cdot)$ is a density which is symmetric about 0, square integrable, and twice continuously differentiable with compact support S. The bandwidth h_n satisfies $h_n \to 0$, $n^{1/2} h_n \to \infty$, and $n^{1/2} h_n^2 \to 0$.*

(d) *$\limsup_{n \to \infty} \sup_{y \in \mathcal{A}_n} \int |\nabla_{ij} f_{\theta_0}(x + y) u_{\theta_0}(x)| dx < \infty$ for $i, j = 1, \ldots, p$, where $\mathcal{A}_n = \{y : y = h_n z, \ z \in S\}$.*

(e) *$n \sup_{t \in S} P(|X_1 - h_n t| > \alpha_n) \to 0$ as $n \to \infty$, for all $\{\alpha_n\}$ as defined above.*

(f) *$(n^{1/2} h_n)^{-1} \int |u_{\theta_0}(x) \chi(|x| \leq \alpha_n)| \to 0$, for all $\{\alpha_n\}$ as defined above.*

(g) *$\sup_{|x| \leq \alpha_n} \sup_{t \in S} \{f_{\theta_0}(x + h_n t)/f_{\theta_0}(x)\} = O(1)$, as $n \to \infty$ for all $\{\alpha_n\}$ as defined above.*

(h) *$A(\delta)$, $A'(\delta)$, $A'(\delta)(\delta + 1)$ and $A''(\delta)(\delta + 1)$ are bounded on $[-1, \infty]$.*

Then, $n^{1/2}(T(G_n) - T(G))$ converges in distribution to $N(0, I^{-1}(\theta_0))$, where $\theta_0 = T(G)$, and T is the minimum distance functional based on ρ_C.

Proof. By condition (c), we have $g_n^*(x) \xrightarrow{\text{a.s.}} f_{\theta_0}(x)$ for every x and

$$\int |g_n^*(x) - f_{\theta_0}(x)| dx \to 0,$$

and hence $T(G_n) \xrightarrow{\mathcal{P}} \theta$. Let us denote $\varrho_n(\theta) = \rho_C(g_n^*, f_\theta)$ and $\hat{\theta} = T(G_n)$. Since $\hat{\theta}$ minimizes $\varrho_n(\cdot)$ over Θ, the Taylor series expansion of $\nabla\varrho_n(\hat{\theta})$ at θ_0 yields

$$0 = n^{1/2}\nabla\varrho_n(\hat{\theta}) = n^{1/2}\nabla\varrho_n(\theta_0) + n^{1/2}\nabla_2\varrho_n(\theta^*)(\hat{\theta} - \theta_0),$$

where θ^* is a point on the line segment joining θ_0 and $\hat{\theta}$. It follows that

$$n^{1/2}(\hat{\theta} - \theta_0) = -[\nabla_2\varrho_n(\theta^*)]^{-1}n^{1/2}\nabla\varrho_n(\theta_0).$$

Therefore it suffices to show that

$$\nabla_2\varrho_n(\theta^*) \xrightarrow{\mathcal{P}} I(\theta_0) \tag{3.25}$$

and

$$-n^{1/2}\nabla\varrho_n(\theta_0) \xrightarrow{\mathcal{D}} Z^* \sim N(0, I(\theta_0)). \tag{3.26}$$

First we prove (3.25). Let $\delta_n(\theta) = g_n^*/f_\theta - 1$. Differentiating with respect to θ, we have

$$\nabla_2\varrho_n(\theta) = -\int A(\delta_n)\nabla_2 f_\theta + \int A'(\delta_n)(\delta_n + 1)u_\theta u_\theta^T f_\theta$$

Let $B_1 = \sup_\delta |A(\delta)|$, $B_2 = \sup_\delta |A'(\delta)(\delta + 1)|$; B_1 and B_2 are finite from condition (h); from condition (a) it follows

$$\left|\int A(\delta_n(\theta^*))(\nabla_2 f_{\theta^*} - \nabla_2 f_{\theta_0})\right| \leq B_1 \int |\nabla_2 f_{\theta^*} - \nabla_2 f_{\theta_0}| \xrightarrow{\mathcal{P}} 0, \tag{3.27}$$

and

$$\left|\int A'(\delta_n(\theta^*))(\delta_n(\theta^*) + 1)(u_{\theta^*}u_{\theta^*}^T f_{\theta^*} - u_{\theta_0}u_{\theta_0}^T f_{\theta_0})\right|$$
$$\leq B_2 \int |u_{\theta^*}u_{\theta^*}^T f_{\theta^*} - u_{\theta_0}u_{\theta_0}^T f_{\theta_0}| \xrightarrow{\mathcal{P}} 0. \tag{3.28}$$

Since $A(0) = 0$ and $\delta_n(\theta^*) \to 0$ as $n \to \infty$, using the dominated convergence theorem we have

$$\int A(\delta_n(\theta^*))\nabla_2 f_{\theta_0} \xrightarrow{\mathcal{P}} 0,$$

and hence by (3.27)

$$\int A(\delta_n(\theta^*))\nabla_2 f_{\theta^*} \xrightarrow{\mathcal{P}} 0.$$

Similarly, since $A'(0) = 1$ and $\delta_n(\theta^*) \to 0$ as $n \to \infty$, we have

$$\int A'(\delta_n(\theta^*))(\delta_n(\theta^*) + 1)u_{\theta_0}u_{\theta_0}^T f_{\theta_0} \xrightarrow{\mathcal{P}} \int u_{\theta_0}u_{\theta_0}^T f_{\theta_0},$$

by the dominated convergence theorem and hence by (3.28)

$$\int A'(\delta_n(\theta^*))(\delta_n(\theta^*) + 1)u_{\theta^*}u_{\theta^*}^T f_{\theta^*} \xrightarrow{\mathcal{P}} \int u_{\theta_0}u_{\theta_0}^T f_{\theta_0}.$$

Therefore we have (3.25). To prove (3.26) note that

$$-n^{1/2}\nabla\varrho_n(\theta_0) = n^{1/2}\int A(\delta_n(\theta_0))\nabla f_{\theta_0},$$

and it follows from a slight extension of the results of Theorem 3.4 that

$$n^{1/2}\int \delta_n(\theta_0)\nabla f_{\theta_0} = n^{1/2}\int (g_n^* - f_{\theta_0})u_{\theta_0} \xrightarrow{\mathcal{D}} Z^* \sim N(0, I(\theta_0)),$$

with u_{θ_0} playing the role of ψ_g in that theorem. The result also follows directly from Theorem 3.3; however, the latter theorem requires stronger conditions than those assumed in the statement of the present theorem.

It is therefore enough to prove that

$$\left| n^{1/2}\int \left\{ A(\delta_n(\theta_0)) - \delta_n(\theta_0) \right\} \nabla f_{\theta_0} \right| \xrightarrow{\mathcal{P}} 0. \qquad (3.29)$$

From condition (h) $A'(\delta)$ and $A''(\delta)(\delta+1)$ are bounded. Thus, using Lemma 2.15, we have a finite B such that

$$\left| A(r^2 - 1) - (r^2 - 1) \right| \le B \times (r - 1)^2$$

for all $r \ge 0$. Thus,

$$\left| A(\delta_n(\theta_0)) - \delta_n(\theta_0) \right| \le B \left[(g_n^*/f_{\theta_0})^{1/2} - 1 \right]^2. \qquad (3.30)$$

Using this and (3.29), we have

$$\left| n^{1/2}\int \left\{ A(\delta_n(\theta_0)) - \delta_n(\theta_0) \right\} \nabla f_{\theta_0} \right| \le B\, n^{1/2}\int \{g_n^{*1/2} - f_{\theta_0}^{1/2}\}^2 |u_{\theta_0}|.$$

Now we consider

$$n^{1/2}\int \{g_n^{*1/2} - f_{\theta_0}^{1/2}\}^2 |u_{\theta_0}|.$$

From the inequality $(a-b)^2 \le (a-c)^2 + (b-c)^2$ for real numbers a, b, c, we get

$$n^{1/2}\int \{g_n^{*1/2} - f_{\theta_0}^{1/2}\}^2 |u_{\theta_0}| \le T_1 + T_2,$$

where

$$T_1 = n^{1/2}\int \{g_n^{*1/2} - \tilde{g}_n^{1/2}\}^2 |u_{\theta_0}|, \quad T_2 = n^{1/2}\int \{\tilde{g}_n^{1/2} - f_{\theta_0}^{1/2}\}^2 |u_{\theta_0}|.$$

The term T_1 is identical to the term $n^{1/2}\tau_{2n}$ of Theorem 3.4 where the convergence of this term to zero has been established. For T_2, note that the integral $\int \{\tilde{g}_n^{1/2} - f_{\theta_0}^{1/2}\}^2 |u_{\theta_0}|$ is essentially of the order of h_n^4, and the given conditions guarantee its convergence to zero. Also see Equation (4.1) of Tamura and Boos (1986), and the succeeding discussion therein. $\qquad\square$

3.4.1 Disparities in This Class

The different conditions on the distances required by Park and Basu (2004) to construct the general structure can now be consolidated as follows:

(a) $C(\cdot)$ is strictly convex and thrice differentiable,

(b) $C(\delta) \geq 0$, with equality only at $\delta = 0$ and $C'(0) = 0$, $C''(0) = 1$. Notice that this implies $A(0) = C'(0) = 0$ and $A'(0) = C''(0) = 1$.

(c) $C'(\delta)$, $A(\delta)$, $A'(\delta)$, $A'(\delta)(\delta + 1)$ and $A''(\delta)(\delta + 1)$ are bounded on $\delta \in [-1, \infty]$.

Although these conditions are quite strong, there is a rich class of distances satisfying the same. We list some of them here. Each family below, except ρ_4, is indexed by a single parameter α. For ρ_4 the tuning parameter is denoted by λ.

$$\rho_1(g, f) = \frac{1}{2} \int \frac{(g - f)^2}{\alpha \, g + \bar{\alpha} \, f}, \qquad \alpha \in (0, 1)$$

$$\rho_2(g, f) = \int \frac{(g - f)^2}{\sqrt{\alpha \, g^2 + \bar{\alpha} \, f^2} + \alpha \, g + \bar{\alpha} \, f}, \qquad \alpha \in (0, 1)$$

$$= \frac{1}{\alpha \bar{\alpha}} \int \left[\sqrt{\alpha \, g^2 + \bar{\alpha} \, f^2} - \alpha g - \bar{\alpha} f \right],$$

$$\rho_3(g, f) = \frac{1}{2} \int \frac{(g - f)^2}{\sqrt{\alpha \, g^2 + \bar{\alpha} \, f^2}}, \qquad \alpha \in (0, 1)$$

$$\rho_4(g, f) = \int \left[\frac{f}{\lambda^2} \{ \exp(\lambda - \lambda \, g/f) - 1 \} + \frac{g - f}{\lambda} \right], \qquad \lambda > 0$$

$$\rho_5(g, f) = \int \left[\frac{4}{\alpha \pi} g \tan \left(\frac{\pi \alpha}{2} \frac{g - f}{g + f} \right) - (g - f) \right], \qquad \alpha \in [0, 1)$$

$$\rho_6(g, f) = \int \left[\frac{2}{\alpha \pi} (g - f) \sin \left(\frac{\pi \alpha}{2} \frac{g - f}{g + f} \right) \right], \qquad \alpha \in [0, 1]$$

$$\rho_7(g, f) = \int \left[\frac{2}{\alpha \pi} (g - f) \tan \left(\frac{\pi \alpha}{2} \frac{g - f}{g + f} \right) \right], \qquad \alpha \in [0, 1)$$

where $\bar{\alpha} = 1 - \alpha$.

The family ρ_1 is the blended weight chi-square (BWCS) family which we have already encountered. The member of this family for $\alpha = 1/2$ is the symmetric chi-square (SCS). The families ρ_2 and ρ_3 are two other variants of ρ_1

TABLE 3.1

The $C(\cdot)$ and the $A(\cdot)$ functions corresponding to the distances presented in this section. Here $\bar{\alpha} = 1 - \alpha$.

Disparity	$C(\delta)$	$A(\delta)$
ρ_1	$\dfrac{\delta^2}{2(\alpha\delta + 1)}$	$\dfrac{\delta}{\alpha\delta + 1} + \dfrac{\bar{\alpha}}{2}\left[\dfrac{\delta}{\alpha\delta + 1}\right]^2$
ρ_2	$\dfrac{\delta^2}{1 + \alpha\delta + \sqrt{\alpha(\delta+1)^2 + \bar{\alpha}}}$	$\dfrac{1}{\alpha} - \dfrac{1}{\alpha\sqrt{\alpha(\delta+1)^2 + \bar{\alpha}}}$
ρ_3	$\dfrac{\delta^2}{2\sqrt{\alpha(\delta+1)^2 + \bar{\alpha}}}$	$\dfrac{\delta}{\sqrt{\alpha(\delta+1)^2 + \bar{\alpha}}} + \dfrac{\bar{\alpha}}{2}\dfrac{\delta^2}{\left[\sqrt{\alpha(\delta+1)^2 + \bar{\alpha}}\right]^3}$
ρ_4	$\dfrac{e^{-\lambda\delta} - 1 + \lambda\delta}{\lambda^2}$	$\dfrac{(\lambda + 1) - [(\lambda + 1) + \lambda\delta]e^{-\lambda\delta}}{\lambda^2}$
ρ_5	$\dfrac{4}{\alpha\pi}(\delta + 1)\tan\left(\dfrac{\alpha\pi}{2}\dfrac{\delta}{\delta + 2}\right) - \delta$	$4\left(\dfrac{\delta + 1}{\delta + 2}\right)^2 \sec^2\left(\dfrac{\alpha\pi}{2}\dfrac{\delta}{\delta + 2}\right) - 1$
ρ_6	$\dfrac{2}{\alpha\pi}\delta \sin\left(\dfrac{\alpha\pi}{2}\dfrac{\delta}{\delta + 2}\right)$	$\dfrac{2\delta(\delta + 1)}{(\delta + 2)^2}\cos\left(\dfrac{\alpha\pi}{2}\dfrac{\delta}{\delta + 2}\right) + \dfrac{2}{\alpha\pi}\sin\left(\dfrac{\alpha\pi}{2}\dfrac{\delta}{\delta + 2}\right)$
ρ_7	$\dfrac{2}{\alpha\pi}\delta \tan\left(\dfrac{\alpha\pi}{2}\dfrac{\delta}{\delta + 2}\right)$	$\dfrac{2\delta(\delta + 1)}{(\delta + 2)^2}\sec^2\left(\dfrac{\alpha\pi}{2}\dfrac{\delta}{\delta + 2}\right) + \dfrac{2}{\alpha\pi}\tan\left(\dfrac{\alpha\pi}{2}\dfrac{\delta}{\delta + 2}\right)$

with similar properties. The family ρ_4 is the generalized negative exponential disparity (Bhandari, Basu and Sarkar, 2006) which includes the negative exponential disparity for $\lambda = 1$. The families ρ_5, ρ_6 and ρ_7 are based on trigonometric functions and also satisfy the properties. Note that each of these three families contain the SCS family as limiting cases of $\alpha \to 0$. The families ρ_5 and ρ_6 have been proposed by Park and Basu (2000). In case of both ρ_6 and ρ_7, although these disparities satisfy conditions (a), (b) and (c), more theoretical and empirical investigations are necessary to get a better idea about the performance of the corresponding estimators. Notice that condition (c) above also implies $C'(\infty)$ and $C(-1)$ are finite, the conditions on the disparity that are necessary for establishing the breakdown results.

3.5 The Basu–Lindsay Approach for Continuous Data

In this section we discuss an alternative approach to minimum distance estimation based on disparities for continuous models. Our discussion in this section follows the work of Basu (1991) and Basu and Lindsay (1994). Suppose that $X_1, \ldots, X_n$ represent an independently and identically distributed random sample of size n from a continuous distribution G which has a corresponding density function g with respect to the Lebesgue measure or some other appropriate dominating measure. This is modeled by the family $\mathcal{F}$, and we wish to estimate the parameter θ which represents the best fitting model distribution. Let G_n denote the empirical distribution function. Because of the discrete nature of the data, a disparity between the data and the model cannot be directly constructed. As we have described earlier in this chapter, Beran (1977) took the approach of constructing a nonparametric kernel density estimate g_n^* from the data, and then minimized the distance between this nonparametric density estimate and the model density over the parameter space Θ. Subsequently, several other authors used this approach.

In Beran's approach, the choice of the sequence of kernels (or, more precisely, the sequence of smoothing parameters) becomes critical. The consistency of the kernel density estimator is very important in this case, and complicated conditions have to be imposed on the kernel to make things work properly. The approach taken by Basu and Lindsay (1994) differs from the above in that it proposes that the model be convoluted with the same kernel as well. To distinguish it from the approach of Beran and others, we will refer to the approach based on model smoothing as the Basu–Lindsay approach. Let us denote the kernel integrated version of the model by f_θ^*. In the Basu–Lindsay approach one then constructs a disparity between g_n^* and f_θ^*, and minimizes it over θ to obtain the corresponding minimum distance estimator.

An intuitive rationale for this procedure is as follows. The real purpose here is to minimize some measure of discrepancy between the data and the model. To make the data continuous, an artificial kernel has to be thrown into the system. However, one needs to ensure – through the imposition of suitable conditions on the kernel function and the smoothing parameter – that the additional smoothing effect due to the kernel vanishes asymptotically; in large samples the constructed disparity should really measure the pure discrepancy between the data and the model with minimal or no impact of the kernel and the smoothing parameter. In the Basu–Lindsay approach, one convolutes the model with the same kernel used on the data. In a sense, this compensates for the distortion due to the imposition of the kernel on the data by imposing the same distortion on the model. It is therefore expected that the kernel will play a less important role in the estimation procedure than it plays in Beran's approach, particularly in small samples. As we will see later in this section, one gets consistent estimators of the parameter θ even when the smoothing parameter is held fixed as the sample size increases to infinity.

For a suitable kernel function $K(x, y, h)$, let $g_n^*(x)$ be the kernel density estimate obtained from the data, and f_θ^* be the kernel smoothed model density. These densities may be defined as

$$g_n^*(x) = \int K(x, y, h) dG_n(y) \tag{3.31}$$

and

$$f_\theta^*(x) = \int K(x, y, h) dF_\theta(y) \tag{3.32}$$

respectively. A typical disparity ρ_C based on a disparity generating function C can now be constructed as

$$\rho_C(g_n^*, f_\theta^*) = \int C(\delta(x)) f_\theta^*(x) dx, \tag{3.33}$$

where the Pearson residual is now defined to be

$$\delta(x) = \frac{g_n^*(x) - f_\theta^*(x)}{f_\theta^*(x)},$$

i.e., it is now a residual between the smoothed data and the smoothed model. The minimum distance estimator corresponding to the above disparity is obtained by minimizing the disparity in (3.33) over the parameter space Θ. Under differentiability of the model, this is achieved by solving the estimating equation

$$-\nabla\rho_C(g_n^*, f_\theta^*) = \int A(\delta(x)) \nabla f_\theta^*(x) dx = 0, \tag{3.34}$$

where ∇ represents the gradient with respect to θ.

Since we are dealing with a kernel smoothed version of the model rather than the model itself, it is necessary to replace the maximum likelihood estimator (MLE) with some different reference point. As a natural analog to the MLE, one may choose the estimator which minimizes the likelihood disparity

$$LD(g_n^*, f_\theta^*) = \int \log(g_n^*(x)/f_\theta^*(x)) g_n^*(x) dx$$

between the smoothed versions of the densities. We call this estimator the MLE*. The subsequent results of this section will establish that all the minimum distance estimators based on disparities presented here are asymptotically equivalent to the MLE* under standard regularity conditions (Theorem 3.19).

The quantities

$$\tilde{u}_\theta(x) = \frac{\nabla f_\theta^*(x)}{f_\theta^*(x)} \tag{3.35}$$

and

$$u_\theta^*(y) = \int \tilde{u}_\theta(x) K(x, y, h) dx \tag{3.36}$$

will be used repeatedly in the rest of the section. We will refer to u_θ^* as the smoothed score function (as opposed to u_θ being the ordinary score function).

In the following, E_θ will represent expectation with respect to the density f_θ.

Lemma 3.9. [Basu and Lindsay (1994, Lemma 3.1)]. *Let g_n^* and f_θ^* be respectively the kernel density estimate obtained from the data and the kernel smoothed model density as defined in (3.31) and (3.32). Then the estimating equation of the MLE* can be written as*

$$-\nabla LD(g_n^*, f_\theta^*) = \frac{1}{n}\sum u_\theta^*(X_i) = 0. \tag{3.37}$$

Further $E_\theta(u_\theta^*(Y)) = \int u_\theta^*(y)f_\theta(y)dy = 0$ for all θ. Thus, Equation (3.37) represents an unbiased estimating equation.

Proof. By taking a derivative of the likelihood disparity $LD(g_n^*, f_\theta^*)$ we get

$$-\nabla LD(g_n^*, f_\theta^*) = \int \frac{g_n^*(x)}{f_\theta^*(x)}\nabla f_\theta^*(x)dx$$

$$= \int \tilde{u}_\theta(x)\left[\int K(x,y,h)dG_n(y)\right]dx.$$

Using Fubini's theorem this becomes

$$-\nabla LD(g_n^*, f_\theta^*) = \int\left[\int \tilde{u}_\theta(x)K(x,y,h)dx\right]dG_n(y)$$

$$= \int u_\theta^*(y)dG_n(y)$$

$$= \frac{1}{n}\sum_{i=1}^n u_\theta^*(X_i).$$

Also,

$$E_\theta[u_\theta^*(Y)] = \int\left[\int \tilde{u}_\theta(x)K(x,y,h)dx\right]f_\theta(y)dy$$

$$= \int \tilde{u}_\theta(x)\left[\int K(x,y,h)f_\theta(y)dy\right]dx$$

$$= \int\left(\frac{\nabla f_\theta^*(x)}{f_\theta^*(x)}\right)f_\theta^*(x)dx$$

$$= \int \nabla f_\theta^*(x)dx = 0.$$

The estimating function of the MLE* is thus an unbiased estimating function. Therefore the standard asymptotic results for unbiased estimating equations may be used to derive the asymptotic properties of our class of minimum distance estimators. Notice, however, that the relation

$$E_\theta[u_\theta^{*2}(X)] = -E_\theta[\nabla u_\theta^*(X)]$$

is not true. $\qquad\qquad\square$

3.5.1   Transparent Kernels

The minimum distance estimators generated through the Basu–Lindsay approach are not automatically first-order efficient. However, sometimes it is possible to choose the kernel (relative to the model) in such a way that no information is lost. Suppose $G = F_\theta$ for some $\theta \in \Theta$, i.e., the true distribution belongs to the model. We will show that in some cases the kernel function can be so chosen that for any minimum distance estimator $\hat\theta$ within the class of disparities, the asymptotic distribution of $n^{1/2}(\hat\theta - \theta)$ is multivariate normal with mean vector zero and asymptotic variance equal to the inverse of the Fisher information matrix; in this sense there is no loss in information. Kernels which allow the generation of first-order efficient estimators as above are called transparent kernels relative to the model. We give a formal mathematical definition of transparent kernels below.

Consider the parametric model $\mathcal{F} = \{F_\theta : \theta \in \Theta \subseteq \mathbb{R}^p\}$. Let $u_\theta^*(x)$ be the smoothed score function and $u_\theta(x) = \nabla \log f_\theta(x)$ be the ordinary score function. Then the kernel $K(x, y, h)$ will be called a transparent kernel for the above family of models provided the following relation holds:

$$Au_\theta(x) + B = u_\theta^*(x). \tag{3.38}$$

Here A is a nonsingular $p \times p$ matrix which may depend on the parameter θ and B is a p dimensional vector. However, since both u and u^* have expectation zero with respect to the density f_θ, we get $B = 0$ and the above relation may be simplified and written as

$$Au_\theta(x) = u_\theta^*(x). \tag{3.39}$$

The following lemma is immediate from an inspection of Equations (3.37) and (3.39). Note that the matrix A in Equation (3.39) is nonsingular.

Lemma 3.10. *Suppose that K is a transparent kernel for the family of models $\mathcal{F}$. Then the estimating equation for the MLE* *is simply the maximum likelihood score equation, and the MLE* *is the ordinary maximum likelihood estimator of θ.*

The above lemma shows that under a transparent kernel, the MLE* is the same as the MLE, and hence all minimum distance estimators, which are asymptotically equivalent to the MLE* are first-order efficient (Theorem 3.19 and Corollary 3.20). We give an example of a transparent kernel below.

Example 3.1. (Normal Model): Suppose that $f_\theta(x)$ is the $N(\mu, \sigma^2)$ density, $\theta = (\mu, \sigma^2)$, and consider K to be the normal kernel with smoothing parameter h (i.e., $K(x, y, h)$ is the $N(y, h^2)$ density at x). Then the smoothed density $f_\theta^*(x)$ is the normal $N(\mu, \sigma^2 + h^2)$ density. We will show that K is a transparent kernel for this family.

Here the ordinary score function and the smoothed score function are given by

$$u_\theta(x) = \begin{pmatrix} \dfrac{x - \mu}{\sigma^2} \\ \dfrac{1}{2\sigma^4}[(x - \mu)^2 - \sigma^2] \end{pmatrix}$$

and

$$u_\theta^*(x) = \begin{pmatrix} \dfrac{x - \mu}{\sigma^2 + h^2} \\ \dfrac{1}{2(\sigma^2 + h^2)^2}[(x - \mu)^2 - \sigma^2] \end{pmatrix}$$

respectively. Therefore $u_\theta^*(x) = A u_\theta(x)$, where A is the matrix

$$A = \begin{pmatrix} \dfrac{\sigma^2}{\sigma^2 + h^2} & 0 \\ 0 & \dfrac{\sigma^4}{(\sigma^2 + h^2)^2} \end{pmatrix},$$

and therefore, by definition K is a transparent kernel for this model.

The same may be verified for the m-variate multivariate normal model $\text{MVN}(\mu, \Sigma)$ with a multivariate normal kernel having covariance matrix $h^2 I$ (i.e., having a smoothing parameter of h across each component) where I represents the m dimensional identity matrix. We demonstrate, from first principles, that the MLE* of $\theta = (\mu, \Sigma)$ is the same as the MLE of θ in this case. Notice that the smoothed model density is the $\text{MVN}(\mu, \Sigma + h^2 I)$ density. Let G_n^* represent the distribution function corresponding to g_n^*. Then the estimating equations for the MLE* are, for the parameters in μ,

$$\int (\Sigma + h^2 I)^{-1}(y - \mu) dG_n^*(y) = 0,$$

and, for the parameters in Σ,

$$\int \{(y - \mu)(y - \mu)^T - E_{f_\theta^*}[(Y - \mu)(Y - \mu)^T]\} dG_n^*(y) = 0.$$

Since the distribution of G_n^* is the convolution of G_n and the $\text{MVN}(0, h^2 I)$ density, the solutions to the above estimating equations are just the ordinary maximum likelihood estimators

$$\hat{\mu} = \bar{X}, \quad \hat{\Sigma} = \frac{1}{n}\sum(X_i - \bar{X})(X_i - \bar{X})^T.$$

Once again, there is no loss of information. ∥

3.5.2 The Influence Function of the Minimum Distance Estimators for the Basu–Lindsay Approach

We will set up some more notation and definitions before we prove the next set of results. Corresponding to $\tilde{u}_\theta(x)$ we will define the following quantities to distinguish between the partials:

$$\tilde{u}_{j\theta}(x) = \nabla_j \log f_\theta^*(x).$$

$$\tilde{u}_{jk\theta}(x) = \nabla_{jk} \log f_\theta^*(x).$$

For similar distinction between successive derivatives in case of the smoothed score function we will write

$$u_{j\theta}^*(y) = \int K(x, y, h)\tilde{u}_{j\theta}(x)dx = \nabla_j \int \log f_\theta^*(x)K(x, y, h)dx.$$

$$u_{jk\theta}^*(y) = \int K(x, y, h)\tilde{u}_{jk\theta}(x)dx = \nabla_{jk} \int \log f_\theta^*(x)K(x, y, h)dx.$$

Definition 3.1. Let $J^*(\theta)$ be the $p \times p$ matrix whose jk-th element is given by $E_\theta[-u_{jk\theta}^*(X)]$. It is easy to see that the matrix $J^*(\theta)$ is nonnegative definite.

For the true, unknown, data generating density $g(x)$, let

$$g^*(x) = \int K(x, y, h)g(y)dy$$

be its kernel smoothed version. For a given disparity measure ρ_C, let $\theta^g = T(G)$ be the minimum distance functional defined by the minimization of $\rho_C(g^*, f_\theta^*)$. Let $A(\cdot)$ be the residual adjustment function associated with the disparity ρ_C. We will define $J^{*g}(\theta^g)$ to be the square matrix of dimension p whose jk-th element is given by

$$\int A'(\delta(x))\tilde{u}_{j\theta^g}(x)\tilde{u}_{k\theta^g}(x)g^*(x)dx - \int A(\delta(x))\nabla_{jk} f_{\theta^g}^*(x)dx \qquad (3.40)$$

and $u_{\theta^g}^{*g}(y)$ to be the p dimensional vector whose j-th component is

$$\int A'(\delta(x))\tilde{u}_{j\theta^g}(x)K(x, y, h)dx - \int A'(\delta(x))\tilde{u}_{j\theta^g}(x)g^*(x)dx. \qquad (3.41)$$

In both of the expressions above, the Pearson residual $\delta(x)$ equals

$$\delta(x) = \frac{g^*(x)}{f_{\theta^g}^*(x)} - 1. \qquad (3.42)$$

Let $G_\epsilon(x) = (1-\epsilon)G(x) + \epsilon\wedge_y(x)$, where $\wedge_y$ is as defined in Section 1.1. Let g_ϵ be the corresponding density, and g_ϵ^* be the associated smoothed density.

We denote the functional $T(G_\epsilon)$ obtained via the minimization of $\rho_C(g_\epsilon^*, f_\theta^*)$ as θ_ϵ. It satisfies

$$\int A(\delta_\epsilon(x))\nabla f_{\theta_\epsilon}^*(x)dx = 0, \tag{3.43}$$

where $\delta_\epsilon(x) = g_\epsilon^*(x)/f_{\theta_\epsilon}^*(x) - 1$. Then the influence function $T'(y)$ of the functional T at the distribution G is the first derivative of θ_ϵ evaluated at $\epsilon = 0$, which may be computed by taking a derivative of both sides of (3.43), and solving for $T'(y)$.

The form of the influence function of our minimum distance estimators is presented in the following theorem. The theorem is easily proved by mimicking the proof of Theorem 2.4.

Theorem 3.11. [Basu and Lindsay (1994, Lemma 5.1)]. *Let $G(x)$ be the true data generating distribution, and let $g(x)$ be the corresponding density; let $g^*(x)$ be the kernel smoothed version of $g(x)$. Also let T represent the minimum distance functional corresponding to a particular disparity measure ρ_C, and let $\theta^g = T(G)$. Let $A(\delta)$ be the corresponding residual adjustment function. In this setup, the influence function of the minimum distance functional is given by*

$$T'(y) = [J^{*g}(\theta^g)]^{-1}u_{\theta^g}^{*g}(y), \tag{3.44}$$

where $J^{*g}(\theta^g)$ and $u_{\theta^g}^{*g}$ are as defined above in Equations (3.40) and (3.41).

Corollary 3.12. *Suppose that $G = F_\theta$ for some $\theta \in \Theta$, so that the true distribution belongs to the model. Then the influence function of the minimum distance functional $T(\cdot)$ given in (3.44) reduces to*

$$T'(y) = [J^*(\theta)]^{-1}u_\theta^*(y). \tag{3.45}$$

If in addition K is a transparent kernel for the model f_θ, the influence function in (3.45) has the simple form

$$T'(y) = I^{-1}(\theta)u_\theta(y) \tag{3.46}$$

where $I(\theta)$ is the Fisher information about θ in f_θ and u_θ is the ordinary score function.

Proof. If $G = F_\theta$, it follows that $\theta = \theta^g$. Replacing this in (3.42), we get $\delta(x) = 0$; thus $A(\delta(x)) = 0$ and $A'(\delta(x)) = 1$. Substituting these in Equation (3.40), $J^{*g}(\theta^g)$ becomes

$$\int \tilde{u}_{j\theta}(x)\tilde{u}_{k\theta}(x)f_\theta^*(x)dx = -\int (\nabla_{jk} \log f_\theta^*(x))f_\theta^*(x)dx = -E_\theta[u_{jk\theta}^*(X)].$$

Again, substituting the above in Equation (3.41), we get

$$\int \tilde{u}_{j\theta}(x)K(x,y,h)dx - \int \tilde{u}_{j\theta}(x)f_\theta^*(x)dx = u_{j\theta}^*(\theta).$$

Thus, Equation (3.45) follows.

If in addition K is a transparent kernel for the family, $u_\theta^*(x) = A u_\theta(x)$, and taking derivatives of both sides of this expression one gets

$$\nabla u_\theta^*(x) = A \nabla u_\theta(x) + u_\theta(x) \nabla A. \tag{3.47}$$

Since $E_\theta(u_\theta(X)) = E_\theta(u_\theta^*(X)) = 0$, taking expectation of both sides of (3.47) one gets

$$E[\nabla u_\theta^*(X)] = -A I(\theta). \tag{3.48}$$

But $E[-\nabla u_\theta^*(X)] = J^*(\theta)$ by definition. Substituting this in Equation (3.48), we get $J^*(\theta) = A I(\theta)$. Thus

$$[J^*(\theta)]^{-1} u_\theta^*(y) = [A I(\theta)]^{-1} A u_\theta(y) = I^{-1}(\theta) A^{-1} A u_\theta(y) = I^{-1}(\theta) u_\theta(y)$$

and relation (3.46) holds. □

Remark 3.3. It is intuitively clear that the smoothed score $u_\theta^*(x)$ converges to the ordinary score function $u_\theta(x)$ when the bandwidth is allowed to go to zero. As a result, the coefficient matrix A defined in Equation (3.39) converges to the p-dimensional identity matrix in this case. This phenomenon may be directly observed in Example 3.1.

3.5.3 The Asymptotic Distribution of the Minimum Distance Estimators

We are now ready to prove the asymptotic distribution of the minimum distance estimators within the class of disparities. Most of the lemmas and theorems of this section are straightforward extensions of the results of Section 2.5. Thus, in the proofs of this section we will generally appeal to the logic of Section 2.5, and only emphasize the additional arguments necessary for the description of the kernel smoothed versions.

We assume that a random sample $X_1, \ldots, X_n$ is drawn from the true distribution G (having density function g), and let $K(x, y, h)$ be the kernel function used in the construction of the smoothed densities. A typical density in the model family is denoted by f_θ, while its kernel smoothed version is represented by f_θ^*; similarly g^* represents the kernel smoothed version of the true density. We are interested in the properties of the minimum distance estimator corresponding to the disparity ρ_C. Let θ^g represent the best fitting parameter which satisfies

$$\rho_C(g^*, f_{\theta^g}^*) = \min_{\theta \in \Theta} \rho_C(g^*, f_\theta^*).$$

The Pearson residuals in this context are defined as $\delta_n(x) = g_n^*(x)/f_\theta^*(x) - 1$ and $\delta_g(x) = g^*(x)/f_\theta^*(x) - 1$.

The first lemma follows in a straightforward manner from the definition in Equation (3.31).

Lemma 3.13. *Provided it exists,* $\text{Var}_g(g_n^*(x)) = \frac{1}{n}\lambda(x)$, *where* $\lambda(x)$ *is given by*

$$\lambda(x) = \int K^2(x, y, h)g(y)dy - [g^*(x)]^2.$$

We will assume that the kernel function K is bounded. From now on let

$$K(x, y, h) \leq M(h) < \infty,$$

where $M(h)$ depends on h, but not on x or y. Then

$$
\begin{aligned}
\lambda(x) &\leq \int K^2(x, y, h)g(y)dy \\
&\leq M(h) \int K(x, y, h)g(y)dy \\
&= M(h)g^*(x). \qquad (3.49)
\end{aligned}
$$

Lemma 3.14. $n^{1/4}(g_n^{*1/2}(x) - g^{1/2}(x)) \to 0$ *with probability 1 if* $\lambda(x) < \infty$.

Proof. One needs an assumption of the finiteness of the variance of $g_n^*(x)$ for the application of the strong law of large numbers. Once that is settled by the assumption of the finiteness of $\lambda(x)$, the rest of the proof is similar to that of Lemma 2.9. □

We will assume that the residual adjustment function $A(\delta)$ is regular in the sense of Definition 2.3 of Chapter 2. As in the case of the discrete models, we will define the Hellinger residuals for the structure under continuous models as

$$\Delta_n(x) = \frac{g_n^{*1/2}(x)}{f_\theta^{*1/2}(x)} - 1$$

$$\Delta_g(x) = \frac{g^{*1/2}(x)}{f_\theta^{*1/2}(x)} - 1.$$

Let $Y_n(x) = n^{1/2}(\Delta_n(x) - \Delta_g(x))^2$.

The following lemma is the analog of Lemma 2.13 and provides the bounds which are useful later in proving the main theorem of the section. The proof essentially retraces the steps of Lemma 2.13. Let $\delta_n(x) = g_n^*(x)/f_\theta^*(x) - 1$ and $\delta_g(x) = g^*(x)/f_\theta^*(x) - 1$.

Lemma 3.15. [Basu and Lindsay (1994, Lemma 6.3)]. *For any* $k \in [0, 2]$ *we have*

(i) $E[Y_n^k(x)] \leq E[|\delta_n - \delta_g|]^k n^{k/2} \leq (\lambda^{1/2}(x)/f_\theta^*(x))^k$.

(ii) $E[|\delta_n - \delta_g|] \leq (\lambda^{1/2}(x)/f_\theta^*(x))$.

Lemma 3.16. [Basu and Lindsay (1994, Lemma 6.4)]. $\lim_n E[Y_n^k(x)] = 0$ *for* $k \in [0, 2)$.

Proof. From Lemma 3.15 (i), $\sup_n E[Y_n^2(x)] < \dfrac{\lambda(x)}{(f_\theta^*(x))^2} < \infty$. The rest of the proof is similar to Lemma 2.14. □

Let $a_n(x) = A(\delta_n(x)) - A(\delta_g(x))$ and $b_n(x) = (\delta_n(x) - \delta_g(x))A'(\delta_g(x))$. We will find the limiting distribution of $S_{1n} = n^{1/2} \int a_n(x)\nabla f_\theta^*(x)dx$ by showing it to be equal to the limiting distribution of $S_{2n} = n^{1/2} \int b_n(x)\nabla f_\theta^*(x)dx$ which is easier to find.

Assumption 3.1. The smoothed version of the true density g satisfies

$$\int g^{*1/2}(x)|\tilde{u}_\theta(x)|dx < \infty. \tag{3.50}$$

.

Lemma 3.17. [Basu and Lindsay (1994, Lemma 6.5)]. *If $A(\cdot)$ is a regular RAF in the sense of Definition 2.3, and if Assumption 3.1 is satisfied, then*

$$\lim_n E\,|S_{1n} - S_{2n}| = 0.$$

Proof. Let $\tau_n(x) = n^{1/2}|a_n(x) - b_n(x)|$. Then by Lemma 2.14,

$$E|S_{1n} - S_{2n}| \le \int E(\tau_n(x))|\nabla f_\theta^*(x)|dx \le B \int E(Y_n(x))|\nabla f_\theta^*(x)|dx. \tag{3.51}$$

But by Lemma 3.15 (i), and Equation (3.49),

$$E(Y_n(x)) \le \frac{\lambda^{1/2}(x)}{f_\theta^*(x)} \le \frac{M^{1/2}(h)g^{*1/2}(x)}{f_\theta^*(x)},$$

so that $BM^{1/2}(h) \int g^{*1/2}(x)|\tilde{u}_\theta(x)|dx$ bounds the integral on the right-hand side of (3.51) which is bounded by Assumption 3.1. The rest of the proof is similar to that of Lemma 2.16. □

By an application of Markov's inequality it follows that $S_{1n} - S_{2n} \to 0$ in probability.

Corollary 3.18. *Suppose that*

$$V = \mathrm{Var}\left[\int K(x, X, h)A'(\delta_g(x))\tilde{u}_\theta(x)dx\right]$$

is finite and Assumption 3.1 holds. Then for a regular RAF,

$$S_{1n} \xrightarrow{D} Z^* \sim N(0, V).$$

Proof. The quantity in question $S_{1n} = n^{1/2} \int a_n(x) \nabla f_\theta^*(x) dx$. The asymptotic distribution of this is the same as that of $S_{2n} = n^{1/2} \int b_n(x) \nabla f_\theta^*(x) dx$ by the previous lemma, which can be written as

$$n^{1/2} \int (\delta_n(x) - \delta_g(x)) A'(\delta_g(x)) \nabla f_\theta^*(x) dx$$

$$= n^{1/2} \int (g_n^*(x) - g^*(x)) A'(\delta_g(x)) \tilde{u}_\theta(x) dx$$

$$= n^{1/2} \frac{1}{n} \sum_{i=1}^{n} \int (K(x, X_i, h) - E(K(x, X_i, h))) A'(\delta_g(x)) \tilde{u}_\theta(x) dx$$

and the result follows by a simple application of the central limit theorem. $\square$

Definition 3.2. We will say that the kernel integrated family of distributions $f_\theta^*(x)$ is smooth if the conditions of Lehmann (1983, p. 409, p. 429) are satisfied with $f_\theta^*(x)$ is place of $f_\theta(x)$.

Suppose $X_1, \ldots X_n$ are n independent and identically distributed observations from a continuous distribution G modeled by $\mathcal{F} = \{F_\theta : \theta \in \Theta \subseteq \mathbb{R}^p\}$. Let g and f_θ represent the corresponding densities, and let g^* and f_θ^* be the corresponding kernel smoothed versions. Consider a disparity $\rho_C(g_n^*, f_\theta^*)$ where g_n^* is the kernel density estimate based on the data. Let $C(\cdot)$ and $A(\cdot)$ be the associated disparity generating and residual adjustment functions respectively, and let θ^g be the best fitting value of the parameter. Let

$$\delta_n^g(x) = \frac{g_n^*(x)}{f_\theta^*(x)} - 1$$

and

$$\delta_g^g(x) = \frac{g^*(x)}{f_\theta^*(x)} - 1.$$

We make the following assumptions for the proof of our main theorem.

(B1) The family $\mathcal{F}$ is identifiable in the sense of Definition 2.2.

(B2) The probability density functions f_θ of the model distributions have common support so that the set $\mathcal{X} = \{x : f_\theta(x) > 0\}$ is independent of θ. Also the true density g is compatible with the model family $\{f_\theta\}$ in the sense of Definition 2.4.

(B3) The family of kernel integrated densities $\{f_\theta^*\}$ is smooth in the sense of Definition 3.2.

(B4) The matrix $J^{*g}(\theta^g)$ as defined in Equation (3.40) is positive definite.

(B5) The quantities

$$\int g^{*1/2}(x) |\tilde{u}_{j\theta}(x)| dx, \int g^{*1/2}(x) |\tilde{u}_{j\theta}(x) \tilde{u}_{k\theta}(x)| dx, \int g^{*1/2}(x) |\tilde{u}_{jk\theta}(x)| dx$$

are bounded for all j and k and all θ in an open neighborhood ω of θ^g.

(B6) For almost all x there exist functions $M_{jkl}(x)$, $M_{jk,l}(x)$, $M_{j,k,l}(x)$ that dominate, in absolute value, $\tilde{u}_{jkl\theta}(x)$, $\tilde{u}_{jk\theta}(x)\tilde{u}_{l\theta}(x)$, $\tilde{u}_{j\theta}(x)\tilde{u}_{k\theta}(x)\tilde{u}_{l\theta}(x)$ for all j, k, l, and that are uniformly bounded in expectation with respect to g^* and f_θ^* for all $\theta \in \omega$.

(B7) The RAF $A(\delta)$ is regular in the sense of Definition 2.3, and K_1 and K_2 represent the bounds for $A'(\delta)$ and $A''(\delta)(1+\delta)$ respectively.

Theorem 3.19. [Basu and Lindsay (1994, Theorem 6.1.)]. *Suppose that conditions (B1)–(B7) hold. Then there exists a consistent sequence θ_n of roots to the minimum disparity estimating equations in (3.34). Also the asymptotic distribution of $n^{1/2}(\theta_n - \theta^g)$ is multivariate normal with mean vector 0 and covariance matrix*

$$[J^{*g}(\theta^g)]^{-1}V_g[J^{*g}(\theta^g)]^{-1}$$

where V_g is the quantity defined in Corollary 3.18, evaluated at $\theta = \theta^g$.

Proof. The proof is essentially the same as the proof of Theorem 2.19, and here we point out just the minor differences.

In case of the linear terms, the difference

$$\left| \int A'(\delta_n^g(x))\nabla_j f_\theta^*(x)dx - \int A'(\delta_g^g(x))\nabla_j f_\theta^*(x)dx \right|$$

is now bounded by

$$K_1 \int |\delta_n^g(x) - \delta_g^g(x)||\nabla_j f_{\theta^g}^*(x)|.$$

But by Lemma 3.15 (ii) and Assumption 3.1,

$$E\left[K_1 \int |\delta_n^g(x) - \delta_g^g(x)| \; |\nabla_j f_{\theta^g}^*(x)|dx \right] \leq K_1 \int \lambda^{1/2}(x)|\tilde{u}_{j\theta^g}(x)|dx$$

$$\leq K_1 M^{1/2}(h) \int g^{*1/2}(x)|u_{j\theta^g}(x)|dx$$

$$< \infty,$$

so the linear term converges as desired.

In case of the quadratic term, the convergences

$$\left| \int (A'(\delta_n^g(x))(1+\delta_n^g) - A'(\delta_g^g(x))(1+\delta_g^g))\tilde{u}_{j\theta^g}(x)\tilde{u}_{k\theta^g}(x)f_{\theta^g}^*(x)dx \right| \to 0$$

and

$$\left| \int A(\delta_n^g)\nabla_{jk}f_{\theta^g}^*(x)dx - \int A(\delta_g^g)\nabla_{jk}f_{\theta^g}^*(x)dx \right| \to 0$$

hold in probability, so that $\nabla_k \int A(\delta_n^g(x))\nabla_j f_\theta^*(x)dx$ converges in probability to

$$-\int A'(\delta_g^g(x))(1+\delta_g^g(x))\tilde{u}_{j\theta^g}(x)\tilde{u}_{k\theta^g}(x)f_{\theta^g}^*(x)dx + \int A(\delta_g^g(x))\nabla_{jk}f_{\theta^g}^*(x)dx,$$

which is the negative of the (j, k)-th term of the matrix $J^{*g}(\theta^g)$. The rest of the proof is similar to Theorem 2.19. $\qquad\square$

Corollary 3.20. *Assume the conditions of Theorem 3.19. In addition, let us suppose that the true distribution belongs to the model ($G = F_\theta$ for some $\theta \in \Theta$) and K is a transparent kernel for the model family. Then the minimum distance estimator θ_n, $n^{1/2}(\theta_n - \theta^g)$ has an asymptotic normal distribution with mean 0 and covariance matrix $I^{-1}(\theta)$, where $I(\theta)$ is the Fisher information matrix about θ in f_θ.*

Proof. If $G = F_\theta$, $V_g = \text{Var}_\theta(u_\theta^*(X))$ and $J^{*g}(\theta^g) = J^*(\theta)$. If in addition K is a transparent kernel then $u_\theta^*(x) = A u_\theta(x)$ and $J^*(\theta) = A I(\theta)$. Then

$$
\begin{aligned}
[J^{*g}(\theta^g)]^{-1} V_g [J^{*g}(\theta^g)]^{-1} &= [J^*(\theta)]^{-1} \text{Var}_\theta(u_\theta^*(X))[J^*(\theta)]^{-1} \\
&= [J^*(\theta)]^{-1} \text{Var}_\theta(u_\theta^*(X))[J^*(\theta)^T]^{-1} \\
&= I^{-1}(\theta) A^{-1} A I(\theta) A^T [A^T]^{-1} I^{-1}(\theta) \\
&= I^{-1}(\theta).
\end{aligned}
$$

$\qquad\square$

Remark 3.4. Theorem 3.19 shows that all minimum distance estimators are asymptotically equivalent to the MLE* for the Basu–Lindsay approach. Corollary 3.20, in addition, shows that when the true distribution belongs to the model and the kernel is a transparent kernel, the relation

$$
\theta_n = \theta + n^{-1/2} I^{-1}(\theta) Z_n(\theta) + o_p(n^{-1/2}) \tag{3.52}
$$

continues to hold where Z_n is as defined in Equation (2.74). Thus, all minimum distance estimators that are asymptotically equivalent to the MLE* are also equivalent to the ordinarily maximum likelihood estimator under these conditions.

3.6 Examples

Example 3.2. This example involves Short's data for the determination of the parallax of the sun, the angle subtended by the earth's radius, as if viewed and measured from the surface of the sun. From this angle and available knowledge of the physical dimensions of the earth, the mean distance from earth to sun can be easily determined. The raw observations are presented in Data Set 2 of Stigler (1977).

To calculate our minimum distance estimators under the normal model, we have used the kernel density function with the Epanechnikov kernel $w(x) = 0.75(1 - x^2)$ for $|x| < 1$. Following Devroye and Györfi (1985, pp. 107–108), the

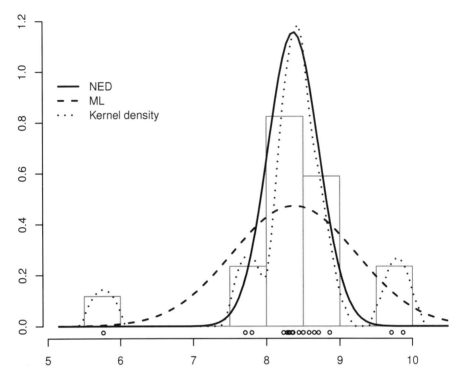

FIGURE 3.1
Normal density fits to Short's data.

mean L_1 criterion with the Epanechnikov kernel and the Gaussian probability density function leads to a bandwidth of the form

$$h_n = (15e)^{1/5}(\pi/32)^{1/10}\sigma n^{-1/5} = 1.66\sigma n^{-1/5}$$

TABLE 3.2
Estimates of the location and scale parameters for Short's data.

	LD	LD+D	HD	$PD_{-0.9}$	PCS
$\hat{\mu}$	8.378	8.541	8.385	8.388	8.198
$\hat{\sigma}$	0.846	0.552	0.348	0.273	1.046

	NED	$BWHD_{1/3}$	SCS	$BWCS_{0.2}$	$GKL_{1/3}$
$\hat{\mu}$	8.369	8.513	8.378	8.572	8.380
$\hat{\sigma}$	0.345	0.611	0.332	0.596	0.347

where σ is the standard deviation. We have used

$$\hat{\sigma} = \text{median}_i\{X_i - \text{median}_j\{X_j\}\}/0.6745$$

in place of σ in the above expression of the bandwidth. The maximum likelihood estimates of location (μ) and scale (σ) are 8.378 and 0.846, respectively. After removing the large outlier at 5.76, the maximum likelihood estimates of the location and scale become 8.541 and 0.552. In Table 3.2, we provide the minimum distance estimates of the location and scale parameters under several minimum distance methods. The LD column in Table 3.2 represents the maximum likelihood estimates for the full data and the LD+D column represents the outlier deleted maximum likelihood estimates after removing the large outlier at 5.76. Among the other minimum distance estimates, except the minimum PCS estimate, all the rest seem of offer some degree of resistance to the outliers. The minimum HD, NED, SCS and $GKL_{1/3}$ estimates are all quite close to each other and neatly downweight the large outliers. The $PD_{-0.9}$ statistic has the strongest downweighting effect. Even the minimum $BWHD_{1/3}$ and the minimum $BWCS_{0.2}$ estimates are significantly improved compared to the maximum likelihood estimator.

In Figure 3.1 we present the histogram of the observed data, on which we superimpose the kernel density estimate, as well as the normal fits corresponding to maximum likelihood and the minimum negative exponential disparity. The actual observations are indicated in the base of the figure. Notice that the maximum likelihood estimate of the scale parameter is widely different from the robust estimates. The outlier deleted maximum likelihood estimate of scale, as presented in Table 3.2, is substantially closer to the robust estimates; however, it is still not in the same ballpark. This is because this estimate is still highly influenced by the two moderate outliers close to 10, whereas the robust estimates model the central part of the data and effectively eliminate these observations as well. ‖

Example 3.3. This example involves Newcomb's light speed data (Stigler, 1977, Table 5). The histogram, a kernel density, and the normal fits using the maximum likelihood estimate and the minimum SCS estimate are presented in Figure 3.2. The data set shows a nice unimodal structure, and the normal model would have provided an excellent fit to the data except for the two large outliers. The minimum SCS estimate automatically discounts these large observations, unlike the maximum likelihood estimate. Table 3.3 provides the values of the different minimum distance estimates of the location and scale parameters. Once again the minimum SCS, the minimum HD, the minimum $GKL_{1/3}$ and the minimum NED estimates are remarkably close to each other. This time the minimum $BWHD_{1/3}$ and $BWCS_{0.2}$ estimates are also in this cluster. In this example, we have used the same kernel and the same formula for the bandwidth as in Example 3.2. Clearly the robust minimum distance estimators automatically discount the effects of the large outliers, unlike the maximum likelihood estimator. ‖

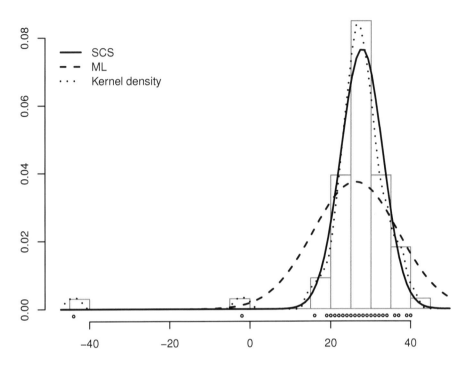

FIGURE 3.2
Normal density fits to the Newcomb data.

TABLE 3.3
Estimates of the location and scale parameters for the Newcomb data.

	LD	LD+D	HD	$PD_{-0.9}$	PCS
$\hat{\mu}$	26.212	27.750	27.738	27.710	14.405
$\hat{\sigma}$	10.664	5.044	5.127	4.746	22.242

	NED	$BWHD_{1/3}$	SCS	$BWCS_{0.2}$	$GKL_{1/3}$
$\hat{\mu}$	27.744	27.744	27.734	27.764	27.741
$\hat{\sigma}$	5.264	5.255	5.204	5.303	5.198

Example 3.4. This example concerns an experiment to test a method of reducing faults on telephone lines (Welch, 1987). Notice again that this data set has a normal structure with one, nonconforming, large outlier. This data set was previously analyzed by Simpson (1989b) using the Hellinger distance. The LD+D column in Table 3.5 represents the maximum likelihood estimate

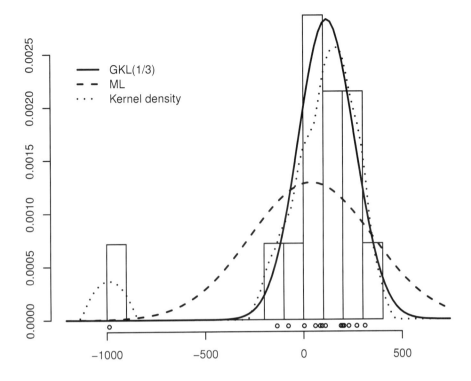

FIGURE 3.3
Normal density fits to the Telephone-line fault data.

after the large outlier (-988) is deleted from the data set. Figure 3.3 presents the kernel density estimate as well as the fits corresponding to the maximum likelihood estimate and the minimum $GKL_{1/3}$ estimate. All the robust estimates again belong to the same cluster and effectively eliminate the outlier. Here we have used the same kernel and the same formula for the bandwidth as in the two previous examples. ‖

All the three examples bear out the same story that has been observed in the examples in Chapter 2. The observations 1–5 in Section 2.6 are all relevant in the above examples. In spite of the fact that the influence function of all our minimum distance estimators are unbounded, a large outlier has no impact on any of our robust estimates which practically eliminates such observations from the data set; this is true for the minimum $BWHD_{1/3}$ and the minimum $BWCS_{0.2}$ estimates as well, although mild outliers produce a somewhat tentative response from these estimates. On the other hand, the maximum likelihood estimator, and, to a greater degree, the minimum PCS estimator, appear to be completely overwhelmed by any small deviation from the assumed model among the observed data.

TABLE 3.4

Estimates of the location and scale parameters for the Telephone-line fault data.

	LD	LD+D	HD	$PD_{-0.9}$	PCS
$\hat{\mu}$	38.929	117.923	116.769	114.646	-21.701
$\hat{\sigma}$	310.232	127.614	137.612	122.678	418.612

	NED	$BWHD_{1/3}$	SCS	$BWCS_{0.2}$	$GKL_{1/3}$
$\hat{\mu}$	119.093	117.600	118.623	119.179	117.794
$\hat{\sigma}$	145.430	142.560	142.553	146.837	141.499

Example 3.5. This example gives some indication of the possible benefits of the effect of model smoothing compared to the case where there is none. For comparison we consider the pseudo random sample of size 40 from the $N(0,1)$ distribution presented by Beran (1977). We used a normal kernel with several different values of the smoothing parameter h_n and determined the minimum Hellinger distance estimates of the location parameter μ and the scale parameter σ using the Basu–Lindsay approach. The obtained values of the estimates are presented in Table 3.5.

For the same data Beran determined the minimum Hellinger distance estimator of the parameters without model smoothing using the Epanechnikov kernel $w(x) = 0.75(1 - x^2)$ for $|x| \leq 1$ with

$$\hat{\mu}^{(0)} = \tilde{\mu} = \text{median}\{X_i\}, \quad \hat{\sigma}^{(0)} = \tilde{\sigma} = (0.674)^{-1}\text{median}\left\{|X_i - \hat{\mu}^{(0)}|\right\}$$

as the initial estimates. His scale estimate s_n was set at $\hat{\sigma}^{(0)}$, and the form of the kernel density estimate is as given in Equation (3.9).

We present the following observations on the basis of the values in Table 3.5, as well as the associated computational experience.

1. For the Basu–Lindsay approach, the estimates $\hat{\mu}$ and $\hat{\sigma}$ tend, respectively, toward the maximum likelihood estimates of the respective parameters, 0.1584 and 1.012, with increasing values of the smoothing parameter. This phenomenon will also be observed later on in other examples and for other distances. It illustrates a general observation for the Basu–Lindsay approach. As the value of the smoothing parameter increases, the minimum distance estimate ultimately settles on the maximum likelihood estimate. Most of the time, however, the parameters change slowly with increasing bandwidth.

2. An equally important observation is the following. While the estimate of the scale parameter in Table 3.5 is fairly stable over h, its rate of

TABLE 3.5

Estimates of the location ($\hat{\mu}$) and scale ($\hat{\sigma}$) parameters for Beran's data under the Basu–Lindsay approach.

h	0.4	0.5	0.6	0.7	0.8	0.9	1.0
$\hat{\mu}$	0.1495	0.1528	0.1551	0.1564	0.1572	0.1577	0.1580
$\hat{\sigma}$	0.9855	0.9903	0.9931	0.9949	0.9960	0.9967	0.9972

TABLE 3.6

Estimates of the location and scale parameters, as reported by Beran (1977).

h_n	0.4	0.5	0.6	0.7	0.8	0.9	1.0
$\hat{\mu}$	0.132	0.137	0.141	0.143	0.146	0.148	0.149
$\hat{\sigma}$	0.962	0.977	0.992	1.007	1.023	1.039	1.056

change with the smoothing parameter is significantly faster for the Beran approach; this can be observed from the results reported by Beran (1977, Table 1); for comparison, these numbers are presented here in Table 3.6. As the value of h_n changes from 0.4 to 1.0, the estimate of the scale parameter changes from 0.962 to 1.056. This is what one should expect. Larger values of the smoothing parameter spread out the density estimate. The estimated value of the scale parameter must show a corresponding increase to match the more spread out density. Thus, in small samples the choice of the smoothing parameter becomes critical. But in the Basu–Lindsay approach there is a corresponding spreading out of the model which compensates for the above. As a result, the variation in the estimate of scale in this approach is restricted between 0.9855 and 0.9972 as the smoothing parameter varies between 0.4 and 1.0.

That the scale estimates presented by him increased with increasing bandwidth was, of course, noticed by Beran who suggested additional considerations, such as closeness to the classical estimators, for choosing the value of the appropriate bandwidth.

3. Another fact related to the above issues which is not reflected directly in the table above but has been observed by the authors in repeated simulations is that smaller values of the smoothing parameter lead to more robust estimates (as opposed to more MLE-like estimates for larger values of the same) for the Basu–Lindsay approach. However, the density estimates do become spikier for smaller values h, and the optimization process becomes relatively more unstable. Typically one would require a larger number of iterations for the corresponding root solving process to converge. ‖

4

Measures of Robustness and Computational
Issues

As we have emphasized in the previous chapters, the primary motivation for developing the minimum distance estimators considered in this book is to generate a class of estimators which combine full asymptotic efficiency with strong robustness features. However, we have also pointed out that the estimators within our class of interest match the maximum likelihood estimator in terms of their influence functions at the model, and hence they are not first-order robust in that sense, although they are first-order efficient. This result notwithstanding, the robustness credentials of the minimum Hellinger distance estimator and many other minimum distance estimators within the class of disparities are undeniable. However, to the followers of the classical robustness theory some explanations about the sources of robustness of these minimum distance estimators are necessary. In this chapter we take on this robustness issue, and discuss in detail several robustness indicators including graphical interpretations, α-influence functions, second-order bias approximations, breakdown analysis and contamination envelopes. This comprehensive description establishes that the minimum distance estimators in this class are genuine competitors to the classical estimators present in the robustness literature. For instance, Tamura and Boos (1986) showed that the affine invariant minimum Hellinger distance estimator for multivariate location and scatter has a breakdown point of at least $\frac{1}{4}$, a quantity independent of the data dimension k; in contrast, the breakdown point of the affine invariant M-estimator is at most $1/(k + 1)$. Donoho and Liu (1988a) showed that the minimum Hellinger distance estimator has the best stability against Hellinger contamination among all Fisher consistent functionals. In real data situations, many of our minimum distance estimators behave like bounded influence estimators for all practical purposes. All in all, as our presentation in this chapter will show, the minimum distance estimators within the class of disparities are solid and useful tools for the applied statistician and researchers in other branches of science.

4.1   The Residual Adjustment Function

We have already introduced the residual adjustment function $A(\delta)$ of the disparity in connection with the minimum disparity estimating equations in Chapter 2; see Equations (2.37), (2.38) and (2.39). In that chapter we used the RAF extensively to describe the estimation procedure based on the minimization of the corresponding measure. In order to emphasize the robustness concept, and to describe the tradeoff between robustness and efficiency, we summarize the following points about the residual adjustment function of a disparity.

1. The description of the minimum distance procedure based on the RAFs of disparities helps to illustrate the adjustment of possible tradeoffs between efficiency and robustness. A very appealing feature of M-estimation in the location model is that one can see directly how the method limits the impact of large observations. Given residuals $\epsilon_i = y_i - \mu$ and an appropriate function ψ, one can solve for μ in the equation $\sum \psi(\epsilon_i) = 0$ to get the M-estimate of the location parameter. While the function $\psi(\epsilon) = \epsilon$ gives the sample mean – the desired estimator for normal theory maximum likelihood – as a solution, other ψ functions with $\psi(\epsilon) \ll |\epsilon|$ for large ϵ will limit the impact of large residuals on the estimator relative to maximum likelihood. Our minimum distance estimators have a very similar form, but with a modified definition of the residual δ. Here we solve an estimating equation which depends solely on a user-selected function $A(\delta)$ such that $A(\delta) = \delta$ gives a solution to the likelihood equations and $A(\delta) \ll \delta$ for large δ implies that large δ observations have a limited impact on the parameter estimate.

2. In spite of the similarity in the form of the estimating equations discussed in item 1 above, the difference in the nature of the residuals leads to different downweighting philosophies. The Pearson residual δ defined in Equation (2.6) is the argument of the residual adjustment function $A(\cdot)$. A large δ observation represents a probabilistic outlier; it depends on the observed relative frequency of a cell and its expectation under the model rather than the geometric distance between the observation and the target parameter. As a result, the procedure based on the residual adjustment function does not downweight a part of the data automatically.

3. The residual adjustment function analysis is distinctly different from the usual influence function approach. Unlike the ψ function in M-estimation, all residual adjustment functions $A(\cdot)$ lead to the same influence function at the model. However, like the ψ function, the choice of the function $A(\cdot)$ has a dramatic effect on robustness.

4. If the residual adjustment function becomes flat very quickly on the positive side of the δ axis, one would expect the corresponding estimator to have good stability properties (see Figures 2.1 and 2.2). The curvature $A_2 = A''(0)$ of the residual adjustment function at zero serves as a local measure of robustness of the minimum distance estimator. As discussed in Chapter 2, A_2 also serves as a measure of the second-order efficiency of the method (Lindsay 1994), with $A_2 = 0$ giving full second-order efficiency in the sense of Rao (1961).

5. The Pearson residuals are bounded below by -1, and we get a value of -1 only when the density estimate is zero at that point. Thus, the value $A(-1)$ reflects the impact – relative to maximum likelihood – of having holes in the data, i.e., empty cells where one would expect observations if the model were correct. Such observations, referred to as inliers in Chapter 2 of this book and the literature, do not usually represent a robustness concern; but, as we will see in Chapter 6, they may distort the small sample efficiency of the estimator and the corresponding test unless they are properly controlled.

   More generally, the term inlier refers to any value x where the observed proportion of data fall below expected levels. For the continuous case, observed and expected levels are described in terms of a nonparametric density estimate and the model density.

6. Suppose that the model has finite Fisher information. For residual adjustment functions satisfying $A(\delta)/\delta^{1/2} = O(1)$ as $\delta \to \infty$ the corresponding minimum distance procedure has a bounded effective influence in the following sense: if the data are contaminated at a fixed level ϵ at a point ξ, then the estimator stays bounded as $\xi \to \infty$. In fact, we will show (Section 4.8) that it converges to the estimator one would get by simply deleting the observation ξ from the sample.

   The above summary indicates that the RAF is a very important component of the description of the robustness of the minimum distance estimator within the class of disparities. Indeed, a description of the RAF is often sufficient to provide a good general idea of the possible behavior of the corresponding estimator in terms of its robustness and efficiency. However, the residual adjustment function has one important limitation. Since it is supported on the infinite set $[-1, \infty)$, it is not possible to graphically describe the behavior of residual adjustment function for very large outlying δ observations. Yet, there are many important and useful disparities for which the RAFs initially rise sharply on the positive side of the δ axis, sometimes growing faster than the residual adjustment function of the likelihood disparity $A_{\text{LD}}(\delta)$, only to eventually stabilize so that the impact of arbitrarily large outliers is entirely discounted by such disparities. This phenomenon, and many other features of the residual adjustment function, cannot be graphically observed. The disparity generating function $C(\cdot)$, which carries much of the same information as the residual adjustment function and can also give indications of the outlier rejection properties of the corresponding estimator, also has the same limitation of an unbounded domain. The weighted likelihood approach of Chapter 7 and Section 4.3 also do not lead to a completely satisfactory graphical representation.

An alternative graphical representation was presented by Park et al. (2002) based on a modified definition of the residual. This definition forces the residual to be constrained in the range $[-1, 1)$, so that neat graphical representations illustrating the features of the estimating equation are now possible. Following Park, Basu and Lindsay (2002), we describe this approach in the next section.

4.2   The Graphical Interpretation of Robustness

Although the discussion in this section is meaningful for both discrete and continuous models, for ease of presentation we illustrate them through discrete models. This will be our general approach throughout this chapter, but wherever appropriate, we will use the continuous model for illustration.

Consider the minimum disparity estimating equation given in (2.39). Under the disparity conditions of Definition 2.1, the residual adjustment function $A(\cdot)$ is strictly increasing. Since $\delta = 0$ implies $A(\delta) = 0$, one may consider the index of summation to be over the nonzero values of δ, and the equation reduces to

$$\sum_{\delta \neq 0} A(\delta(x)) \nabla f_\theta(x) = 0.$$

The above equation can be further rewritten as

$$\sum_{\delta \neq 0} w(\delta(x)) \delta(x) \nabla f_\theta(x) = 0, \tag{4.1}$$

where the weight function is given by

$$w(\delta(x)) = \frac{A(\delta(x))}{\delta(x)}. \tag{4.2}$$

Note that the left-hand side in (4.1) reduces to the ordinary maximum likelihood score function (or the estimating function for the likelihood disparity) when $w(\delta) \equiv 1$, identically. The weight function $w(\cdot)$ may be interpreted as a quantification of the amount by which the estimating function downweights (or magnifies) the effect of a particular observation relative to maximum likelihood. This facilitates a straightforward comparison with the latter method. However, under the disparity conditions in Definition 2.1, the weight function in (4.2) still operates on the unbounded domain $(0, \infty)$. We now introduce a modified version of the residual which eliminates this problem.

In analogy to the definition of the Pearson residual, consider the *Neyman residual* defined by

$$\delta_N(x) = \frac{d_n(x) - f_\theta(x)}{d_n(x)}. \tag{4.3}$$

To distinguish between these residuals, we will denote the Pearson residual defined in (2.6) by $\delta_P(x)$. Notice that the Neyman residual takes values in $(-\infty, 1]$, while the Pearson residual takes values in $[-1, \infty)$.

We now define a *combined residual* $\delta_c(x)$ as

$$\delta_c(x) = \begin{cases} \delta_P(x) & : & d_n \le f_\theta \\ \delta_N(x) & : & d_n > f_\theta \end{cases}.$$

It should be noted that this combined residual has a bounded domain given by $[-1, 1)$.

Since

$$\delta_P(x) = \frac{\delta_N(x)}{1 - \delta_N(x)},$$

the weight function $w(\delta_P)$, written in terms of the Neyman residuals, equals

$$\frac{1 - \delta_N}{\delta_N} A\left(\frac{\delta_N}{1 - \delta_N}\right).$$

Using the combined residuals, we rewrite the weight function as

$$w_c(\delta_c) = \begin{cases} \dfrac{A(\delta_c)}{\delta_c} & : & -1 \le \delta_c < 0 \\ A'(0) & : & \delta_c = 0 \\ \dfrac{1 - \delta_c}{\delta_c} A\left(\dfrac{\delta_c}{1 - \delta_c}\right) & : & 0 < \delta_c < 1 \\ A'(\infty) & : & \delta_c = 1 \end{cases}, \tag{4.4}$$

where $A'(\infty) = \lim_{x \to \infty} \frac{A(x)}{x}$. In the case $0 < \delta_c < 1$, the above weight function $w_c(\delta_c)$ can be interpreted as the weight function in (4.2) of Pearson residuals but in the Neyman scale.

Although the domain of δ_c values is a right open interval, we define $w_c(1) = A'(\infty)$ for the sake of continuity which gives completeness and ease of interpretation. By doing so, we have a smooth weight function $w_c(\cdot)$ defined over a bounded interval $[-1, 1]$ which satisfies the following:

1. It is differentiable at zero (the point where the Neyman and Pearson residuals are merged along the δ-axis).

2. For the likelihood disparity (LD), we have $w_c(\delta_c) \equiv 1$ identically.

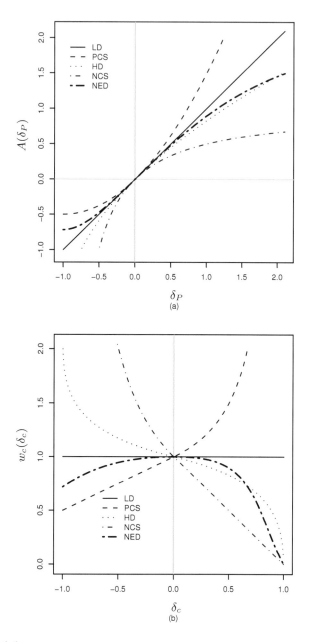

FIGURE 4.1
RAF functions, $A(\delta_P)$, and weight functions, $w_c(\delta_c)$, for LD, PCS, HD, NCS and NED.

3. Wherever the disparity downweights the corresponding values relative to maximum likelihood, the graph lies below the horizontal line $w_c(\delta_c) = 1$. This is true for the case of inliers (negative δ) as well as outliers (positive δ).

Having thus described the construction of the weight function, we now proceed to give graphical illustrations of its usefulness. For this purpose, we present, in Figure 4.1, the residual adjustment functions of the likelihood disparity (LD), the Hellinger distance (HD), the Pearson's chi-square (PCS), the Neyman's chi-square (NCS) and the negative exponential disparity (NED), respectively, together with the weight functions for these disparities. The disparity generating function and the residual adjustment function of these disparities are given in Table 2.1. Some important features of the weight functions become directly visible in the graphs of the weight functions in Figure 4.1. In what follows, we discuss these features of the weight function in relation to the figure, together with the correspondence of their properties with the features of the residual adjustment function.

1. It is clear that the weight functions in (4.4) as represented pictorially in Figure 4.1 cover the entire spectrum of possibilities in terms of the different values of the residual and its treatment by the corresponding residual adjustment function and the disparity. The graph shows the behavior of the weight function near $\delta_c = 0$, as well as around the boundary values $+1$ and -1.

2. Consider the behavior of the weight functions near $\delta_c = 0$. The requirement that $A'(0) = 1$ guarantees that the weight function satisfies $w(\delta_c) = 1$ at $\delta_c = 0$. Thus, unless the derivative of the weight function of a given disparity equals zero at $\delta_c = 0$, the weight function crosses the horizontal line $w_c(\delta_c) = 1$ at $\delta_c = 0$.

3. The derivative $w'_c(\delta_c)$ at $\delta_c = 0$ equals $A''(0)/2$; thus for disparities with negative values of the curvature parameter $A''(0)$, the weight function dips locally below the horizontal line $w_c(\delta_c) = 1$ as δ_c moves away from zero in the positive direction and rises locally above the same if $A''(0)$ is positive.

4. If $A''(0) = 0$, as in the case of the NED, the weight function is tangent to the horizontal line $w_c(\delta_c) = 1$ at $\delta_c = 0$. As it follows from Lindsay (1994), $A''(0) = 0$ (or equivalently $w'_c(0) = 0$) implies second-order efficiency in the sense of Rao (1961); more generally, curvature at zero is a measure of the second-order efficiency of the procedure. When $A''(0) = 0$, the second derivative of $w_c(\delta_c)$ exists at $\delta_c = 0$ and equals $A'''(0)/3$.

5. Since we are now able to view the behavior of the weight function over the entire domain of values of the residual, we can now observe the limiting behavior of the function when the combined residual becomes

arbitrarily close to the limiting value of $+1$. Values of δ_c close to 1 indicate that data are observed at these points at significantly higher proportions than predicted by the model. Figure 4.1 shows that $w_c(\delta_c)$ converges to zero as $\delta_c \to 1$ for each of the distances NED, NCS and HD. Thus, when the residual for an observation increases to its limiting value on the positive side, the estimating equations for these distances behave as if the observation is simply deleted from the data set.

On the other hand, the weight function for the PCS goes to ∞ as $\delta_c \to 1$. Instead of downweighting outlying observations, the PCS helps to magnify their effect. Thus, under the presence of observations for which residuals δ_c become arbitrarily close to 1, the procedure based on the PCS will break down completely. The maximum likelihood estimator and the likelihood disparity-based methods meet essentially the same fate, as they also fail to control the effect of large outliers. Note that for controlling large outliers, the weight function must converge to zero as $\delta_c \to 1$; converging to a finite constant strictly greater than zero is not enough.

As the weight function $w_c(\delta_c)$ for the likelihood disparity is identically equal to 1, the likelihood based methods do not satisfy this robustness requirement.

6. The limiting behavior of the weight function as $\delta_c \to -1$ is also important in establishing the overall performance of the corresponding minimum distance procedure, although it is usually not a robustness concern. Cells leading to large negative values of the combined residual (also called inliers) have less data than predicted under the model and appear to have a significant impact on the small sample efficiency of the estimators. The maximum likelihood estimator, although not outlier robust, seems to deal with inliers satisfactorily, and weight functions which stay around or below the horizontal line $w_c(\delta_c) = 1$ on the negative side of the δ_c axis seem to perform better in terms of inlier control; the weight function need not converge to zero as $\delta_c \to -1$ for proper inlier control. From Figure 4.1 we see that the PCS, the NED and the likelihood disparity control the inliers effectively, but the HD and the NCS fail to do so. For the NCS, the weight function actually diverges to infinity as $\delta_c \to -1$, signifying that the NCS is not even defined if there is a single empty cell.

Illustrative examples concerning the effects of downweighting outliers and inliers are given in Park, Basu and Lindsay (2002). As outlier robust disparities rarely treat inliers satisfactorily, modifications which improve the handling of inliers without affecting the treatment of outliers are of great practical value. An interesting approach for tackling this problem is the imposition of an empty cell penalty (Harris and Basu 1994, Basu, Harris and Basu 1996, Basu and Basu 1998, Mandal, Basu and Pardo 2010). Another inlier controlling approach involves the use of combined disparities – see Park, Basu and

Basu (1995) and Mandal, Bhandari and Basu (2011). More recently, other approaches presented in Patra, Mandal and Basu (2008) and Mandal and Basu (2010a,b) have provided additional interesting options for inlier control strategies. We will discuss all such methods in detail in Chapter 6.

In Figure 4.2, we investigate the weight function in (4.4) graphically for each of the following six families of disparities:

(a) the power divergence (PD) family: (Cressie and Read, 1984),

(b) the blended weight Hellinger distance (BWHD) family: (Lindsay, 1994; Basu and Lindsay, 1994; Basu and Sarkar, 1994b),

(c) the blended weight chi-square (BWCS) family: (Lindsay, 1994; Basu and Sarkar, 1994b),

(d) the generalized negative exponential disparity (GNED) family: Bhandari, Basu and Sarkar (2006),

(e) the generalized Kullback–Leibler (GKL) divergence family: (Park and Basu, 2003),

(f) the (one parameter) robustified likelihood disparity (RLD) family: (Chakraborty, Basu and Sarkar, 2001).

In the graphs of Figure 4.2, we focused on the tuning parameters over the ranges where they seem to matter most. Thus, we chose λ in $[-2, 1]$ for the PD, α in $[0, 1]$ for the BWHD and BWCS families, λ in $[0, 2]$ for the GNED, τ in $[0, 1]$ for the GKL family, and α in $[0, 1]$ for the RLD. For these families there is little practical reason for considering members outside these ranges. Note that the standard distributions generated by specific tuning parameters within these families are provided in Table 2.1.

For comparison, we have presented the second-order efficient member of each of these families in the corresponding graphs. Notice that for the PD and the GKL families the likelihood disparity itself is the second order efficient member. For the GNED the ordinary negative exponential disparity is the second-order efficient member. For the BWHD and the BWCS families the member corresponding to $\alpha = 1/3$ represent the second-order efficient element. For the BWHD family, in fact, the second-order efficient member is also third-order efficient (in the sense that $A'''(0) = 0$), so the corresponding weight function has a second order contact with the $w_c(\delta_c) = 1$ line. For the RLD family, all members are second-order efficient.

The graphs in Figure 4.2 show some important characteristics which cannot be graphically described using the residual adjustment function or the disparity generating function. Members of the BWHD and BWCS families with small positive values of α and members of the GNED family with small positive values of λ will appear to be highly nonrobust when viewed through their residual adjustment functions; such members have positive curvature parameters and inflate the weights of outliers locally around $\delta_c = 0$. However, the weight functions in Figure 4.2 show that the weights of these members eventually drop down to zero as $\delta_c \to 1$, and hence arbitrarily large outliers fail to unduly affect the inference procedure for these disparities. In terms of the big picture, therefore, the behavior of the minimum $\text{BWCS}_{0.2}$ estimator in Tables 2.2 and 2.3 is, after all, expected.

It can in fact be verified that for the BWHD and BWCS families, all members corresponding to $\alpha > 0$ eventually generate a zero weight as $\delta_c \to 1$. The same is true for all members corresponding to $\lambda > 0$ within the GNED family, and all members corresponding to $\tau > 0$ within the GKL family; the RLD members also have the same property as long as α is a positive finite constant. Within the PD family, in contrast, the weight functions converge to infinity as $\delta_c \to 1$ for all values of $\lambda > 0$. Thus, all members within this family beyond the second-order efficient likelihood disparity in the λ scale allow the weights to run off to infinity as the outliers become arbitrarily large.

Different families treat inliers differently. The GNED family, and to some extent the BWCS family, have a certain amount of natural protection against the inliers. For the GNED family, there is no serious inlier problem even for the highly robust members of the family. For the BWCS family, robust members such as the SCS (symmetric chi-square, corresponding to $\alpha = 1/2$) appear to be fairly stable against inliers, although the inlier problem does eventually show up within this family for larger values of α. For the RLD family, there is never any overweighting of either inliers or outliers. In fact, the two parameter RLD family can downweight inliers appropriately as well as outliers.

Clearly, the PD, the BWHD and the GKL families have a problem with inliers in the sense that the robust members of these families (those with negative values of the estimation curvature) inflate the weights of the inliers. Inlier controlling strategies are useful in the context of these families and this will be discussed in Chapter 6.

On the basis of our discussion in this section and given the background of the combined residuals and the graphical interpretation, we present a general definition of a robust disparity as follows.

Definition 4.1. A disparity function satisfying the conditions in Definition 2.1 will be called a "robust disparity" if the corresponding weight function $w_c(\delta_c)$ converges to zero as $\delta_c \to 1$.

Notice that in the above definition we have not specifically required that the estimation curvature of the disparity be negative (i.e., $w'(\delta_c) < 0$ at $\delta_c = 0$), so as to include the disparities such as the $\text{BWHD}_{0.2}$, $\text{BWCS}_{0.1}$ and $\text{NED}_{0.2}$, as presented in Figure 4.2 within our class of robust distances. In this sense we are primarily focusing on the manner in which the disparities treat the extreme outliers. Depending on the situation we may consider more restricted definitions of robust disparities which may require conditions on the estimation curvature, or, in a more restricted sense, require that $w_c(\delta_c) \leq 1$ for all $\delta_c > 0$.

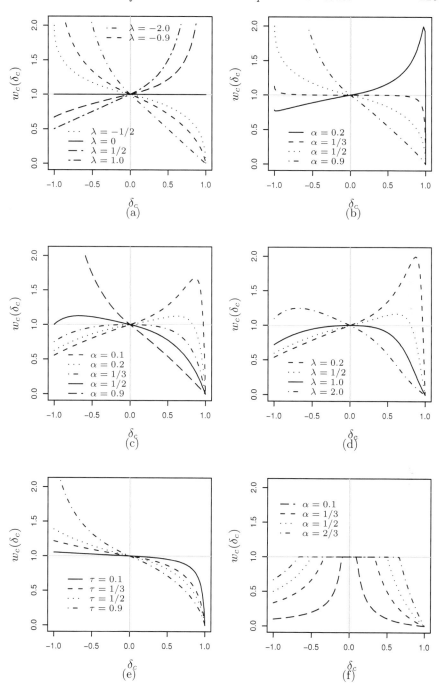

FIGURE 4.2
Weight function $w_c(\delta_c)$ for several classes of disparities. (a) PD; (b) BWHD; (c) BWCS; (d) GNED; (e) GKL; (f) RLD.

4.3   The Generalized Hellinger Distance

While most of our discussion in this chapter will relate to a section of disparities satisfying certain robustness conditions, in this section we will focus on a particular subclass of disparities where the robustness potential can be immediately recognized from the mathematical structure of the disparities in this subclass. The Hellinger distance is a particular member of this subclass. We use the discrete model for illustration. Suppose that we have a random sample $X_1, \ldots, X_n$, from a discrete distribution G modeled by a parametric family $\mathcal{F}$ as defined in Section 1.1. Let f_θ represent the density function of F_θ. Let $\mathcal{X} = \{0, 1, \ldots\}$ be the sample space, and let $d_n(x)$ be the relative frequency defined in Section 2.3.1. Notice that the minimum Hellinger distance estimator of θ minimizes

$$2 \sum_x (d_n^{1/2}(x) - f_\theta^{1/2}(x))^2 = 4(1 - \sum_x d_n^{1/2}(x) f_\theta^{1/2}(x)). \qquad (4.5)$$

Alternatively, therefore, the MHDE maximizes $\phi_{n,\theta} = \sum_x d_n^{1/2}(x) f_\theta^{1/2}(x)$. Under differentiability of the model, the corresponding (standardized) estimating equation can be expressed as

$$\phi_{n,\theta}^{-1} \sum_x d_n^{1/2}(x) f_\theta^{1/2}(x) u_\theta(x) = 0. \qquad (4.6)$$

In contrast, the ordinary likelihood equation has the form

$$\sum_x d_n(x) u_\theta(x) = 0, \qquad (4.7)$$

where $u_\theta(x)$ is the score function. It is immediately observed that while the estimating equations agree in the limit under the model, there are important differences as well. The effect of a large deviation from the model at a point x (reflected by a substantially higher value of $d_n(x)$ compared to what is predicted under the model f_θ) is severely downweighted, since the expectation in (4.6) is with respect to the density $\phi_{n,\theta}^{-1} d_n^{1/2}(x) f_\theta^{1/2}(x)$, unlike (4.7) where the expectation is with respect to the empirical density $d_n(x)$. Thus, while the model density has no role in the weighting of the individual score in the likelihood equation, it has a substantial impact in the case of the Hellinger distance, and as a result the effect of large outliers is often sufficiently muted in the estimating equation of the minimum Hellinger distance estimator (see Simpson, 1987; Basu, Basu and Chaudhuri, 1997).

An obvious generalization of the form in Equation (4.5) produces the generalized family of distances

$$D_\alpha(d_n, f_\theta) = K_\alpha(1 - \sum d_n^\alpha(x) f_\theta^{1-\alpha}(x)), \quad \alpha \in (0,1), \qquad (4.8)$$

where K_α is an appropriate nonnegative standardizing constant. The minimizer of $D_\alpha(d_n, f_\theta)$ does not depend on the value of K_α. The minimum Hellinger distance estimator corresponds to $\alpha = 1/2$. We denote the family defined in (4.8) as the generalized Hellinger distance family GHD_α, indexed by the tuning parameter α. Let $\phi_{n,\theta,\alpha} = \sum_x d_n^\alpha(x) f_\theta^{1-\alpha}(x)$. The *minimum generalized Hellinger distance estimator* with index α (MGHDE$_\alpha$) maximizes $\phi_{n,\theta,\alpha}$ over θ in Θ. Under differentiability of the model, the estimating equation for the corresponding estimator has the form

$$\phi_{n,\theta,\alpha}^{-1} \sum_x d_n^\alpha(x) f_\theta^{1-\alpha}(x) u_\theta(x) = 0, \tag{4.9}$$

so that the smaller the value of α, the higher is the degree of downweighting applied to an observation inconsistent with the model; for values of α smaller than $1/2$ a stronger downweighting effect relative to the minimum Hellinger distance estimator is exerted on such observations (see Simpson 1987, and Basu, Basu and Chaudhuri, 1997). The limiting case $\alpha = 1$ actually corresponds to maximum likelihood.

By writing $d_n^\alpha(x) f^{1-\alpha}(x) = (f_\theta(x)/d_n(x))^{1-\alpha} d_n(x) = w(x) d_n(x)$, and by rewriting the sum over x as a sum over the sample index i, (4.9) can be expressed in the equivalent weighted likelihood equation form

$$\frac{1}{n} \sum_i w(X_i) u_\theta(X_i) = 0 \tag{4.10}$$

where $w(x) = (f_\theta(x)/d_n(x))^{1-\alpha}$. For a cell with an unusually large frequency relative to the model, larger downweighting will be provided for smaller values of α. One can then use the final values of the fitted weights as diagnostics for the aberrant cells. On the other hand, the weights all converge to 1 as $n \to \infty$ when the model is correct. Thus, the weighted likelihood estimating equation has the property that it downweights observations discrepant with the model, but asymptotically behaves like the maximum likelihood score equation when the model is correct. Other weighted likelihood estimating equations having the same property and a somewhat broader scope have been discussed in Markatou, Basu and Lindsay (1997, 1998) and in Chapter 7 of this book, where the weights are based on the residual adjustment functions defined by Lindsay (1994). The weights in (4.10) are slightly different from the "disparity weights," although they share the same spirit.

The weighted likelihood estimating equations in (4.10) can themselves be solved by using an iteratively reweighted fixed point algorithm similar to the iteratively reweighted least squares. One can start with an initial approximation to the unknown θ, and create weights w; the next approximation to θ can then be obtained by solving the estimating equation in (4.10) by treating the weights w as fixed constants. This process can be continued till convergence.

It is easy to see that the residual adjustment function of the generalized

TABLE 4.1
The MGHDE$_\alpha$ of θ for several values of the tuning parameter in the range $\alpha \in (0, 1)$.

| α | Parameter Estimate | |
	First Experimental Run	Second Experimental Run
0.10	0.068	0.276
0.20	0.088	0.317
0.30	0.101	0.339
0.40	0.112	0.353
0.50	0.125	0.364
0.60	0.147	0.372
0.70	0.177	0.379
0.80	0.227	0.384
0.90	0.291	0.389
0.99	0.351	0.448
0.999	0.357	2.511
0.9999	0.357	3.003
0.99999	0.357	3.053
0.999999	0.357	3.058

Hellinger distance estimator has the form

$$A_\alpha(\delta) = \frac{(\delta + 1)^\alpha - 1}{\alpha}. \tag{4.11}$$

Example 4.1. We use different members of the generalized Hellinger distance family to estimate the parameter θ under a Poisson model for the data considered in Example 2.2.

For each of the two experimental runs the estimate $\hat{\theta}$ is presented in Table 4.1 for several values of the tuning parameter α ranging from 0.1 to 1. Notice that for the first experimental run, there is a slow degeneration in the parameter as α moves toward 1, but in the second case there is an abrupt "breakdown" in the parameter estimate around 0.999. Remarkably, the parameter estimates for the second case remain stable even when the tuning parameter α is in the neighborhood of 0.99. Thus, the outlier stability property within this family is not compromised by many disparities which are only minimally separated from the likelihood disparity in the α scale. ‖

4.3.1   Connection with Other Distances

Some routine algebra shows that the generalized Hellinger distance can be written as

$$-K_\alpha \sum_x d_n(x) \left[\left(\frac{d_n(x)}{f_\theta(x)} \right)^{\alpha-1} - 1 \right], \tag{4.12}$$

and a comparison with the form of the Cressie–Read family in (2.10) shows that the generalized Hellinger distance family is a subclass of the Cressie–Read family with $\alpha = \lambda + 1$, and $K_\alpha = [\alpha(1-\alpha)]^{-1}$. Thus, the family defined by Equation (4.8) represents the Cressie–Read family restricted to $\lambda \in (-1, 0)$.

The generalized Hellinger distance family has a very convenient mathematical form which is helpful in providing a nice description of the robustness properties of the corresponding estimators and tests, and we will heavily exploit this form in the rest of this chapter, as well as in the next.

The connection with the generalized Hellinger distance, and the Rényi divergences (see Section 2.3.2) are also immediately obvious. Comparing (2.31) and (4.8), we have

$$\mathrm{GHD}_\alpha(d_n, f_\theta) = \frac{1}{\alpha(1-\alpha)} \left[1 - e^{\alpha(\alpha-1)\mathrm{RD}_\alpha(d_n, f_\theta)} \right], \quad \alpha \in (0,1).$$

In the following theorem, we present the influence function of the minimum GHD_α estimators. These forms will be useful to us later on.

Theorem 4.1. *For the minimum generalized Hellinger distance functional* $T_\alpha(\cdot)$ *with parameter* α, *the influence function has the representation* $T'_\alpha(y) = D_\alpha^{-1} N_\alpha$, *where*

$$N_\alpha = \alpha u_{\theta^g}(y) g^{\alpha-1}(\xi) f_{\theta^g}^{1-\alpha}(\xi)$$

$$D_\alpha = -[(1-\alpha) \sum_x g^\alpha(x) f_{\theta^g}^{1-\alpha}(x) u_{\theta^g}(x) u_{\theta^g}^T(x) + \sum_x g^\alpha(x) f_{\theta^g}^{1-\alpha}(x) \nabla u_{\theta^g}(x)],$$

and $\theta^g = T_\alpha(G)$ *is the minimum generalized Hellinger distance functional at the true distribution* G. *Also,* $\nabla u_{\theta^g}(x)$ *is the* $p \times p$ *matrix of second partial derivatives of* $\log f_\theta(x)$ *evaluated at* θ^g. *The dimensions of* N_α *and* D_α *are* $p \times 1$ *and* $p \times p$ *respectively.*

4.4   Higher Order Influence Analysis

We have seen earlier in Section 2.3.7 that the influence function of all minimum distance estimators satisfying the disparity conditions is identical with that of the maximum likelihood estimator at the model. This is in fact necessary if the estimators are to have full asymptotic efficiency at the model. Yet, many of these estimators have very strong robustness features which are in significant contrast with the properties of the maximum likelihood estimator. This shows that the influence function, a tool extensively used in classical robustness literature, is not useful in describing the robustness of the minimum distance estimators within the class of disparities; on the basis of the influence function analysis, one would expect the other minimum distance estimators within the class of disparities to be no more robust than the maximum likelihood estimator when considering the sensitivity of the estimator to an infinitesimal contamination of a model element. In this section we will show that a second-order analysis (in comparison to the first-order analysis represented by the classical influence function approach) may be more meaningful in describing the robustness of these estimators, because it can often give a more accurate representation of the bias under contamination. The comparison of the behaviors of the first-order and the second-order approximations in the influence function approach was suggested by Lindsay (1994), and is a helpful tool in characterizing the robustness of these minimum distance estimators.

Let G represent the true distribution, and let $T(G_\epsilon) = T\left((1 - \epsilon)G + \epsilon \wedge_y\right)$ be the value of the functional T of interest evaluated at the contaminated distribution G_ϵ; here ϵ is the contaminating proportion, y is the contaminating point, and $\wedge_y$ is a degenerate distribution with all its mass on the point y. The influence function of the functional $T(\cdot)$ is given by

$$T'(y) = \frac{\partial T(G_\epsilon)}{\partial \epsilon}\Big|_{\epsilon=0} = \frac{\partial T((1 - \epsilon)G + \epsilon \wedge_y)}{\partial \epsilon}\Big|_{\epsilon=0}. \qquad (4.13)$$

As described earlier, the influence function gives a description of the effect of an infinitesimal contamination on the functional of interest.

Viewed as a function of ϵ, $\Delta T(\epsilon) = T(G_\epsilon) - T(G)$ quantifies the amount of bias and describes how the functional changes with contamination. Consider the first-order Taylor series expansion

$$\Delta T(\epsilon) = T(G_\epsilon) - T(G) \approx \epsilon T'(y). \qquad (4.14)$$

This approximation shows that the predicted bias up to the first order will be the same for all functionals having the same influence function. A generalization of this approximation shows that

$$T(G_n) - T(G) = \frac{1}{n}\sum T'(X_i) + o_p(n^{-1/2}) \qquad (4.15)$$

(von Mises, 1947; Fernholz, 1983) where $X_1, \ldots, X_n$ represents a random sample from the true distribution G modeled by $\{F_\theta\}$ and G_n is the empirical distribution function. This implies that under the true model any estimator with the same influence function as the maximum likelihood estimator has the same variance as the latter and is therefore asymptotically optimal provided the approximation in (4.15) is valid.

The approximations in (4.14) and (4.15) suggest that whenever two functionals have identical influence functions, they will be theoretically equivalent in terms of their first-order efficiency, as well in terms of their first-order robustness properties. In actual practice, the comparison of the maximum likelihood estimator with the other minimum distance estimators based on the above approximation turns out to be highly inaccurate in the description of the robustness properties. In the following, we present an example which illustrates the limitation of the influence function in capturing the robustness of the minimum Hellinger distance estimator. The example will show that the influence function approximation can drive the predicted bias of the minimum Hellinger distance estimator to its maximum possible value, but the actual bias is of a substantially smaller magnitude.

Example 4.2. To present the study in a simple context, we consider the estimation of the mean value parameter μ in a one parameter exponential family model. In this case the influence function of the maximum likelihood estimator equals

$$T'_{\mathrm{ML}}(y) = y - \mu \tag{4.16}$$

and since the maximum likelihood estimator is just the sample mean in this case, the approximation (4.14) is exact for the maximum likelihood estimator. Any other first-order efficient estimator including the minimum Hellinger distance estimator has the same influence function as the maximum likelihood estimator at the model as well as the same bias approximation as given in (4.14); this example will illustrate that while the influence function accurately describes the maximum likelihood estimator, it provides a poor and unacceptable approximation for for some of our other robust estimators.

The binomial $(10, \theta)$ model is chosen for illustration, with true value of θ equal to $1/2$. The contamination is mass ϵ at $y = 10$. The parameter θ of interest is a scalar multiple of the mean parameter, so that the analysis can be carried out in the spirit of the approximation in (4.16). As this approximation is exact for the maximum likelihood estimator, the plot of $\Delta T_{\mathrm{ML}}(\epsilon)$ is exactly linear. Together with $\Delta T_{\mathrm{ML}}(\epsilon)$, the bias plots $\Delta T_{\mathrm{HD}}(\epsilon)$ of the minimum Hellinger distance estimator and $\Delta T_{\mathrm{SCS}}(\epsilon)$ of the minimum symmetric chi-square estimator are presented in Figure 4.3 (a). Although the first-order predicted bias of all the three estimators are the same, the figure shows that there are huge differences in the actual bias plots. The bias plot for the maximum likelihood estimator is the diagonal straight line. Both the other two plots are practically flat till $\epsilon = 0.5$, indicating that there is virtually no bias for these estimators at this level; around $\epsilon = 0.5$ the estimated values (and hence their biases) of both these estimators shoot up to a point close to their maximum possible value. As a result, these curves have an entirely different structure than the smoothly and continuously increasing bias curve of the maximum likelihood estimator.

Although it is not recognizable from Figure 4.3 (a), if one were to blow up the lower left-hand portion of this figure one would see that the slope of the T_{HD} and T_{SCS} curves at $\epsilon = 0$ are actually the same as the slope of the T_{ML} curve at that point, as it should be from the influence function analysis.

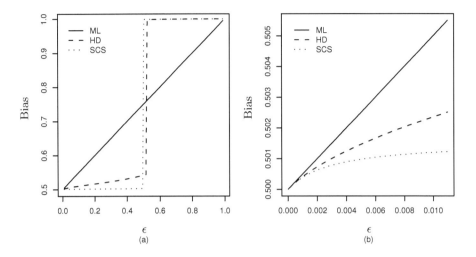

FIGURE 4.3
Bias plots for the ML, HD and SCS methods for the binomial example; the contaminating point is $y = 10$.

This is demonstrated in Figure 4.3 (b). In the case of the maximum likelihood curve, the linear form is exact and there is no additional impact of higher order terms; but in the case of the Hellinger distance and the symmetric chi-square, the impact of the higher order terms is substantial, which forces the T_{HD} and T_{SCS} curves to shrink very fast as ϵ increases. Thus, except for extremely small values of ϵ, the predicted bias from the linear form will be nowhere near the true bias for these two estimators. This provides ample motivation to search for alternatives to the first order influence function analysis when studying the robustness of the minimum distance estimators.   ‖

In the following, we will look at the second-order prediction of the bias curve by extending the above approach. In particular, we will look at the ratio of the two approximations and try to determine the range of values of ϵ where the first-order approximation is particularly poor compared to the second. As a natural generalization to the notation in Equation (4.13), let $T''(y)$ be the second derivative of $T(G_\epsilon)$ evaluated at $\epsilon = 0$. The second-order approximation to bias is given by

$$\Delta T(\epsilon) = \epsilon T'(y) + \frac{\epsilon^2}{2} T''(y),$$

and when we look at the ratio of the second-order (quadratic) approximation to the first (linear) as a simple measure of adequacy of the first-order approximation, we get

$$\frac{\text{quadratic approximation}}{\text{linear approximation}} = 1 + \frac{[T''(y)/T'(y)]\epsilon}{2}.$$

This indicates that if ϵ is larger than

$$\epsilon_{\text{crit}} = \left| \frac{T'(y)}{T''(y)} \right|,$$

the second-order approximation will differ by more than 50% compared to the first-order approximation. As in our other second-order results in the previous chapters, we will show that the estimation curvature A_2 is again one of the key items in the second-order analysis that we are going to present in this section.

In the next theorem we present the results of our second-order analysis in connection with a model involving a scalar parameter. Let G be the true distribution, and let $\theta = T(G)$.

Theorem 4.2. [Lindsay (1994, Proposition 4)]. *Let $I(\theta)$ be the Fisher information about a scalar parameter θ in model $\{f_\theta\}$. Assume that the true distribution belongs to the model. For an estimator defined by an estimating function of the form (2.39), we get, in the notation described in this section,*

$$T''(y) = T'(y)I^{-1}(\theta)\big[m_1(y) + A_2 m_2(y)\big],$$

where

$$m_1(y) = 2\nabla u_\theta(y) - 2E_\theta[\nabla u_\theta(X)] + T'(y)E_\theta[\nabla_2 u_\theta(X)]$$

and

$$m_2(y) = \frac{I(\theta)}{f_\theta(y)} + E_\theta[u_\theta^3(X)]\frac{u_\theta(y)}{I(\theta)} - 2u_\theta^2(y).$$

Here ∇ represents gradient with respect to θ, and A_2 is the estimation curvature of the disparity corresponding to the estimating Equation (2.39).

Proof. From Equation (2.49), the derivative of the functional $T(G_\epsilon)$ as defined in Section 2.3.7 with respect to ϵ has the representation

$$\frac{\partial}{\partial \epsilon}T(G_\epsilon) = \frac{A'(\delta_\epsilon(y))u_{\theta_\epsilon}(y) - \sum A'(\delta_\epsilon(x))u_{\theta_\epsilon}(x)g(x)}{\sum A'(\delta_\epsilon(x))u_{\theta_\epsilon}^2(x)g_\epsilon(x) - \sum A(\delta_\epsilon(x))\nabla_2 f_{\theta_\epsilon}(x)}, \qquad (4.17)$$

where $\theta_\epsilon = T(G_\epsilon)$ and $\delta_\epsilon(x) = g_\epsilon(x)/f_{\theta_\epsilon}(x) - 1$. Now suppose that the true distribution belongs to the model for some θ, and denote

$$g_\epsilon(x) = f_{\theta,\epsilon}(x) = (1 - \epsilon)f_\theta(x) + \epsilon\chi_y(x),$$

and let $G_\epsilon(x) = F_{\theta,\epsilon}(x)$. To find $T''(y)$, we need to take another derivative of the quantity on the right-hand side of Equation (4.17) and evaluate it at $\epsilon = 0$. We will, one by one, find the derivatives of the numerator and denominator in Equation (4.17) under $g_\epsilon(x) = f_{\theta,\epsilon}(x)$, and evaluate them at $\epsilon = 0$.

$$\text{Derivative of Numerator} = A''(\delta_\epsilon(y))\delta'_\epsilon(y)u_{\theta_\epsilon}(y)$$
$$+ A'(\delta_\epsilon(y))\nabla u_{\theta_\epsilon}(y)\frac{\partial}{\partial\epsilon}T(F_{\theta,\epsilon})$$
$$- \sum A''(\delta_\epsilon(x))\delta'_\epsilon(x)u_{\theta_\epsilon}(x)f_\theta(x)$$
$$- \sum A'(\delta_\epsilon(x))f_\theta(x)\nabla u_{\theta_\epsilon}(x)\frac{\partial}{\partial\epsilon}T(F_{\theta,\epsilon}).$$

Evaluating $\delta'_\epsilon(x)\big|_{\epsilon=0}$ (the derivative of $\delta_\epsilon(x)$ with respect to ϵ at $\epsilon = 0$), we get,

$$\delta'_\epsilon(x)\Big|_{\epsilon=0} = \frac{\partial\delta_\epsilon(x)}{\partial\epsilon}\Big|_{\epsilon=0} = -1 + \frac{\chi_y(x)}{f_\theta(x)} - u_\theta(x)T'(y). \qquad (4.18)$$

Substituting this in the above equation and evaluating at $\epsilon = 0$, we get,

$$\text{Derivative of Numerator}|_{\epsilon=0} = A_2\left[-u_\theta(y) + \frac{u_\theta(y)}{f_\theta(y)} - T'(y)u_\theta^2(y)\right]$$
$$+ T'(y)\nabla u_\theta(y) + T'(y)I(\theta).$$

Similarly,

$$\text{Derivative of Denominator}|_{\epsilon=0} = A_2\left[u_\theta^2(y) - I(\theta) - T'(y)\sum u_\theta^3(x)f_\theta(x)\right]$$
$$- I(\theta) - \nabla u_\theta(y) - T'(y)\sum u_\theta(x)\nabla_2 f_\theta(x).$$

Also, the value of the numerator and denominator evaluated at $\epsilon = 0$ are equal to $u_\theta(y)$ and $I(\theta)$ respectively. Then substituting these values in the derivative of (4.17), the desired result holds. □

Straightforward calculations and substitutions based on the above theorem lead to the following corollary.

Corollary 4.3. *Suppose that the model is a one parameter exponential family, and θ is the mean value parameter. In this case we get $m_1(y) = 0$, so that*

$$T''(y) = A_2 T'(y)Q(y), \quad \text{and} \quad \epsilon_{\text{crit}} = |A_2 Q(y)|^{-1},$$

where

$$Q(y) = \frac{1}{f_\theta(y)} + (y - \theta)\frac{E(Y - \theta)^3}{[E(Y - \theta)^2]^2} - 2\frac{(y - \theta)^2}{E(Y - \theta)^2}.$$

In particular, if y represents an observation with very small probability, the leading term of $Q(y)$ becomes dominant, and we get $Q(y) \approx 1/f_\theta(y)$, so that the quadratic approximation to the bias caused by a point mass contamination will predict a bias which is smaller by 50% or more compared to the linear approximation of the bias whenever ϵ is of the order of magnitude $\epsilon_{\text{crit}} \approx$

TABLE 4.2
$Q(y)$ for different possible contaminations for the binomial $(10, \theta)$ model in Example 4.2 for true value of $\theta = 0.5$
.

y	$Q(y)$	$1/Q(y)$
10	1004.00	0.001
9	89.600	0.011
8	17.156	0.058
7	5.933	0.168
6	4.419	0.226
5	4.349	0.230
4	4.419	0.226
3	5.933	0.168
2	17.156	0.058
1	89.600	0.011
0	1004.00	0.001

$f_\theta(y)/|A_2|$ or greater, provided A_2 is negative. On the other hand, for positive A_2, the quadratic approximation to the bias will predict *more* bias compared to the linear approximation. Once again the curvature parameter A_2 appears in the role of an indicator for robustness.

Example 4.3 (Continuation of Example 4.2). Going back to our binomial example, we compute the exact values of $Q(y)$ for each possible value of $y = 0, 1, \ldots, 10$. These numbers are presented in Table 4.2. Notice that the curvature parameter A_2 equals $-1/2$ for both the Hellinger distance and the symmetric chi-square measures. Thus, a point mass contamination at an extreme point like 10 or 0 will produce a predicted bias for the quadratic approximation which is smaller by 50% or more compared to the predicted bias for linear approximation whenever ϵ is greater than $2 \times 0.001 = 0.002$. Even a point mass contamination at the central value of 5 will lead to similar discrepancies between the two approximations for any contamination proportion higher than $2 \times 0.230 = 0.460$. Our calculations show that for a point mass contamination at $y = 10$, the actual values of ϵ beyond which the quadratic approximation to linear approximation differs by more than 50% in this example are approximately $\epsilon = 0.005$ for SCS and $\epsilon = 0.01$ for HD. ‖

4.5   Higher Order Influence Analysis: Continuous Models

Here we reproduce calculations similar to Section 4.4 for continuous models. We choose the Basu–Lindsay approach to illustrate the application of this technique in this case. Under the notation of Sections 3.5 and 4.4, we have the following analog of Theorem 4.2.

Theorem 4.4. *Let the parametric model be represented by the family of densities* $\{f_\theta\}$. *Let* θ *be a scalar parameter and* $I(\theta)$ *be the corresponding Fisher information which is assumed to be finite. Assume that the true distribution belongs to the model. For an estimator defined by an estimating function of the form (3.43), we get,*

$$T''(y) = T'(y)I^{-1}(\theta)[m_1(y) + A_2 m_2(y)],$$

where

$$m_1(y) = 2\nabla u_\theta^*(y) - 2E_\theta[\nabla u_\theta^*(X)] + T'(y)E_\theta[\nabla_2 u_\theta^*(X)] \qquad (4.19)$$

and

$$
m_2(y) = \frac{\int \tilde{u}_\theta^2(x) f_\theta^*(x) dx}{u_\theta^*(y)} \int K^2(x, y, h)\tilde{u}_\theta(x)\frac{1}{f_\theta^*(x)}dx
$$
$$
-2\int \tilde{u}_\theta^2(x)K(x, y, h)dx + T'(y)\int \tilde{u}_\theta^3(x) f_\theta^*(x)dx. \qquad (4.20)
$$

In addition, when the model is a one parameter exponential family model with θ being the mean value parameter, and K is a transparent kernel for the model, we get, from Equation (4.19)

$$m_1(y) = A\left\{2\nabla u_\theta(y) - 2E_\theta[\nabla u_\theta(X)] + T'(y)E_\theta[\nabla_2 u_\theta(X)]\right\},$$

where $u_\theta(x)$ represents the ordinary score function. Hence $m_1(y)$ equals zero from Corollary 4.3. Here A is the matrix in Equation (3.39) defining the transparent kernel.

  An interpretation similar to $m_2(y) \approx 1/f_\theta(y)$ as in the discrete case is more difficult to achieve here because of the presence of the kernel function, the bandwidth, the smoothed densities and the associated integrals. However, actual calculations show that the quantity $m_2(y)$ has behavior similar to the discrete case; it assumes a larger positive value at a more unlikely observation under the model. For example, at the $N(\mu, 1)$ model, a normal kernel with bandwidth $h = 0.05$ gives $m_2(2) = 96.336$ and $m_2(4) = 41732.330$, for the true value of $\mu = 0$. The bandwidth does have a substantial effect on the value of $m_2(y)$ however, and under $h = 0.5$, the same quantities are now $m_2(2) = 3.325$ and $m_2(4) = 1794.768$. This is consistent with our previous observation that smaller h leads to greater robustness and larger h leads to more MLE-like solutions. See Basu (1991) and Basu and Lindsay (1994) for more details of such calculations.

The description in the following section is general, relating to both discrete and continuous models.

4.6   Asymptotic Breakdown Properties

The breakdown point of the estimator (Hampel, 1971) is another of the accepted measures of its robustness. The paper by Donoho and Huber (1983) is still one of the most comprehensive descriptions of the notion of breakdown. The breakdown point of a statistical functional is roughly the smallest fraction of contamination in the data that may cause an arbitrarily extreme value in the estimate.

Many of the minimum distance estimators within the class of disparities are remarkably robust in terms of their breakdown properties. In this section we will describe the breakdown issue in the context of these estimators. The breakdown properties of the minimum Hellinger distance estimator again takes historical precedence.

4.6.1   Breakdown Point of the Minimum Hellinger Distance Estimator

For the minimum Hellinger distance estimator, we will present the derivation of the breakdown point bound at a generic distribution G. For this purpose, we will consider the contaminated sequence of distributions

$$H_{\epsilon,n} = (1 - \epsilon)G + \epsilon K_n, \qquad (4.21)$$

where $\{K_n\}$ is some appropriate sequence of contaminating distributions and ϵ is the contaminating proportion. Our description in this section will follow the work of Simpson (1987). Given two arbitrary distributions H and K, we will define, in the spirit of the above work, the corresponding affinity measure

$$s(h, k) = \int h^{1/2} k^{1/2},$$

where h and k are the density functions for the distributions H and K respectively. We state the breakdown point result for the minimum Hellinger distance functional in the following theorem.

Theorem 4.5. [Simpson (1987, Theorem 3)]. *Let $\hat{s} = \max s(g, f_t)$, $t \in \Theta$ and suppose the maximum occurs in the interior of Θ. Let $s^* = \lim\limits_{M \to \infty} \sup\limits_{|t| > M} s(g, f_t)$.*

If

$$\epsilon < (\hat{s} - s^*)^2/[1 + (\hat{s} - s^*)^2] \tag{4.22}$$

then there is no sequence of the form (4.21) for which

$$|T(H_{\epsilon,n}) - T(G)| \to \infty$$

as $n \to \infty$.

Proof. Suppose $|T(H_{\epsilon,n}) - T(G)| \to \infty$, where $H_{\epsilon,n}$ is as in (4.21). The proof proceeds by showing that this must imply that ϵ exceeds the bound in (4.22). Let $\theta = T(G)$ be the maximizer of $s(g, f_t)$ over $t \in \Theta$. Since $|T(H_{\epsilon,n}) - T(G)| \to \infty$, there must be a sequence $\{\theta_n\}$ with $|\theta_n| \to \infty$ for which $s(h_{\epsilon,n}, f_{\theta_n}) > s(h_{\epsilon,n}, f_\theta)$ infinitely often. The quantity $\{\theta_n\}$ is essentially representing the functional $T(H_{\epsilon,n})$ at the contaminated distribution under breakdown. Now

$$
\begin{aligned}
s(h_{\epsilon,n}, f_\theta) &= \int h_{\epsilon,n}^{1/2} f_\theta^{1/2} \\
&= \int [(1-\epsilon)g + \epsilon k_n]^{1/2} f_\theta^{1/2} \\
&\geq (1-\epsilon)^{1/2} \int g^{1/2} f_\theta^{1/2} = (1-\epsilon)^{1/2} \hat{s}.
\end{aligned}
$$

On the other hand,

$$
\begin{aligned}
s(h_{\epsilon,n}, f_{\theta_n}) &= \int h_{\epsilon,n}^{1/2} f_{\theta_n}^{1/2} \\
&= \int [(1-\epsilon)g + \epsilon k_n]^{1/2} f_{\theta_n}^{1/2} \\
&\leq (1-\epsilon)^{1/2} \int g^{1/2} f_{\theta_n}^{1/2} + \epsilon^{1/2} \int k_n^{1/2} f_{\theta_n}^{1/2} \\
&\leq (1-\epsilon)^{1/2} s(g, f_{\theta_n}) + \epsilon^{1/2}.
\end{aligned}
$$

But, by definition

$$s(g, f_{\theta_n}) \leq s^* + \gamma$$

eventually, for every $\gamma > 0$. Thus, $s(h_n, f_{\theta_n}) > s(h_n, f_\theta)$ infinitely often implies that

$$(1-\epsilon)^{1/2} \hat{s} \leq (1-\epsilon)^{1/2} s^* + \epsilon^{1/2}.$$

It is then a simple matter to check that this implies

$$\epsilon \geq (\hat{s} - s^*)^2/[1 + (\hat{s} - s^*)^2].$$

This completes the proof.   □

Very often, $s^* = 0$, in which case the breakdown bound depends solely on the value of $\hat{s}$; the latter value quantifies the fit of the model to the distribution G, with $\hat{s} = 1$ when a perfect fit is achieved. This is the case when G belongs to the model family. Thus, for example, if $\mathcal{F}$ represents the Poisson model and G is a specific Poisson distribution, $\hat{s} = 1$ and $s^* = 0$, so that the asymptotic breakdown point as given by (4.22) is $\frac{1}{2}$.

4.6.2   The Breakdown Point for the Power Divergence Family

As it has been demonstrated in Section 4.3, the generalized Hellinger distance is essentially the subclass of the power divergences of Cressie and Read (1984) restricted to the range $\lambda \in (-1, 0)$. The primary reason for expressing the distance in the particular form as has been done in Section 4.3 is to exploit the corresponding affinity representation

$$s_\alpha(g, f) = \int g^\alpha f^{1-\alpha}, \tag{4.23}$$

which appears to be extremely useful in demonstrating the robustness potential of these distances. Working with the above form of the affinity, one can easily extend the breakdown results of Theorem 4.5 to the case of the generalized Hellinger distance. From the robustness point of view, this is the segment of the power divergence family which primarily concerns us. Values of λ equal to or larger than zero have little or nothing to offer in terms of outlier stability. Values of λ which are equal to or smaller than -1 generate distances which are not defined even if there is a single empty cell; hence their practical use is limited (although their outlier resistant properties are not in question). In this sense the following Theorem relates to the members of the power divergence family that are of real interest to us.

Theorem 4.6. *Let s_α be as defined in (4.23), and consider the contaminated model given in (4.21). Let $\hat{s}_\alpha = \max s_\alpha(g, f_t), t \in \Theta$, and suppose that the maximum occurs in the interior of Θ. Let $s_\alpha^* = \lim_{M \to \infty} \sup_{|t| > M} s_\alpha(g, f_t)$. If*

$$\epsilon < (\hat{s}_\alpha - s_\alpha^*)^{1/\alpha} / \left[1 + (\hat{s}_\alpha - s_\alpha^*)^{1/\alpha} \right] \tag{4.24}$$

then there is no sequence of the form (4.21) for which $|T(H_{\epsilon,n}) - T(G)| \to \infty$ as $n \to \infty$.

Proof. Note that

$$s_\alpha(h_{\epsilon,n}, f_\theta) \geq (1 - \epsilon)^\alpha \hat{s}_\alpha$$

and

$$s_\alpha(h_{\epsilon,n}, f_{\theta_n}) \leq (1 - \epsilon)^\alpha s_\alpha(g, f_{\theta_n}) + \epsilon^\alpha,$$

where $\theta = T_\alpha(G)$ and $\theta_n = T_\alpha(H_{\epsilon,n})$. The result is then a simple modification of Theorem 4.5.   □

The implications of the above theorem are essentially the same as that of Theorem 4.5. Thus, for example, if G is a specific Poisson distribution and $\mathcal{F}$ represents the Poisson model, the asymptotic breakdown point of the MGHD$_\alpha$ functional is $\frac{1}{2}$. Indeed the same result holds for most parametric models. Thus, Simpson's (1987) result for the breakdown of the minimum Hellinger distance estimator extends to the members of the power divergence family for $\lambda \in (-1, 0)$. See also Simpson (1989a) and Basu, Basu and Chaudhuri (1997).

Example 4.4. In this example, we contrast some minimum distance methods with an outlier screen used by Woodruff et al. (1984). A similar comparison was presented by Simpson (1987). The comparison amounts to a study of the breakdown points of the two procedures.

The sequential outlier screen of Woodruff et al. (1984) attempts to delete erroneous large counts and excludes the largest count at any stage by comparing it to a probability generated by the Poisson model. Let $F_\theta(\cdot)$ be the Poisson distribution function corresponding to parameter θ. Let $\bar{X}_n$ be the sample mean, and $X_{(n)}$ be the largest count at a given stage involving n observations. According to the outlier screen strategy, one uses $\bar{X}_n$ to estimate the Poisson mean parameter θ if

$$[F_{\bar{X}_n}(X_{(n)})]^n \leq 1 - \alpha_0$$

for some prespecified small probability α_0 (such as 0.01). Otherwise one discards $X_{(n)}$, and repeats the procedure by setting $n = n - 1$.

Consider the effect of adding k large contaminants at the value c to augment the given data. The mean of the augmented sample is $m(c, k) = (n\bar{X}_n + kc)/(n + k)$. One would fail to detect and eliminate one or more such outliers, i.e., masking would occur, when

$$[F_{m(c,k)}(c)]^{n+j} \leq 1 - \alpha_0$$

for some $j \in \{1, \ldots, k\}$. As a specific example of the effect of masking, consider 10 observations with frequencies 4, 3, 2 and 1 at the values 0, 1, 2 and 3 respectively. The ordinary mean is $\bar{X}_{10} = 1$. Consider the effect of adding additional "sevens" to the data, and let $\alpha_0 = 0.01$. Then the outlier screen fails to detect the effect of more than one contaminating values, and the estimated mean registers a jump from 1 (at $k = 1$) to 2 (at $k = 2$). The results are presented in Table 4.3. The three estimators are the outlier screened maximum likelihood estimator obtained through the above method and the minimum distance estimators based on the Hellinger distance (HD) the symmetric chi-square (SCS). The table shows that the effect of the additional contaminants on the last two estimators is negligible.

Simpson (1987) shows that a breakdown in the above outlier screen procedure does not occur for fixed n and k. All outliers will be eventually excluded if they are sufficiently large. However, in the asymptotic setup where $n \to \infty$ and $k/(n + k) \to \epsilon \in (0, 1)$, Simpson demonstrates the existence of a sequence for which the outlier screen eventually fails, so that the asymptotic breakdown point of the outlier screen is zero.   ‖

TABLE 4.3
Example of masking in the outlier screen of Woodruff et al. (1984)

			$\hat{\theta}$		
k	$m(c,k)$	$[F_{m(c,k)}(c)]^{n+k}$	OML	HD	SCS
0	1.0	0.9999	1.0	0.9338	0.9872
1	17/11	0.9977	1.0	0.9614	0.9938
2	2.0	0.9869	2.0	0.9741	0.9978

4.6.3 A General Form of the Breakdown Point

The derivations of the previous section made use of the specific forms of the disparities. Here we establish the breakdown point of the minimum distance functional $T(G)$ under general conditions on the disparity. We also expand below the scope of our assumptions to deal with the case that is of real interest. Consider the contamination model.

$$H_{\epsilon,n} = (1 - \epsilon)G + \epsilon K_n,$$

where $\{K_n\}$ is a sequence of contaminating distributions. Let $h_{\epsilon,n}$, g and k_n be the corresponding densities. Following Simpson (1987), we say that there is breakdown in T for ϵ level contamination if there exists a sequence K_n such that $|T(H_{\epsilon,n}) - T(G)| \to \infty$ as $n \to \infty$.

We write below $\theta_n = T(H_{\epsilon,n})$. We make the following assumptions for the breakdown point analysis. The first three assumptions presented here intuitively reflect the worst possible choice of the contamination scenario and hence describe the situation of real interest. Assumption A4 below denotes the specific assumption on the disparity.

Assumption 4.1. The contaminating sequence of densities $\{k_n\}$, the truth $g(x)$ and the model $f_\theta(x)$ satisfy the following:

A1. $\int \min\{g(x), k_n(x)\} \to 0$ as $n \to \infty$. That is, the contamination distribution becomes asymptotically singular to the true distribution.

A2. $\int \min\{f_\theta(x), k_n(x)\} \to 0$ as $n \to \infty$ uniformly for $|\theta| \le c$ for any fixed c. That is, the contamination distribution is asymptotically singular to specified models.

A3. $\int \min\{g(x), f_{\theta_n}(x)\} \to 0$ as $n \to \infty$ if $|\theta_n| \to \infty$ as $n \to \infty$. That is, large values of the parameter θ give distributions which become asymptotically singular to the true distribution.

A4. $C(-1)$ and $C'(\infty)$ are finite for the disparity generating function $C(\cdot)$ associated with the chosen disparity ρ_C. Here $C'(\infty)$ is as defined in Lemma 2.1.

Theorem 4.7. [Park and Basu (Theorem 4.1, 2004)]. *Under the assumptions A1–A4 above, the asymptotic breakdown point ϵ^* of the minimum distance functional is at least $\frac{1}{2}$ at the model.*

Proof. Let θ_n denote the minimizer of $\rho_C(h_{\epsilon,n}, f_\theta)$ and ϵ denote a level of contamination. If breakdown occurs, there exists a sequence $\{K_n\}$ such that $|\theta_n| \to \infty$, where $\theta_n = T(H_{\epsilon,n})$. Thus, we have

$$\rho(h_{\epsilon,n}, f_{\theta_n}) = \int_{A_n} D(h_{\epsilon,n}(x), f_{\theta_n}(x)) + \int_{A_n^c} D(h_{\epsilon,n}(x), f_{\theta_n}(x)), \qquad (4.25)$$

where $A_n = \{x : g(x) > \max(k_n(x), f_{\theta_n}(x))\}$ and $D(g,f) = C(g/f - 1)f$. From A1, $\int_{A_n} k_n(x) \to 0$, and from A3, $\int_{A_n} f_{\theta_n}(x) \to 0$, so under $k_n(\cdot)$ and $f_{\theta_n}(\cdot)$ the set A_n converges to a set of zero probability. Thus, on A_n, $D(h_{\epsilon,n}(x), f_{\theta_n}(x)) \to D((1-\epsilon)g(x), 0)$ as $n \to \infty$, so

$$\left| \int_{A_n} D(h_{\epsilon,n}(x), f_{\theta_n}(x)) - \int_{A_n} D((1-\epsilon)g(x), 0) \right| \to 0 \qquad (4.26)$$

by dominated convergence theorem, and by A1 and A3, we have

$$\left| \int_{A_n} D((1-\epsilon)g(x), 0) - \int_{g>0} D((1-\epsilon)g(x), 0) \right| \to 0. \qquad (4.27)$$

Using (4.26) and (4.27)

$$\left| \int_{A_n} D(h_{\epsilon,n}(x), f_{\theta_n}(x)) - \int_{g>0} D((1-\epsilon)g(x), 0) \right| \to 0.$$

Notice that for $g > 0$

$$D((1-\epsilon)g(x), 0) = \lim_{f \to 0} D((1-\epsilon)g(x), f) = (1-\epsilon)C'(\infty)g(x)$$

and

$$\int_{g>0} D((1-\epsilon)g(x), 0) = \int D((1-\epsilon)g(x), 0) = (1-\epsilon)C'(\infty).$$

Thus, we have

$$\left| \int_{A_n} D(h_{\epsilon,n}(x), f_{\theta_n}(x)) - \int D((1-\epsilon)g(x), 0) \right| \to 0,$$

and hence

$$\int_{A_n} D(h_{\epsilon,n}(x), f_{\theta_n}(x)) \to (1-\epsilon)C'(\infty). \qquad (4.28)$$

From A1 and A3, $\int_{A_n^c} g(x) \to 0$ as $n \to \infty$, so under $g(\cdot)$, the set A_n^c converges to a set of zero probability. By similar arguments, we get

$$\left| \int_{A_n^c} D(h_{\epsilon,n}(x), f_{\theta_n}(x)) - \int D(\epsilon k_n(x), f_{\theta_n}(x)) \right| \to 0. \qquad (4.29)$$

Notice that $\int D(\epsilon k_n(x), f_{\theta_n}(x)) \geq C(\epsilon - 1)$ by Jensen's inequality. Using (4.25), (4.28) and (4.29), it follows that

$$\liminf_{n \to \infty} \rho_C(h_{\epsilon,n}, f_{\theta_n}) \geq C(\epsilon - 1) + (1 - \epsilon)C'(\infty) = a_1(\epsilon), \qquad (4.30)$$

say. We will have a contradiction to our assumption that $\{k_n\}$ is a sequence for which breakdown occurs if we can show that there exists a constant value θ^* in the parameter space such that for the same sequence $\{k_n\}$,

$$\limsup_{n \to \infty} \rho_C(h_{\epsilon,n}, f_{\theta^*}) < a_1(\epsilon) \qquad (4.31)$$

as then the $\{\theta_n\}$ sequence above could not minimize $\rho_C(h_{\epsilon,n}, f_{\theta^*})$ for every n.

We will now show that Equation (4.31) is true for all $\epsilon < 1/2$ under the model when we choose θ^* to be the minimizer of $\int D((1 - \epsilon)g(x), f_\theta(x))$. For any fixed θ, let $B_n = \{x : k_n(x) > \max(g(x), f_\theta(x))\}$. From A1, $\int_{B_n} g(x) \to 0$ and from A2 $\int_{B_n} f_\theta(x) \to 0$. Similarly from A1 and A2 $\int_{B_n^c} k_n(x) \to 0$ as $n \to \infty$. Thus, under k_n, the set B_n^c converges to a set of zero probability, while under g and f_θ, the set B_n converges to a set of zero probability. Thus, on B_n, $D(h_{\epsilon,n}, f_\theta(x)) \to D(\epsilon k_n(x), 0)$ as $n \to \infty$ and

$$\left| \int_{B_n} D(h_{\epsilon,n}(x), f_\theta(x)) - \int_{k_n > 0} D(\epsilon k_n(x), 0) \right| \to 0,$$

by dominated convergence theorem. As in the derivation of (4.28)

$$D(\epsilon k_n(x), 0) = \lim_{f \to 0} D(\epsilon k_n, f) = \epsilon C'(\infty) k_n(x)$$

and

$$\int_{k_n > 0} D(\epsilon k_n(x), 0) = \int D(\epsilon k_n(x), 0) = \epsilon C'(\infty).$$

Similarly we have

$$\left| \int_{B_n^c} D(h_{\epsilon,n}(x), f_\theta(x)) - \int D((1 - \epsilon)g(x), f_\theta(x)) \right| \to 0.$$

Hence, we have

$$\lim_{n \to \infty} \rho_c(h_{\epsilon,n}, f_\theta) = \epsilon C'(\infty) + \int D((1 - \epsilon)g(x), f_\theta(x))$$

$$\geq \epsilon C'(\infty) + \inf_\theta \int D((1 - \epsilon)g(x), f_\theta(x)) \qquad (4.32)$$

with equality for $\theta = \theta^*$. Let $a_2(\epsilon) = \epsilon C'(\infty) + \int D((1-\epsilon)g(x), f_{\theta^*}(x))$. Notice from (4.32) that among all fixed θ the disparity $\rho_C(h_{\epsilon,n}, f_\theta)$ is minimized in the limit by θ^*.

If $g(\cdot) = f_{\theta_t}$, that is, the true distribution belongs to the model

$$\int D\Big((1-\epsilon)f_{\theta_t}(x), f_{\theta}(x)\Big) \geq C(-\epsilon)$$

and $C(-\epsilon)$ is also the lower bound over $\theta \in \Theta$ for $\int D((1-\epsilon)f_{\theta_t}(x), f_{\theta}(x))$. Thus, in this case $\theta^* = \theta_t$, and from (4.32)

$$\lim_{n\to\infty} \rho_C(h_{\epsilon,n}, f_{\theta^*}) = C(-\epsilon) + \epsilon C'(\infty). \tag{4.33}$$

As a result, asymptotically there is no breakdown for ϵ level contamination when $a_3(\epsilon) < a_1(\epsilon)$, where $a_3(\epsilon)$ is the right-hand side of Equation (4.33). Notice that $a_1(\epsilon)$ and $a_3(\epsilon)$ are strictly decreasing and increasing respectively in ϵ and $a_1(1/2) = a_3(1/2)$, so that asymptotically there is no breakdown and $\limsup_{n\to\infty} |T(H_{\epsilon,n})| < \infty$ for $\epsilon < 1/2$. □

See Park and Basu (2004) for further discussion on this issue and related topics.

4.6.4   Breakdown Point for Multivariate Location and Covariance Estimation

In this section we consider the breakdown point of the minimum Hellinger distance functional in the case of the estimation of multivariate location and covariance. This discussion follows Tamura and Boos (1986). The setup here is the following. Let X represent the original data set of a given size n; let Y be a contaminating data set of size m $(m \leq n)$. The estimator $\hat{\theta}$ will be said to break down if, through proper choice of the elements of the data set Y, the difference $\hat{\theta}(X \cup Y) - \hat{\theta}(X)$ can be made arbitrarily large. If m^* is the smallest number of the contaminating values for which the estimator breaks down, then the breakdown point of the corresponding estimator at X is $m^*/(m+n)$. In the case of multivariate location and covariance, the breakdown point of the joint estimation of location and covariance has been defined by Donoho (1982) through the following measure of discrepancy

$$B(\theta_1, \theta_2) = \mathrm{tr}(\Sigma_1\Sigma_2^{-1} + \Sigma_1^{-1}\Sigma_2) + ||\mu_1 - \mu_2||^2 \tag{4.34}$$

between parameter values $\theta_1 = (\mu_1, \Sigma_1)$ and $\theta_2 = (\mu_2, \Sigma_2)$, where tr represents the trace of a matrix and $|| \cdot ||$ represents the Euclidean norm. The joint estimate of multivariate location and covariance will break down when the supremum of the discrepancy, as given in the above equation, between the pure data estimate at X and the contaminated data estimate at $X \cup Y$, is infinite.

To prove the breakdown results for the minimum Hellinger distance estimator in connection with multivariate location and covariance estimation, we first present the two following lemmas. Let f_1, f_2 and f_3 represent three densities from continuous distributions. The first lemma, which is an easy consequence of the relation

$$\text{HD}(g, f) = 4 \left\{ 1 - \int g^{1/2} f^{1/2} \right\},$$

is a convexity property of the distance.

Lemma 4.8. [Tamura and Boos (1986, Lemma 5.1)]. *For $0 \le t \le 1$,*

$$\text{HD}((1 - t)f_1 + t f_2, f_3) \le (1 - t)\text{HD}(f_1, f_3) + t\text{HD}(f_2, f_3).$$

The next lemma establishes a triangle type inequality for the HD measure.

Lemma 4.9. [Tamura and Boos (1986, Lemma 5.2)]. *For the densities f_1, f_2 and f_3*

$$\text{HD}(f_1, f_2) \le 2 \left[\text{HD}(f_1, f_3) + \text{HD}(f_2, f_3) \right].$$

Proof. By triangle inequality

$$\left[\int (f_1^{1/2} - f_2^{1/2})^2 \right]^{1/2} \le \left[\int (f_1^{1/2} - f_3^{1/2})^2 \right]^{1/2} + \left[\int (f_2^{1/2} - f_3^{1/2})^2 \right]^{1/2}.$$

Squaring both sides of the above equation, and using the relation $a^2 + b^2 \ge 2ab$, we get

$$\int (f_1^{1/2} - f_2^{1/2})^2 \le 2 \left[\int (f_1^{1/2} - f_3^{1/2})^2 + \int (f_2^{1/2} - f_3^{1/2})^2 \right],$$

so that the result follows. $\qquad\square$

Let

$$\vartheta^* = \liminf_{\theta_1, \theta_2} \text{HD}(f_{\theta_1}, f_{\theta_2}),$$

where the limit is taken as $B(\theta_1, \theta_2) \to \infty$. Let $g_n^*(c_n) = g_n^*(c_n, x)$ be the kernel density estimate at x as defined in Equation (3.15) based on the bandwidth c_n, and let $f_{\theta_n(c_n)}$ be the model density nearest to the above kernel density estimate in terms of the Hellinger distance. Let $a_{n,m} = \text{HD}(g_n^*(c_{n+m}), f_{\theta_n(c_n)})$. Then the breakdown point of the minimum Hellinger distance estimator $\theta_n(c_n)$ is as given in the following theorem.

Theorem 4.10. [Tamura and Boos (1986, Theorem 5.1)]. *Under the above definitions, the breakdown point $\epsilon^*(\theta_n(c_n))$ of $\theta_n(c_n)$ satisfies*

$$\epsilon^*(\theta_n(c_n)) \ge (\vartheta^*/4 - a_{n,m})/(4 - a_{n,m}).$$

Proof. From the definition of the functional, and by Lemma 4.9,

$$
\begin{aligned}
2\text{HD}(g_{n+m}^*(c_{n+m}), f_{\theta_n(c_n)}) &\ge 2\text{HD}(g_{n+m}^*(c_{n+m}), f_{\theta_{n+m}(c_{n+m})}) \\
&\ge \text{HD}(f_{\theta_{n+m}(c_{n+m})}, f_{\theta_n(c_n)}) \\
&\quad - 2\text{HD}(g_{n+m}^*(c_{n+m}), f_{\theta_n(c_n)})
\end{aligned}
$$

implying

$$4\mathrm{HD}(g_{n+m}^*(c_{n+m}), f_{\theta_n(c_n)}) \geq \mathrm{HD}(f_{\theta_{n+m}(c_{n+m})}, f_{\theta_n(c_n)}).$$

The occurrence of breakdown must imply, by definition,

$$\mathrm{HD}(f_{\theta_{n+m}(c_{n+m})}, f_{\theta_n(c_n)}) \geq \vartheta^*,$$

and hence

$$\mathrm{HD}(g_{n+m}^*(c_{n+m}), f_{\theta_n(c_n)}) \geq \vartheta^*/4. \tag{4.35}$$

We denote the empirical distributions of X, Y and $X \cup Y$ by G_n, G_m and G_{n+m} respectively. Since

$$G_{n+m} = \frac{n}{n+m}G_n + \frac{m}{n+m}G_m,$$

applying the same to Equation (3.16), we get

$$g_{n+m}^*(c_{n+m}, x)) = \frac{n}{n+m}g_n^*(c_{n+m}, x) + \frac{m}{n+m}g_m^*(c_{n+m}, x).$$

Using Lemma 4.8, we then get

$$\mathrm{HD}(g_{n+m}^*(c_{n+m}), f_{\theta_n(c_n)}) \leq \left[1 - \frac{m}{n+m}\right]\mathrm{HD}(g_n^*(c_{n+m}), f_{\theta_n(c_n)})$$
$$+ \left[\frac{m}{n+m}\right]\mathrm{HD}(g_m^*(c_{n+m}), f_{\theta_n(c_n)}). \tag{4.36}$$

Using Equations (4.35), (4.36), and the fact that $\mathrm{HD}(g_m^*(c_{n+m}), f_{\theta_n(c_n)}) \leq 4$, we finally get

$$\vartheta^*/4 \leq \left[1 - \frac{m}{n+m}\right]\mathrm{HD}(g_n^*(c_{n+m}), f_{\theta_n(c_n)}) + \left[\frac{4m}{n+m}\right]. \tag{4.37}$$

Collecting the terms in $m/(n+m)$, which stands for the proportion of data contamination causing breakdown, the desired result follows. □

Remark 4.1. For the multivariate normal family $\vartheta^* = 4$ (Tamura, 1984). Let g be the true density; if

$$\mathrm{HD}(g, f_{\theta_n(c_n)}) \to a$$

and

$$g_n^*(c_{n+m}) \to g(x) \text{ for each } x,$$

we get, $a_{n,m} \to a$ almost surely, and Theorem 4.10 yields

$$\lim_{n\to\infty} \inf \epsilon^*(\theta_n(c_n)) \geq \frac{\vartheta^*/4 - a}{4 - a}. \tag{4.38}$$

If the true distribution belongs to the model so that $g = f_\theta$ for some $\theta \in \Theta$, we get $a = 0$ and the right-hand side of Equation (4.38) is $1/4$. Thus, breakdown cannot occur for the minimum Hellinger distance estimator in this case for $\epsilon < 1/4$.

4.7   The α-Influence Function

In the traditional approach to robustness, boundedness of the influence function is often assumed to be a primary requirement. As the influence function of the maximum likelihood functional is unbounded for most common models, the boundedness requirement of the influence function necessarily makes the corresponding functional deficient at the model in relation to the maximum likelihood estimator. At one time, this was considered inevitable, and the loss in efficiency was viewed as the price to be paid for robustness.

Yet, many members of our class of minimum distance estimators based on disparities have strong robustness features in spite of having the same influence function as the maximum likelihood estimator. For a more extensive exploration of this issue, we have considered, in one of the previous sections, the extension of the influence function approach to a second-order approximation following Lindsay (1994). In this section we present the α-influence function technique following Beran (1977), which further strengthens the robustness credentials of the minimum distance estimators based on disparities.

Beran (1977) appears to be the first to argue and demonstrate that there is no intrinsic conflict between the robustness of an estimator and optimal model efficiency. We now describe the setup under which Beran's result may be described. Let $\gamma_y(x)$ denote the uniform density on the interval $(y - \eta, y + \eta)$, where $\eta > 0$ is very small, and let $f_{\theta,\alpha,y}(x) = (1 - \alpha)f_\theta(x) + \alpha\gamma_y(x)$ for $\theta \in \Theta$, $\alpha \in [0,1)$, and real y. The density $f_{\theta,\alpha,y}$ represents a contamination of the models with $100\alpha\%$ gross errors located near y. Let $T(\cdot)$ denote the minimum Hellinger distance functional, so that $T(F_{\theta,\alpha,y})$ and $T(F_\theta) = \theta$ represent the functionals evaluated at the indicated distributions. For every $\alpha \in (0,1)$, Beran considered the difference quotient (or the α-influence function) $\alpha^{-1}[T(F_{\theta,\alpha,y}) - \theta]$, and demonstrated that it is a bounded continuous function of y such that $\lim_{y\to\infty}[T(F_{\theta,\alpha,y}) - \theta] = 0$ for any $\alpha \in (0,1)$. Hence the functional T is robust against $100\alpha\%$ contamination by gross errors at any arbitrary real y; the boundedness (or lack of it) of the influence function of the functional $T(\cdot)$ has no role in the above.

We formally present the α-influence function result in the following theorem.

Theorem 4.11. *For every* $\alpha \in (0,1)$, *every* $\theta \in \Theta$, *and under the assumptions of Lemma 2.6,* $T(F_{\theta,\alpha,y})$ *is a continuous bounded function of* y *such that*

$$\lim_{y\to\infty} T(F_{\theta,\alpha,y}) = \theta. \tag{4.39}$$

If $f_\theta(x)$ *is a positive density continuous in* x *and if the conclusions of Theorem 3.1 hold for* $g = f_\theta$, *then, under the notation of Theorem 3.1,*

$$\lim_{\alpha\to 0} \alpha^{-1}[T(F_{\theta,\alpha,y}) - \theta] = \int [2s_\theta(x)]^{-1} \vartheta_{f_\theta}(x)\gamma_y(x)dx \tag{4.40}$$

for every real y.

Proof. We will first prove (4.39). We denote $\theta_y = T(F_{\theta,\alpha,y})$, $s_t = f_t^{1/2}$ and $s_{t,\alpha,y} = f_{t,\alpha,y}^{1/2}$. Also, let

$$m_y(t) = \int s_t(x) s_{\theta,\alpha,y}(x) dx$$

and

$$k_y(t) = \int s_t(x) \left\{ (1-\alpha)^{1/2} s_\theta(x) + \alpha^{1/2} \gamma_y^{1/2}(x) \right\} dx.$$

Now

$$|k_y(t) - m_y(t)| = \left| \int s_t(x)[(1-\alpha)^{1/2} s_\theta(x) + \alpha^{1/2} \gamma_y^{1/2}(x) - s_{\theta,\alpha,y}(x)] dx \right|$$

$$\leq \left[\int \left\{ (1-\alpha)^{1/2} s_\theta(x) + \alpha^{1/2} \gamma_y^{1/2}(x) - s_{\theta,\alpha,y}(x) \right\}^2 dx \right]^{1/2}.$$

$$(4.41)$$

Notice that the right-hand side of (4.41) is independent of t; also for a fixed θ, f_θ and γ_y eventually become singular as $y \to \infty$. Thus

$$\lim_{y\to\infty} \sup_{t\in\Theta} |k_y(t) - m_y(t)| = 0.$$

We will show that this necessarily implies that $\theta_y \to \theta$ as $y \to \infty$. If not, there exists a subsequence $\{\theta_z\} \subset \{\theta_y\}$ such that $\theta_z \to \theta_1 \neq \theta$. But

$$\lim_{y\to\infty} m_y(\theta_y) = \lim_{y\to\infty} k_y(\theta_y) = (1-\alpha)^{1/2} \int s_{\theta_1}(x) s_\theta(x) dx$$

$$< (1-\alpha)^{1/2} = \lim_{y\to\infty} m_y(\theta) \qquad (4.42)$$

On the other hand, since $t = \theta_y$ maximizes $m_y(t)$ over Θ,

$$\lim_{y\to\infty} m_y(\theta_y) \geq \lim_{y\to\infty} m_y(\theta),$$

which contradicts (4.42). Thus, $\theta_y \to \theta$. The continuity follows from the Hellinger continuity of the functional (Lemma 2.6), and the boundedness follows from (4.39).

To prove (4.40) it suffices, using relation (3.6) in Theorem 3.1, to show that for every $\sigma \in L_2$, $\sigma \perp s_\theta$,

$$\lim_{\alpha\to0} \alpha^{-1} \int \sigma(x)[s_{\theta,\alpha,y}(x) - s_\theta(x)] dx = \int [2s_\theta(x)]^{-1} \sigma(x) \gamma_y(x) dx.$$

By taking a derivative of $s_{\theta,\alpha,y}$ with respect to α and evaluating at $\alpha = 0$, we get, over the set $\{x : |x - y| \geq \eta\}$,

$$\lim_{\alpha \to 0} \alpha^{-1} \int_{|x-y| \geq \eta} \sigma(x)[s_{\theta,\alpha,y}(x) - s_\theta(x)]dx = -2^{-1} \int_{|x-y| \geq \eta} \sigma(x)s_\theta(x)dx.$$
(4.43)

On the other hand, using the relation $(a^{1/2} - b^{1/2}) = (a - b)/(a^{1/2} + b^{1/2})$ for positive numbers a and b, we get,

$$\lim_{\alpha \to 0} \alpha^{-1} \int_{|z-y| < \eta} \sigma(x)[s_{\theta,\alpha,y}(x) - s_\theta(x)]dx$$

$$= \lim_{\alpha \to 0} \int_{|x-y| < \eta} [s_{\theta,\alpha,y}(x) + s_\theta(x)]^{-1} \sigma(x)(\gamma_y(x) - f_\theta(x))dx$$

$$= \int_{|x-y| < \eta} [2s_\theta(x)]^{-1} \sigma(x)\gamma_y(x)dx - 2^{-1} \int_{|x-y| < \eta} \sigma(x)s_\theta(x)dx. \quad (4.44)$$

Combining Equations (4.43) and (4.44), we get

$$\lim_{\alpha \to 0} \alpha^{-1} \int \sigma(x)[s_{\theta,\alpha,y}(x) - s_\theta(x)]dx$$

$$= \int_{|x-y| < \eta} [2s_\theta(x)]^{-1} \sigma(x)\gamma_y(x)dx - 2^{-1} \int \sigma(x)s_\theta(x)dx.$$

But $\gamma_y(x) = 0$ outside the set $\{x : |x - y| < \eta\}$, and σ is $\perp$ to s_θ. Therefore the required result is established. □

Remark 4.2. Let us expand the scope of the above theorem to the generalized Hellinger distance family for values of the tuning parameter in $(0, 1)$. It is easily seen, by considering the generalized objective function

$$\int f_t^a(x) f_{\theta,\alpha,y}^{1-a}(x)dx, \quad a \in (0, 1), \quad \alpha \in (0, 1)$$

that the α-influence function of the corresponding functional is a continuous bounded function, for each value of $\alpha, a \in (0, 1)$, even when the ordinary influence function is not. Also see Bhandari, Basu and Sarkar (2006) for a corresponding result for the class of generalized negative exponential disparities.

4.8   Outlier Stability of Minimum Distance Estimators

Here we present an outlier stability analysis of minimum distance estimators based on disparities following Lindsay (1994). In this section we will refer to the model in terms of the density functions $\{f_\theta\}$. The presentation here assumes a discrete model. But similar arguments work for continuous models; see e.g., Markatou, Basu and Lindsay (1998).

Consider a fixed model $\{f_\theta\}$ and let $d_n(x)$ be the observed relative frequencies obtained from a random sample generated by the unknown true distribution. Also let $\{\xi_j : j = 1, 2, \ldots, \}$ be a sequence of elements of the sample space and ϵ be the contamination proportion. Consider the ϵ contaminated data

$$d_j(x) = (1 - \epsilon)d_n(x) + \epsilon\chi_{\xi_j}(x),$$

where $\chi_y(x)$ is the indicator function at y, and let $\delta_j(x) = d_j(x)/f_\theta(x) - 1$ denote the Pearson residual for the contaminated data.

Definition 4.2. We will say that ξ_j constitutes an outlier sequence for the model f_θ and data d_n if $\delta_j(\xi_j) \to \infty$ and $d_n(\xi_j) \to 0$ as $j \to \infty$.

Lemma 9 of Lindsay (1994) shows that ξ_j constitutes an outlier sequence if and only if $d_n(\xi_j) \to 0$ and $f_\theta(\xi_j) \to 0$ as $j \to \infty$. The result is intuitive, and is easy to prove. The quantities $d_j(x)$ and $\delta_j(x)$ are also functions of n, but we keep that implicit.

Let us consider the limiting behavior of the disparity measure $\rho_C(d_n, f_\theta)$ under contamination through an outlier sequence $\{\xi_j\}$. Let

$$d_\epsilon^*(x) = (1 - \epsilon)d_n(x).$$

While $d_\epsilon^*(x)$ is not a density function, one can formally calculate $\rho_C(d_\epsilon^*, f_\theta)$. Following Lindsay (1994), we note that

$$\rho_C(d_\epsilon^*, f_\theta) \to \rho_C(d_n, f_\theta) \text{ as } \epsilon \to 0 \tag{4.45}$$

under standard regularity conditions and the use of dominated convergence. If in addition

$$\rho_C(d_j, f_\theta) \to \rho_C(d_\epsilon^*, f_\theta) \text{ as } j \to \infty \tag{4.46}$$

then, for extreme outliers and small contaminating fractions ϵ, the distance between the contaminated data d_j and f_θ is close to the distance obtained by simply deleting the outlier from the sample. Equation (4.45) exhibits a continuity property of the distance measure that is closely related to the notion of qualitative robustness. Equation (4.46) represents a key stability property of the distance which demonstrates its outlier rejection capability under an outlier sequence.

In order to precisely describe the conditions under which the minimum distance estimators have the desired robustness properties, we assume that the disparity generating function $C(\delta)$ satisfies the following condition.

Assumption 4.2. $C(-1)$ is finite, and $C(\delta)/\delta \to 0$ as $\delta \to \infty$.

Remark 4.3. Note that the condition $C(\delta)/\delta \to 0$ as $\delta \to \infty$ could actually have been relaxed to $C(\delta)/\delta \to m$ as $\delta \to \infty$, where m is a finite constant. The disparity generating function may be normalized to

$$C^*(\delta) = C(\delta) - m\delta$$

where the function C^* generates the same disparity ρ_C as C and has the property $C^*(\delta)/\delta \to 0$ as $\delta \to \infty$. Hence the condition $C(\delta)/\delta \to 0$ as $\delta \to \infty$ essentially encompasses all disparities with $C(\delta)/\delta \to m$ as $\delta \to \infty$ for some finite m.

Lemma 4.12. *Suppose that the disparity ρ_C satisfies Assumption 4.2. Then the disparity generating function $C(\delta)$ is decreasing.*

Proof. Since C is a convex function, C' is increasing. By L'Hospital's rule, the stability conditions in Assumption 4.2 lead to the result $C'(\delta) \to 0$ as $\delta \to \infty$. Since $C'(\delta)$ is increasing to zero, it must be negative for all finite δ. Therefore C must be a decreasing function. $\qquad\square$

Lemma 4.13. [Lindsay (1994, Proposition 12)]. *Under Assumption 4.2, the convergence in (4.46) holds for the disparity measure determined by $C(\cdot)$.*

Proof. Let $\delta_\epsilon^* = d_\epsilon^*/f_\theta - 1$. We have

$$
\begin{aligned}
\rho_C(d_j, f_\theta) &= \sum C(\delta_j(x))f_\theta(x) \\
&= \sum C(\delta_\epsilon^*(x))f_\theta(x) + I_{1j} - I_{2j} \\
&= \rho_C(d_\epsilon^*, f_\theta) + I_{1j} - I_{2j}
\end{aligned}
$$

where $I_{1j} = C(\delta_j(\xi_j))f_\theta(\xi_j)$, and $I_{2j} = C(\delta_\epsilon^*(\xi_j))f_\theta(\xi_j)$, so that the convergence in (4.46) will hold if $I_{1j} - I_{2j} \to 0$ as $j \to 0$. We will, in fact, show that each of the terms I_{1j} and I_{2j} individually tend to zero as $j \to \infty$. Since $f_\theta(x) \le |1/\delta_j(x)|$, we get

$$|I_{1j}| = |C(\delta_j(\xi_j))f_\theta(\xi_j)| \le |C(\delta_j(\xi_j))/\delta_j(\xi_j)|$$

which tends to zero as $j \to \infty$ from Assumption 4.2. To tackle I_{2j}, note that

$$|C(\delta_\epsilon^*(\xi_j))| \le \max\{C(-1), |C(\delta_j(\xi_j))|\},$$

so that the convergence of this term to 0 follows from Assumption 4.2 and a simple extension of the argument applied for the convergence of I_{1j}. $\qquad\square$

As an application, notice that the stability conditions of Assumption 4.2 are satisfied by all members of the PD_λ family with $\lambda \in (-1, 0)$, and all members of the BWHD_α and BWCS_α families with $\alpha \in (0, 1)$. Notice that unlike the PD_λ family of Cressie and Read, several members of the BWCS_α and BWHD_α families with positive values of the curvature parameter A_2 satisfy the stability conditions in Assumption 4.2.

For $\alpha \in (0, 1)$, the minimization of the generalized Hellinger distance $\mathrm{GHD}_\alpha(d_n, f_\theta)$ is equivalent to the maximization of

$$S_\alpha(d_n, f_\theta) = \sum d_n^\alpha(x) f_\theta^{1-\alpha}(x),$$

and conditions (4.45) and (4.46) may be stated in terms of the convergence of S_α's. Notice that

$$S_\alpha(d_\epsilon^*, f_\theta) = (1 - \epsilon)^\alpha \sum d_n^\alpha(x) f_\theta^{1-\alpha}(x) = (1 - \epsilon)^\alpha S_\alpha(d_n, f_\theta),$$

so that the maximizers of $S_\alpha(d_n, f_\theta)$ and $S_\alpha(d_\epsilon^*, f_\theta)$ (or the minimizers of $\mathrm{GHD}_\alpha(d_n, f_\theta)$ and $\mathrm{GHD}_\alpha(d_\epsilon^*, f_\theta)$) are one and the same. Thus, the convergence in (4.46) will guarantee that under an outlier sequence the minimum GHD_α estimate of θ will eventually settle down on the corresponding minimum distance estimator for the pure data.

4.8.1   Outlier Stability of the Estimating Functions

There is a corresponding outlier stability property of the minimum distance estimating equations themselves. We will denote a residual adjustment function $A(\delta)$ to be outlier stable if

$$\sum A(\delta_j(x)) \nabla f_\theta(x) \to \sum A(\delta_\epsilon^*(x)) \nabla f_\theta(x) \tag{4.47}$$

as $j \to \infty$ under an outlier sequence $\{\xi_j\}$.

Theorem 4.14. [Lindsay (1994, Proposition 14)]. *Let $u_\theta(\cdot)$ be the maximum likelihood score function for the model. If for some k greater than 1, $E_\theta[|u_\theta(X)|^k]$ is finite for all θ, $A(-1)$ is finite, and $A(\delta) = O(\delta^{(k-1)/k})$ as $\delta \to \infty$, then $A(\cdot)$ is outlier stable for the model f_θ.*

Proof. Since

$$E_\theta[|u_\theta(X)|^k] = \sum |u_\theta(x)|^k f_\theta(x) = \sum f_\theta(x)^{(1-k)} |\nabla f_\theta(x)|^k < \infty,$$

the finiteness of the sum implies that the terms in the summation representation converge to zero. Therefore

$$f_\theta(\xi_j)^{(1-k)/k} |\nabla f_\theta(\xi_j)| \to 0 \ \text{ as } \ j \to \infty.$$

Now

$$\sum A(\delta_j(x)) \nabla f_\theta(x) \ = \ \sum A(\delta_\epsilon^*(x)) \nabla f_\theta(x) + K_{1j} - K_{2j}$$

so that the result will be proved if

$$K_{1j} = A(\delta_j(\xi_j)) \nabla f_\theta(\xi_j) \ \text{ and } \ K_{2j} = A(\delta_\epsilon^*(\xi_j)) \nabla f_\theta(\xi_j)$$

both tend to zero individually. Notice that both $A(\delta_j(\xi_j))$ and $A(\delta_\epsilon^*(\xi_j))$ are bounded in absolute value by $M_j = \max\{|A(-1)|, A(1/f_\theta(\xi_j))\}$. Therefore each of the terms K_{1j} and K_{2j} are bounded, in absolute value by $U_j \times V_j$, where

$$U_j = M_j \times f_\theta(\xi_j)^{(k-1)/k}, \quad V_j = f_\theta(\xi_j)^{(1-k)/k}|\nabla f_\theta(\xi_j)|.$$

The given conditions ensure that U_j is bounded and $V_j \to 0$ as $j \to \infty$. Thus, the required result is proved.   □

Consider the estimating equation of the generalized Hellinger distance family.

$$\sum_x A_\alpha(\delta_\epsilon^*(x))\nabla f_\theta(x) = 0,$$

$\alpha \in (0, 1)$, where A_α is as given in (4.11). The above equation reduces to

$$(1 - \epsilon)^\alpha \sum \left(\frac{d_n(x)}{f_\theta(x)}\right)^\alpha \nabla f_\theta(x) = 0,$$

so that the solution to the limiting equation is exactly the same as the one that would be obtained if the outlier were simply deleted from the data set.

4.8.2   Robustness of the Estimator

The final item in this sequence is the robustness of the estimator itself. This is a more difficult problem than the other items discussed here. The distance measures may have multiple local minima, so the convergence of the estimating equations alone do not provide the full picture. So we further explore the convergence of the disparity measures in (4.46) and try to determine the conditions under which the global minimum based on the contaminated data d_j converges to the estimate under the reduced data d_ϵ^*. These results will also help us address the discontinuities of the bias curves observed in Figure 4.3 (a).

Definition 4.3. The strong breakdown point of the minimum distance functional corresponding to the observed density d_n will be the supremum of those ϵ for which the global minima of $\rho_C(d_j, f_\theta)$ converges to the global minimum of $\rho_C(d_\epsilon^*, f_\theta)$ for any outlier sequence.

We have seen that under Assumption 4.2, the convergence in (4.46) holds. It seems reasonable to expect that under stronger conditions the infima of the left-hand terms will converge to the infimum of the right. Yet, such a result can hardly be expected to hold in complete generality; when more than 50% of the data are being moved over to infinity, it will be unfair to expect such a stability result will generally apply. At the end of our discussion we will show that the strong breakdown point is 0.5 under the right conditions, and for the robust distances under consideration the observed outcome in Figure 4.3 is as one would expect.

Assumption 4.3. We assume the following:

(i) The distances $\rho_C(d_j, f_\theta)$ and $\rho_C(d_\epsilon^*, f_\theta)$ are continuous in θ, and the latter has a unique minimum at θ^*.

(ii) The convergence in (4.46) is uniform in θ for any compact set B of parameter values containing θ^*.

Note that it follows from the proof of Lemma 4.13 that Assumption 4.3 (ii) holds whenever $\sup_{\theta \in B} f_\theta(\xi_j) \to 0$ as $j \to \infty$. The convergence holds for the exponential family with an infinite sample space.

Let B be a compact set of parameter values containing θ^*. Let $m(\epsilon, B)$ be the minimized value of the distance $\rho_C(d_\epsilon^*, f_\theta)$ over $\theta \in B$. Under Assumption 4.3 (ii) it is easy to see that the sequence t_j of values of θ that minimize $\rho_C(d_j, f_\theta)$ over B does converge to θ^*, and the minimized value of the distances converges to $m(\epsilon, B)$. We will show that there exists a lower bound $L(\epsilon, B^c)$ for the values of $\rho_C(d_j, f_\theta)$ which is valid for all large j and all $\theta \in B^c$. If we have the relation

$$m(\epsilon, B) < L(\epsilon, B^c), \tag{4.48}$$

for some ϵ, the global minimum of $\rho_C(d_j, f_\theta)$ will eventually lie within B, and breakdown cannot occur for such a value of ϵ.

For the derivation of the lower bound above, we make the following assumption.

Assumption 4.4. It is assumed that for each $0 < \gamma < 1$, there exists a subset S of the sample space such that

(i) $d_n(S) = \sum_{x \in S} d_n(x) \geq 1 - \gamma$ and

(ii) $W = \{\theta : f_\theta(S) \geq \gamma\}$ is a compact set.

Lemma 4.15. [Lindsay (1994, Lemma 20)]. *Under Assumptions 4.2, 4.3 and 4.4, for every $\alpha > 0$, there exists a compact parameter set B_α containing θ^* such that*

$$\lim_{j \to \infty} \inf_{\theta \in B_\alpha^c} \rho_C(d_j, f_\theta) \geq C(\epsilon - 1) - \alpha.$$

Proof. The proof of this theorem is quite technical, and we refer the reader to Lindsay (1994). Together with the relation (4.48), the results of the above lemma imply that there can be no breakdown for ϵ satisfying $C(\epsilon - 1) - \alpha > m(\epsilon, B_\alpha)$.

In particular, let us now choose $d_n(x) = f_{\theta_0}(x)$, i.e., the density $d_n(x)$ is now a model density. Let $f_j(x) = (1 - \epsilon)f_{\theta_0}(x) + \epsilon \chi_{\xi_j}(x)$, and $f_\epsilon^* = (1 - \epsilon)f_{\theta_0}$. From the strict convexity of the function $C(\cdot)$ and Jensen's inequality, we have

$$\rho_C(d_\epsilon^*, f_\theta) = \rho_C(f_\epsilon^*, f_\theta) \geq C((1 - \epsilon) - 1) = C(-\epsilon), \tag{4.49}$$

and thus the minimum of the function $\rho_C(f_\epsilon^*, f_\theta)$ is obtained at $\theta^* = \theta_0$; from the identifiability of the model family, $\theta^* = \theta_0$ is also the unique minimum. $\square$

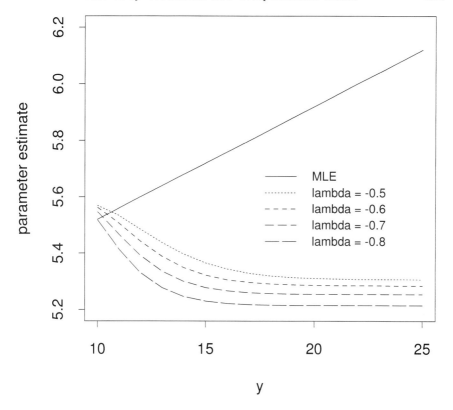

FIGURE 4.4
The change in the minimum distance estimators as a function of the maximum value y as it is moved farther and farther away from the main body of the data.

We are now ready to state the concluding result on strong breakdown.

Theorem 4.16. [Lindsay (1994, Proposition 22)]. *For disparities satisfying Assumptions 4.2–4.4, the strong breakdown point of the corresponding minimum distance estimator at a model point f_{θ_0} is at least 0.5.*

Proof. Choose $\epsilon < 0.5$. Since $\theta_0 = \theta^* \in B_\alpha$, we have $m(\epsilon, B_\alpha) = C(-\epsilon)$ from Equation (4.49). Since $C(\cdot)$ is a decreasing function, $C(-\epsilon) < C(\epsilon - 1)$ for such an ϵ. Choose α such that $C(-\epsilon) < C(\epsilon - 1) - \alpha$. By Lemma 4.15, the minimum of $\rho_C(f_j, f_\theta)$ over $\theta \in B_\alpha$ is eventually smaller than the distance for any $\theta \in B_\alpha^c$. Thus, breakdown cannot occur for any $\epsilon < 0.5$. $\qquad\square$

Example 4.5. To illustrate the effect of an outlier sequence on the estimator within the power divergence family, we present a small numerical study. A sample of size 25 was generated randomly from a Poisson distribution with mean 5. Estimates of θ, under a Poisson (θ) model, were obtained using the minimum distance estimators corresponding to $\lambda = 0$ (LD), $\lambda = -0.5$ (HD), $\lambda = -0.6$, $\lambda = -0.7$ and $\lambda = -0.8$ within the power divergence family. Subsequently, the largest sampled value y (in our sample 10) was increased successively in steps of 1, and all the estimates were recalculated. This was repeated for all values of y up to 25, and all the estimates are presented in Figure 4.4 as a function of y. The observations are self explanatory. The maximum likelihood estimate shows a steady increase as a function of y, while our robust estimates quickly discount its effect. In fact, the values at which the parameter estimates settle down for the robust distances as $y \rightarrow \infty$ are the estimates that one would get when the outlier is simply deleted from the sample.   ‖

4.9   Contamination Envelopes

In this section we construct, as functions of $\epsilon \in (0, 1)$, upper and lower contamination envelopes for the values of distances satisfying disparity conditions when $100\epsilon\%$ of the data comes from a contaminating distribution and not the target data distribution. The envelopes are in fact tight, in the sense that they are actually attained for particular contaminating distributions. The results presented here provide a general idea about the stability of disparity-based distances under data contamination. Among other uses, these envelopes can be used to construct the power-breakdown bounds for the disparity difference tests. Our approach provides a comprehensive framework for studying disparity difference tests under contamination, and generalizes and improves the results of Simpson (1989b), Bhandari, Basu and Sarkar (2006), Dey and Basu (2007), and others.

The results presented here will generally apply to all distances within the class of disparities satisfying certain regularity conditions; however, following Lindsay and Basu (2010), we will use the Hellinger distance (HD), the negative exponential disparity (NED) and the symmetric chi-square (SCS) as our candidate distances on which the results are illustrated. Given two densities g and f, these distances may be expressed in the form

$$\mathrm{HD}(g, f) = 2 \int (g^{1/2} - f^{1/2})^2, \quad \mathrm{NED}(g, f) = \int (e^{-\delta} - 1)f,$$

and

$$\mathrm{SCS}(g, f) = \int \frac{(g - f)^2}{g + f},$$

respectively, where $\delta = g/f - 1$.

We would like to determine, depending on the true distribution G and the model $\mathcal{F}$, the upper and lower limits U_ϵ and L_ϵ of the disparity between an ϵ contaminated version of the true density g and its nearest model element, by varying the contamination density v over its domain and choosing the worst case.

One could develop this approach in its most general setting if one could determine the worst cases for the upper and lower contamination envelopes by varying them over the true density and the model density (subject to a fixed initial distance between the true density and the model density "nearest" to it). While the general upper bound U_ϵ seems tractable under such an approach, finding the general lower bound seems to be a very difficult problem in practice. For given problems, we will derive the lower bound L_ϵ specific to the given model and the true density.

We begin with the following important assumption.

Assumption 4.5. (Robustness Assumption). The $C(\cdot)$ functions generating the disparities under consideration satisfy the following robustness considerations.

$$C(-1) \text{ is finite and } \sup_\delta C'(\delta) = M < \infty. \tag{4.50}$$

Notice that since $C''(\delta) \geq 0$, $C'(\delta)$ is an increasing function. Thus, we must have $\lim_{\delta \to \infty} C'(\delta) = M$. For disparities satisfying the above assumption, we will put the function $C(\delta)$ in its canonical form by redefining it as $C(\delta) - M\delta$. This does not change the disparity, but in this form $C'(\delta) \leq 0$ for all δ, and so the canonical $C(\delta)$ is a decreasing function. Because of this $C(-1)$ is an upper bound on $\rho_C(g, f)$, so the distance is bounded. The above assumption also indicates that if for a fixed $g(x)$ we let $f(x)$ go to zero, then $C(\delta(x))f(x) \to 0$ as well, where $\delta(x) = g(x)/f(x) - 1$.

In finding the contamination envelopes between the contaminated version of the density g and a model density f, we will make heavy use of the ratio of the densities g and f, which we will denote by γ. As $\gamma(x) = g(x)/f(x) = \delta(x) + 1$, the key quantity for us under this setup will be the function $C(\gamma - 1) = C(\delta)$. For our three illustrative cases – Hellinger distance, negative exponential disparity, and symmetric chi-square – the corresponding forms of the C functions which satisfy Assumption 4.5 have the form

$$C_{\mathrm{HD}}(\gamma - 1) = 4(1 - \gamma^{1/2}), \quad C_{\mathrm{NED}}(\gamma - 1) = e^{1-\gamma} - 1,$$

and

$$C_{\mathrm{SCS}}(\gamma - 1) = \frac{4}{\gamma + 1} - 2$$

respectively. The likelihood disparity does not meet the robustness conditions; although $C(-1) = 1$, $\sup_\delta C'(\delta) = \infty$.

Given fixed densities g and f and a fixed contamination proportion ϵ, the upper envelope is represented by the maximum of the disparity

$$\rho(h, f) = \int C\left(\frac{h(x)}{f(x)} - 1\right) f(x) = \int C\left(\frac{\bar{\epsilon}g(x) + \epsilon v(x)}{f(x)} - 1\right) f(x) \tag{4.51}$$

over the contaminating density v; here $h(x) = \bar{\epsilon}g(x) + \epsilon v(x)$ and $\bar{\epsilon} = 1 - \epsilon$. Here and elsewhere the integrals will be assumed to be with respect to the appropriate dominating measure and $v(x)$ is a density with respect to the same measure. The following results are based on Lindsay and Basu (2010).

Theorem 4.17. *For fixed densities g and f and fixed contamination proportion ϵ, the maximum increase in the disparity $\rho(h, f)$, $h = \bar{\epsilon}g + \epsilon v$ through varying v occurs when v is singular to f.*

Proof. The disparity $\rho(h, f)$ is as given in Equation (4.51). The integrand on the right-most side of the above equation is restricted to the support of $f(\cdot)$ (the integrand is zero elsewhere from Assumption 4.5). However, since C is decreasing,

$$C\left(\frac{\bar{\epsilon}g(x) + \epsilon v(x)}{f(x)} - 1\right) \leq C\left(\frac{\bar{\epsilon}g(x)}{f(x)} - 1\right),$$

for all x, so that

$$\rho(h, f) = \int C\left(\frac{\bar{\epsilon}g(x) + \epsilon v(x)}{f(x)} - 1\right) f(x) \leq \int C\left(\frac{\bar{\epsilon}g(x)}{f(x)} - 1\right) f(x) = \rho(\bar{\epsilon}g, f),$$

irrespective of the choice of v. Since the right-hand side of the last equation represents the disparity $\rho(h, f)$ when v is singular to f, the result is proved. □

We can, under an analytic assumption, make an important generalization by considering how much an initial distance $\rho(g, f)$ could be perturbed upward by contamination when we only know the distance

$$\rho(g, f) = \int C\left(\frac{g(x)}{f(x)} - 1\right) f(x) = c_1 \tag{4.52}$$

and not necessarily the actual densities g and f. That is, we consider the worst case envelope curve by solving the maximization problem

$$\max_{g,f} \int C\left(\bar{\epsilon}\frac{g(x)}{f(x)} - 1\right) f(x) \tag{4.53}$$

subject to

$$\int C\left(\frac{g(x)}{f(x)} - 1\right) f(x) = c_1, \quad \int \left(\frac{g(x)}{f(x)}\right) f(x) = 1 \text{ and } \int f(x) = 1. \tag{4.54}$$

This worst case maximization result, proved by a rather lengthy treatment using Tchebycheff functions in Lindsay and Basu (2010), is presented below.

Theorem 4.18. *Let ρ_C be a distance satisfying the disparity conditions and Assumption 4.5, and let c_1 be the initial distance as given in Equation (4.52) between the true density g and the model density f. Then for any fixed $\epsilon \in (0, 1)$, the worst case upper contamination envelope in (4.53) is given by*

$$U_\epsilon = C(\bar{\epsilon}\gamma_1 - 1),$$

where $\bar{\epsilon} = 1 - \epsilon$, and γ_1 is obtained by solving the equation $C(\gamma_1 - 1) = c_1$.

We now turn our attention to the lower envelope. However, unlike the upper envelope problem, the general worst case solution seems to be an extremely difficult one, so that we work with the minimization results with varying v over fixed f and g. Consider the following optimization problem

$$\min_{v(x)} \int C\left(\bar{\epsilon}\frac{g(x)}{f(x)} + \epsilon\frac{v(x)}{f(x)} - 1\right)f(x). \tag{4.55}$$

To describe the solution, define the sets

$$S_\gamma = \{x : g(x)/f(x) < \gamma\}, S_\gamma^* = \{x : g(x)/f(x) \le \gamma\}, \ \gamma > 0,$$

and $S_0 = S_0^* = \{x : g(x) = 0\}$. Define the functions

$$K(\gamma) = \frac{\bar{\epsilon}\int_{S_\gamma} g(x) + \epsilon}{\int_{S_\gamma} f(x)}$$

and

$$K^*(\gamma) = \frac{\bar{\epsilon}\int_{S_\gamma^*} g(x) + \epsilon}{\int_{S_\gamma^*} f(x)}.$$

Let

$$\gamma_0 = \sup\{\gamma \in (0, \infty) : K(\gamma) \ge \bar{\epsilon}\gamma\}.$$

Finally, we present the lower bound in the following theorem.

Theorem 4.19. *Assume that $C''(\gamma - 1)$ exists and is strictly positive for all $\gamma \in (0, \infty)$. Then there exists a minimizer $\hat{v}(x)$ to (4.55) that satisfies, for $x \in S_{\gamma_0}^*$,*

$$\bar{\epsilon}\frac{g(x)}{f(x)} + \epsilon\frac{\hat{v}(x)}{f(x)} = K^*(\gamma_0) \tag{4.56}$$

and satisfies $\hat{v}(x) = 0$ for $x \in S_{\gamma_0}^{*c}$. The attained minimum is

$$L_\epsilon = C(K^*(\gamma_0) - 1)\int_{S_{\gamma_0}^*} f(x) + \int_{S_{\gamma_0}^{*c}} C(\bar{\epsilon}(g(x)/f(x) - 1))f(x).$$

Proof. The reader is referred to Lindsay and Basu (2010) for a proof of the above results. □

The above set of theorems provide general perturbation bounds for the class of disparities. Specific applications of these results in case of power breakdown of tests is presented in Chapter 5.

4.10   The Iteratively Reweighted Least Squares (IRLS)

While the density-based minimum distance estimators provide attractive alternatives to the maximum likelihood estimator, the defining equations of these estimators are usually nonlinear and require numerical techniques such the Newton–Raphson algorithm in order to handle them. The numerical difficulty involved in the solution of the above estimators increases greatly with the number of parameters. In the estimation of (μ, Σ) in a d-dimensional multivariate normal model, for example, the number of unknown parameters equals $p = d + d(d + 1)/2 = d(d + 3)/2$. For this problem, each step of the Newton–Raphson algorithm requires $(p + 1)(p + 2)/2$ numerical integrations and the inversion of a p dimensional Hessian matrix.

In this section we will present a method with the aim to reduce the numerical effort necessary to solve the estimating equations of these minimum distance estimators. This method was systematically developed by Basu and Lindsay (2004), and is similar in spirit to the iteratively reweighted least squares algorithm used frequently in robust regression. Also see Basu (1991). The method, called the *iteratively reweighted estimating equations* (IREE) is easily motivated by the substantial gain in programming effort. In case of the d dimensional normal, this method requires $(p + 2)$ numerical integrations and no matrix inversion per step. Thus, if $d = 3$ (so that $p = 9$), each Newton–Raphson step requires 55 numerical integrations and the inversion of a 9×9 matrix, while the new method requires only 11 numerical integrations per step. While the price one might expect to pay for this is a decrease from quadratic to linear convergence, a striking fact is that by a careful selection of weights one can make the method competitive in speed with the Newton–Raphson algorithm even in the univariate model (where $p = 2$). In fact, a simple scalar adjustment makes the method quadratically convergent when the data exactly fit the model.

4.10.1   Development of the Algorithm

We first discuss the *iteratively reweighted least squares* (IRLS) algorithm, and then present the development of the IREE along those lines. The IRLS is an algorithm often used in determining the parameter estimates in robust regression. This algorithm is generally attributed to Beaton and Tukey (1974), and is far simpler to apply than the Newton–Raphson method. Holland and Welsch (1977), McCullagh and Nelder (1989) and Green (1984) are good general references. Byrd and Pyne (1979) and Birch (1980) discuss convergence results for this algorithm, and Del Pino (1989) provides an extensive bibliography.

Consider the standard regression model

$$Y_{n \times 1} = X_{n \times p} \beta_{p \times 1} + \epsilon_{n \times 1}.$$

A robust estimate $\hat{\beta}$ of β is found by minimizing

$$\sum_{i=1}^{n} \rho\left(\frac{Y_i - X_i\beta}{\sigma}\right)$$

where σ is a known or previously estimated scale parameter and X_i is the i-th row of X.

Let ψ represent the first derivative of ρ. Then $\hat{\beta}$ satisfies the estimating equation

$$\sum_{i=1}^{n} x_{ij}\psi\left(\frac{Y_i - X_i\hat{\beta}}{\sigma}\right) = 0 \tag{4.57}$$

for $j = 1, 2, \ldots, p$. Here x_{ij} is the j-th component of X_i. Solving this set of equations directly typically requires the application of numerical methods.

Define the weight function $w(r)$ as $w(r) = \psi(r)/r$. Equation (4.57) can be written as

$$\sum_{i=1}^{n} \left(\frac{Y_i - X_i\hat{\beta}}{\sigma}\right) w\left(\frac{Y_i - X_i\hat{\beta}}{\sigma}\right) x_{ij} = 0. \tag{4.58}$$

One can solve (4.58) iteratively using a weighted least-squares algorithm. Let W_β be the $n \times n$ diagonal matrix whose i-th diagonal element is $w((Y_i - X_i\beta)/\sigma)$. Then for a given starting value β_0, the first iteration yields

$$\beta_1 = (X^T W_{\beta_0} X)^{-1} X^T W_{\beta_0} Y. \tag{4.59}$$

This iteration scheme is continued till convergence is achieved. Note that if Y exactly fits the model, in the sense $Y = X\hat{\beta}$, then (4.58) converges in one step.

In numerical analysis terms, the IRLS is using a fixed point method to solve (4.58). To illustrate, we begin with the one variable case. The fixed point method is a simple algorithm to determine the roots of the equation $f(x) = 0$ in some interval (a, b). The conditions leading to its convergence are well known results of numerical analysis; among others, Ralston and Rabinowitz (1978), and Ortega (1990) are sources of detailed discussions on this subject. However, as our development of the method described herein depends critically on the convergence mechanism of the fixed point algorithm, we provide a very brief description of the same in the following. Consider a target value α, a starting value $x^{(0)}$, and a sequence $\{x^{(i)} : i = 1, 2, \ldots\}$ in $\mathbb{R}^p$. We will say that $x^{(i)}$ converges linearly to α if for a starting value sufficiently close to the target value there exists a constant $c \in (0, 1)$ such that

$$||x^{(i+1)} - \alpha|| \leq c||x^{(i)} - \alpha||$$

where $||\cdot||$ denotes the Euclidean norm. The sequence converges quadratically if for a sufficiently close starting value there exists a constant c such that

$$||x^{(i+1)} - \alpha|| \leq c||x^{(i)} - \alpha||^2.$$

The fixed point iteration method can be used to determine the root of an equation $f(x) = 0$, when the equation can be rewritten in the alternative form $x = F(x)$. First consider the univariate case where f is a real valued function of a single real variable x. Let $x = F(x)$ be the fixed point formulation of the equation $f(x) = 0$. In this case we start with an initial approximation $x^{(0)}$ and at the i-th stage perform the next iteration as $x^{(i+1)} = F(x^{(i)})$. Given that $x^{(i)}$ is the value at the i-th stage, we assume that $F(x)$ has a continuous derivative in the closed interval bounded by $x^{(i)}$ and the true solution α. Since $\alpha = F(\alpha)$, it follows that

$$x^{(i+1)} - \alpha = F(x^{(i)}) - F(\alpha) = (x^{(i)} - \alpha)F'(\xi^{(i)})$$

where $\xi^{(i)}$ lies between $x^{(i)}$ and α. When the iteration converges $x^{(i)} \to \alpha$, and $F'(\xi^{(i)}) \to F'(\alpha)$. Thus, we get $(x^{(i+1)} - \alpha) \sim (x^{(i)} - \alpha)F'(\alpha)$, and hence also $(x^{(i+1)} - \alpha) \sim A[F'(\alpha)]^i$ for a constant A. Thus, $|F'(\alpha)| < 1$ is a necessary condition for the iteration to be asymptotically stable. When $|F'(\alpha)| < 1$, and the initial value is sufficiently close to α, the sequence $x^{(i)}$ will converge to α. Since $x^{(i+1)} - \alpha = (x^{(i)} - \alpha)F'(\xi^{(i)})$, sufficient closeness of $x^{(i)}$ to α and the continuity of $F'(x)$ at α will mean that

$$|x^{(i+1)} - \alpha| = |(x^{(i)} - \alpha)||F'(\xi^{(i)})|$$

and so $x^{(i)}$ converges to α at a linear rate.

It is well known that if $F'(\alpha) = 0$, then it leads to quadratic convergence for the fixed point method; in this case

$$
\begin{aligned}
(x^{(i+1)} - \alpha) &= F(x^{(i)}) - F(\alpha) \\
&= (x^{(i)} - \alpha)F'(\alpha) + \frac{1}{2}(x^{(i)} - \alpha)^2 F''(\xi^{(i)}) \\
&= \frac{1}{2}(x^{(i)} - \alpha)^2 F''(\xi^{(i)}),
\end{aligned}
$$

where $\xi^{(i)}$ is between $x^{(i)}$ and α. So if the method converges, the error in $x^{(i+1)}$ tends to be proportional to the square of the error in $x^{(i)}$.

When the number of variables is p, the fixed point formulation $x_i = F_i(x_1, x_2, \ldots, x_p)$, $i = 1, \ldots, p$, has to be solved in p unknowns. Let α be the true solution and let $x^{(j)} = (x_1^{(j)}, \ldots, x_p^{(j)})$ be the value at the j-th stage. For a suitably close starting value $x^{(0)}$,

$$(x^{(j)} - \alpha) \sim D^j z$$

where $D = F'(\alpha)$, F' being the Jacobian matrix, and z a constant vector. Let $\lambda_1, \lambda_2, \ldots, \lambda_p$ be the eigenvalues of D. The necessary condition for convergence now is that the spectral radius of D given by $\rho(D) = \max_i |\lambda_i|$ is less than 1. When the method converges, the rate of convergence is linear. However, as in the scalar case, the rate of convergence becomes quadratic when the matrix D is a null matrix or a nilpotent matrix (a matrix is nilpotent if some power of it is the null matrix).

4.10.2   The Standard IREE

Let us now examine how the above ideas can be used to solve for the roots of the minimum disparity estimating equations. Assume that $\Theta \in \mathbb{R}^p$. For simplicity, we illustrate this with the discrete model first. Here we are solving the estimating equation

$$\sum_x A(\delta(x))\nabla f_\theta(x) = 0.$$

Assuming that $\sum_x f_\theta(x)$ can be differentiated within the summation sign, we can write

$$\sum_x (A(\delta(x)) - \lambda)f_\theta(x)\frac{\nabla f_\theta(x)}{f_\theta(x)} = 0$$

for any constant λ, or

$$\sum_x w(x)\frac{\nabla f_\theta(x)}{f_\theta(x)} = 0 \qquad (4.60)$$

where

$$w(x) = (A(\delta(x)) - \lambda)f_\theta(x). \qquad (4.61)$$

This is a weighted version of the estimating equation of the likelihood disparity, just as (4.58) is a weighted version of the ordinary least squares. Let $\theta = (\theta_1, \ldots, \theta_p)^T$, and let ∇_i be the gradient with respect to θ_i. If $f_\theta(x)$ is in the exponential family, a relationship of the form

$$\frac{\nabla_i f_\theta(x)}{f_\theta(x)} = K(\theta)[S_i(x, \theta) - \theta_i]) \qquad (4.62)$$

is often found to be true. It is always true if θ represents the set of natural parameters. The function S_i may depend on θ. Assuming that we have a relationship of the form (4.62), we can write the i-th equation of (4.60) as

$$\sum_x w(x)[S_i(x, \theta) - \theta_i] = 0$$

or

$$\theta_i = \frac{\sum w(x)S_i(x, \theta)}{\sum w(x)} \qquad (4.63)$$

and hence we arrive at the fixed point equation $\theta = F(\theta)$, where F is a function from $\mathbb{R}^p$ to $\mathbb{R}^p$. The iteration will be carried on till convergence to a specific level of tolerance is reached. As described earlier, this algorithm converges linearly.

   This IREE method does not require the evaluation of the second partial derivatives, and the inversion of the Hessian matrix. The discrete case is similar with integrals replaced with summations. In Theorems 4.20 and 4.21, we will show how to improve the rate of convergence of the algorithm when the functions S_i are independent of θ.

Since the RAF $A(\delta)$ is increasing on $[-1, \infty)$, the weights $w(x)$ in (4.61) will be non-negative if we use $\lambda = A(-1)$. We will refer to this case as the standard IREE (or IREE with standard weights). To illustrate the use of this method, let us look at the one parameter exponential family. Let $\mu = \theta$ be the mean and V be the variance for the model f_θ. Then

$$\frac{\nabla f_\theta(x)}{f_\theta(x)} = \frac{x - \mu}{V}$$

where ∇ represents the gradient with respect to the mean value parameter. The IREE will solve the equation

$$\sum w(x) \left\{ \frac{(x - \mu)}{V} \right\} = 0$$

for μ as

$$\mu = F(\mu) = \frac{\sum x w(x)}{\sum w(x)}.$$

In general, for a univariate parameter θ one gets,

$$F(\theta) = \frac{\sum S(x, \theta) w(x)}{\sum w(x)}, \tag{4.64}$$

if $\nabla f_\theta(x) / f_\theta(x) = K(\theta)[S(x, \theta) - \theta]$.

4.10.3   Optimally Weighted IREE

The iteratively reweighted estimating equation described above is a useful algorithm. As with all algorithms, its rate of convergence is critical for its practical applicability. Being a linearly convergent algorithm, one could improve its rate of convergence using the Aitken acceleration (e.g., Ralston and Rabinowitz, 1978). However, a more effective and simple modification, described below, often works better in practice.

Consider a scalar parameter θ, and let the weight function w of the IREE be as defined in Equation (4.61). The standard IREE is obtained by replacing λ with $A(-1)$. This keeps the weights nonnegative. The convergence properties of the algorithm can be greatly enhanced if we allow negative weights. As discussed in Section 4.10.1, the convergence of the fixed point algorithm applied to the fixed point formulation (4.64) depends on the derivative $F'(\alpha)$ at the solution, and the rate of convergence is quadratic if this derivative is zero. If $S(x, \theta) = S(x)$ is independent of θ, direct differentiation of (4.64), combined with the result that at the solution $\theta = F(\theta) = \sum w(x) S(x) / \sum w(x)$, gives

$$F'(\theta) = \frac{\sum w'(x)(S(x) - \theta)}{\sum w(x)} \tag{4.65}$$

at the solution, where $w'(x) = \nabla w(x)$. Thus, the form of the data will determine convergence properties. An important special case occurs when the data fit the model well.

Theorem 4.20. [Basu and Lindsay (2004, Theorem 3.1)]. *Suppose that* $d_n(x) = f_\theta(x)$ *and* $\nabla f_\theta(x)/f_\theta(x) = K(\theta)[S(x) - \theta]$ *where* $S(x)$ *is independent of* θ. *Then for* $\lambda = -1$, *we get* $F'(\theta) = 0$ *at the solution and thus the IREE converges at a quadratic rate.*

Proof. In this case the derivative $F'(\theta)$ at the solution is as in Equation (4.65). By direct differentiation,

$$w'(x) = \frac{\partial A(\delta(x))}{\partial \delta(x)} \nabla \delta(x) f_\theta(x) + (A(\delta(x)) - \lambda) \nabla f_\theta(x)$$

and when $d_n(x) = f_\theta(x)$, we get $\delta(x) = 0$, $A(\delta(x)) = 0$, $\partial A(\delta(x))/\partial \delta(x) = 1$, and $\partial \delta(x)/\partial \theta = -\nabla f_\theta(x)/f_\theta(x)$, so that $w'(x) = -(1 + \lambda) \nabla f_\theta(x)$, which vanishes for $\lambda = -1$. As a result, the right-hand side of Equation (4.65) vanishes as well, implying $F'(\theta) = 0$ at the solution if $\lambda = -1$. □

Thus, under the conditions of the theorem, the IREE converges quadratically.

In particular for the mean value parameter $\theta = \mu$, we get, at the solution,

$$F'(\mu) = \frac{\sum w'(x)(x - \mu)}{\sum w(x)}. \tag{4.66}$$

For the conditions of the above theorem, we get $w(x) = -\lambda f_\theta(x)$ and

$$w'(x) = -\left\{\frac{(x - \mu)}{V}\right\} f_\theta(x) - \lambda \left\{\frac{(x - \mu)}{V}\right\} f_\theta(x)$$

and replacing these values in (4.66) we obtain $F'(\mu) = 1 + 1/\lambda$. At $\lambda = -1$, $F'(\mu) = 0$.

We will refer to the case where $\lambda = -1$ is used in the weight function $w(x)$ as the optimal IREE (or the IREE with optimal weights).

Theorem 4.21. [Basu and Lindsay (2004, Theorem 3.2)]. *Suppose that θ is p-dimensional. Assume that the quantities $S_i(x, \theta)$ used in Equation (4.62) are independent of θ for each $i, i = 1, \ldots, p$. In such cases the IREE will converge quadratically at the model $(d_n(x) = f_\theta(x))$ if we use $\lambda = -1$.*

Proof. Equation (4.63) can now be represented as

$$\theta_i = \frac{\sum w(x) S_i(x)}{\sum w(x)}$$

where $S_i(x)$ depends on x only. The (i, j)-th element of the Jacobian matrix D at the solution is

$$\frac{\sum \nabla_j w(x)(S_i(x) - \theta_i)}{\sum w(x)}$$

where ∇_j represents the gradient with respect to θ_j. As in the unidimensional case, the above expression is 0 at the model when $\lambda = -1$, and this is true for all i and j, making the Jacobian matrix at the solution a null matrix. In this case, therefore, the optimally weighted IREE will converge quadratically. □

4.10.4   Step by Step Implementation

In this section we present a step by step guideline for the implementation of the Iteratively Reweighted Estimating Equation algorithm. For illustration we consider the continuous case under the Basu–Lindsay approach. The discrete case, as also the continuous case under the Beran approach, can be handled similarly with appropriate changes to represent the difference in the model and the nature of smoothing.

1. Select the parametric model. While the algorithm can be implemented in other situations as well, the biggest benefit of the method will come when the likelihood score function of the smoothed model may be represented in the form (4.62).

2. Create the smoothed empirical density by choosing an appropriate kernel function. For the multivariate normal model, choose the multivariate normal kernel. Choose the bandwidth to be a multiple of a robust equivariant scale estimator.

3. Choose a robust starting value $\theta^{(0)}$. For the univariate normal model one can choose

$$\hat{\mu}^{(0)} = median\{X_i\}, \text{ and } \hat{\sigma}^{(0)} = (0.674)^{-1}median\{|X_i - \hat{\mu}^{(0)}|\}$$

as the starting values.

4. Create the smoothed model density and construct the Pearson residuals δ. Use the same kernel applied in Step 2 above to construct the smoothed model density.

5. Choose an appropriate RAF to create the weight functions. The Hellinger distance is a generic choice for most applications, but other robust choices may also be useful, particularly when the sample size is small.

6. Choose the tuning parameter λ. Notice that $\lambda = -1$ is algorithmically optimal, but conservative choices of λ closer to $A(-1)$ may provide better convergence properties, particularly when the sample size is small.

7. Create weights $w(x)$ as in (4.61) and solve the corresponding weighted likelihood estimating equation assuming the weights to be fixed constants to get the updated value for the next stage.

8. Repeat steps 4–7 with the current parameter estimate until the convergence criterion at the desired level of tolerance is satisfied. The form of the RAF used and the tuning parameter λ do not change from iteration to iteration.

5

The Hypothesis Testing Problem

The hypothesis testing problem is a fundamental paradigm in the theory of statistical inference. A popular and useful statistical tool for the hypothesis testing scenario is the likelihood ratio test. This is one of the oldest techniques in statistical literature (see, e.g., Neyman and Pearson, 1928; Wilks, 1938) and is widely used by practitioners. Under standard regularity conditions this method can be routinely applied to many hypothesis testing problems and also enjoys certain optimality properties.

Yet, just like the robustness problems faced by the maximum likelihood estimator, the likelihood ratio test can also be very significantly affected by model misspecifications and the presence of outliers. One possible alternative in this scenario is to use tests based on distance functions such as disparities. We will show that just as the maximum likelihood estimator belongs to a large class of estimators including other minimum distance estimators, so also the likelihood ratio test belongs to a larger class of tests of the "disparity difference" type, many of whom possess strong robustness features while being equivalent to the likelihood ratio test under the null and contiguous alternatives.

As in the case of estimation, the methods based on the Hellinger distance have a special place in the class of robust tests of hypotheses based on the minimum distance idea, and historically preceded the other tests. We will begin with a description of the disparity difference test (DDT) based on the Hellinger distance.

5.1 Disparity Difference Test: Hellinger Distance Case

Our development in this section follows that of Simpson (1989b). We assume that we have a parametric model $\mathcal{F}$ as defined in Section 1.1 and an independent and identically distributed random sample $X_1, \ldots, X_n$ is available from the true distribution G. Consider testing the null hypothesis

$$H_0 : \theta \in \Theta_0 \quad \text{versus} \quad H_1 : \theta \in \Theta \setminus \Theta_0, \tag{5.1}$$

where Θ_0 is a proper subset of Θ. The likelihood ratio test statistic (LRT) is one of the most common tests that may be employed in this situation. In order to express this in terms of a general, unified, notation, we will also refer to this as the Disparity Difference Test based on the likelihood disparity. The test statistic in this case equals twice the negative log likelihood ratio, and can be expressed, as a function of the observed data vector d_n as

$$
\begin{aligned}
\mathrm{DDT}_{\mathrm{LD}}(d_n) &= \mathrm{LRT}(d_n) \\
&= 2\left[\log\left(\prod_{i=1}^{n} f_{\hat{\theta}^{\mathrm{ML}}}(X_i)\right) - \log\left(\prod_{i=1}^{n} f_{\hat{\theta}_0^{\mathrm{ML}}}(X_i)\right)\right] \quad (5.2) \\
&= 2n\left[\mathrm{LD}(d_n, f_{\hat{\theta}_0^{\mathrm{ML}}}) - \mathrm{LD}(d_n, f_{\hat{\theta}^{\mathrm{ML}}})\right], \quad (5.3)
\end{aligned}
$$

where $\hat{\theta}^{\mathrm{ML}}$ and $\hat{\theta}_0^{\mathrm{ML}}$ are the unrestricted maximum likelihood estimator and the maximum likelihood estimator under the null hypothesis respectively, and LD is the likelihood disparity.

As an analog of the likelihood ratio test, we consider the disparity difference test based on the Hellinger distance (Simpson, 1989b) to test the hypothesis given in Equation (5.1). Note that the statistic in Equation (5.3) (and hence, equivalently, in Equation (5.2)) may be viewed as $2n$ times the difference between the minimized value of the likelihood disparity under the null and the unrestricted minimum of the likelihood disparity. In the same spirit, one may define the corresponding disparity difference test based on the Hellinger distance as

$$
\mathrm{DDT}_{\mathrm{HD}}(d_n) = 2n\left[\mathrm{HD}(d_n, f_{\hat{\theta}_0}) - \mathrm{HD}(d_n, f_{\hat{\theta}})\right], \quad (5.4)
$$

where $\hat{\theta}$ and $\hat{\theta}_0$ are the unrestricted minimum Hellinger distance estimator and the minimum Hellinger distance estimator over the null space respectively. We will show that under appropriate regularity conditions, the distribution of the disparity difference test statistic based on the Hellinger distance has the same asymptotic limit as that of the likelihood ratio test under the null hypothesis. As in Section 1.4.1, let us suppose that Θ_0 is defined by a set of $r \le p$ restrictions on Θ defined by $R_i(\theta) = 0$, $1 \le i \le r$. We assume that the parameter space under H_0 can be described through a parameter $\gamma = (\gamma_1, \ldots, \gamma_{p-r})^T$, with $p - r$ independent components, i.e., H_0 specifies that there exists a function $b : \mathbb{R}^{p-r} \to \mathbb{R}^p$ where $\theta = b(\gamma)$, $\gamma \in \Gamma \subseteq \mathbb{R}^{p-r}$. The function b is assumed to have continuous derivatives $\dot{b}(\gamma)$ of order $p \times (p - r)$ with rank $p - r$. Then $\hat{\theta}_0 = b(\hat{\gamma})$, where $\hat{\gamma}$ is the minimum Hellinger distance estimator of the parameter in the γ formulation of the model. Let $G = F_\theta$ be the true distribution which belongs to the model, and θ be the true parameter. Under H_0, let γ be the true value of the reduced parameter, so that $\theta = b(\gamma)$ in this case. When the null hypothesis is true it is easily shown – under standard regularity conditions – that $\hat{\gamma}$ and $\hat{\theta}_0$ are consistent and first-order efficient estimators of γ and θ respectively (e.g., Theorem 2.19). Let $J(\gamma)$ be the information matrix for the model under the $f_{b(\gamma)}$ formulation, and let $I(\theta)$ be the information matrix under the f_θ formulation respectively. Then $J(\gamma) = \dot{b}(\gamma)^T I(b(\gamma))\dot{b}(\gamma)$ where $\theta = b(\gamma)$. It is well known that the asymptotic distribution of the likelihood ratio statistic in Equation (5.2) is a chi-square with r degrees of freedom when the null hypothesis is true.

For the continuous case we describe the disparity difference test based on the Hellinger distance in the same way as in Equation (5.3) by first constructing a nonparametric kernel density estimate of the true unknown density. Thus, if g_n^* is a kernel density estimate as obtained in Chapter 3, the disparity difference test based on the Hellinger distance in this case will be given by

$$\text{DDT}_{\text{HD}}(g_n^*) = 2n \left[\text{HD}(g_n^*, f_{\hat{\theta}_0}) - \text{HD}(g_n^*, f_{\hat{\theta}}) \right]. \tag{5.5}$$

In this case also we will show that this test statistic has the same null distribution as that of the LRT under appropriate conditions. One could, of course, construct the disparity difference test based on likelihood disparity by substituting d_n with the kernel density estimate g_n^* in Equation (5.3), but that is not our usual likelihood ratio test.

In terms of exhibiting the equivalence of the null distributions of the test statistics, we will actually show a little bit more. It will, in fact, be proved that the disparity difference test statistic DDT_{HD} based on the Hellinger distance is equivalent to the likelihood ratio test under the null as well as under local alternatives. On the other hand, we will demonstrate that under data contamination the DDT_{HD} statistic is substantially more stable compared to the likelihood ratio test and is far more reliable in terms of holding its level and power.

We now present some technical conditions to complete the setup for the derivation of the asymptotic properties of the disparity difference test based on the Hellinger distance. The resulting theorem will be presented in the notation corresponding to a continuous distribution, but the same result holds for the discrete case with the observed proportions d_n replacing the smoothed density g_n^*. Suppose that the mapping from θ to $f_\theta^{1/2}$ is twice differentiable in L_2; that is, suppose that for θ in the interior of Θ, there are p dimensional functions ψ_θ and $p \times p$ matrix valued functions $\dot{\psi}_\theta$ with components in L_2 such that

$$\left\{ \int \left(f_{\theta+t}^{1/2} - f_\theta^{1/2} - t^T \psi_\theta \right)^2 \right\}^{1/2} = o(||t||) \tag{5.6}$$

and

$$\psi_{\theta+t} - \psi_\theta - t^T \dot{\psi}_\theta = o(||t||) \tag{5.7}$$

component-wise in L_2 as $||t|| \to 0$. Under these assumptions, the Hellinger distance between the smoothed data density g_n^* (or any other density g with respect to the dominating measure) and the model density f_θ is twice differentiable in θ and the expansion

$$\text{HD}(g, f_{\theta+t}) = \text{HD}(g, f_\theta) + t^T \nabla \text{HD}(g, f_\theta) + \frac{1}{2} t^T \nabla_2 \text{HD}(g, f_\theta) t + o(||t^2||) \tag{5.8}$$

holds uniformly in g, where ∇ is the gradient with respect to θ. We need the approximation

$$-n^{1/2} \nabla \text{HD}(g_n^*, f_\theta) = Z_n(\theta) + o_p(1) \tag{5.9}$$

under f_θ, where $Z_n(\theta)$ is as in Equation (2.74). This is true under standard regularity assumptions for the class of minimum distance estimators based on disparities including the minimum Hellinger distance estimator. Under f_θ, the assumption $\mathrm{HD}(g_n^*, f_\theta) \to 0$ as $n \to \infty$ (i.e., the kernel density estimator is consistent for the true density in the Hellinger metric) immediately yields

$$\nabla_2 \mathrm{HD}(g_n^*, f_\theta) = I(\theta) + o_p(1) \tag{5.10}$$

through an easy modification of Lemma 2.11.

The minimum Hellinger distance estimator is first-order efficient, and the unrestricted estimator satisfies

$$\hat\theta = \theta + n^{-1/2} I^{-1}(\theta) Z_n(\theta) + o_p(n^{-1/2}). \tag{5.11}$$

See, e.g., Equation (2.84). The constrained estimator $\hat\theta_0$ satisfies $\hat\theta_0 = b(\hat\gamma)$, where $\hat\gamma$ is first-order efficient for the true parameter γ_0. Thus,

$$\hat\gamma = \gamma_0 + n^{-1/2} \left[\dot b(\gamma_0)^T I(b(\gamma_0)) \dot b(\gamma_0) \right]^{-1} \dot b(\gamma_0)^T Z_n(b(\gamma_0)) + o_p(n^{-1/2}) \tag{5.12}$$

under $f_{b(\gamma)}$.

We now provide the conditions formally imposed by Simpson (1989b) for the statement and proof of the series of results that describe the asymptotic distribution of the $\mathrm{DDT}_{\mathrm{HD}}$ test statistic under the null and under local alternatives respectively.

(i) The density estimate g_n^* converges to the density f_θ in the Hellinger metric under f_θ; notice that this convergence is implied by the L_1 convergence under f_θ.

(ii) The mapping from θ to $f_\theta^{1/2}$ is twice differentiable in L_2, i.e., (5.6) and (5.7) hold and $I(\theta)$ is positive definite.

(iii) The unconstrained estimators $\hat\theta$ and $\hat\theta^{\mathrm{ML}}$ are first-order efficient in the sense of (5.11).

(iv) Either H_0 is simple and $\Theta_0 = \{\theta_0\}$ where θ_0 is in the interior of Θ, or H_0 is composite and $\Theta_0 = \{b(\gamma) : \gamma \in \Gamma \subseteq \mathbb{R}^{p-r}\}$; b has a continuous derivative $\dot b_{p \times (p-r)}$ of rank $(p-r)$.

(v) If H_0 is composite then the constrained estimators satisfy $\hat\theta_0 = b(\hat\gamma)$ and $\hat\theta_0^{\mathrm{ML}} = b(\hat\gamma^{\mathrm{ML}})$ where $\hat\gamma$ and $\hat\gamma^{\mathrm{ML}}$ are first-order efficient in the sense of (5.12).

(vi) Approximation (5.9) holds.

There may be several approaches for proving the main result of this section. However, we follow Simpson's approach and first present two auxiliary lemmas. The result in Lemma 5.1 is standard, but is presented under integrated smoothness conditions.

Lemma 5.1. [Simpson (1989b, Lemma 1)]. *Suppose that the conditions (i)–(vi) hold, θ is in the interior of Θ and θ^* is first-order efficient in the sense of (5.11). Then*

$$2\log\left\{\frac{L_n(\theta^*)}{L_n(\theta)}\right\} - Z_n^T(\theta)I^{-1}(\theta)Z_n(\theta) \to 0 \tag{5.13}$$

in probability under f_θ where $L_n(\theta) = \prod_{i=1}^{n} f_\theta(X_i)$.

Proof. Define $t_n = n^{1/2}(\theta^* - \theta)$. From Equation (5.11), this quantity is equal to $I^{-1}(\theta)Z_n(\theta) + o_p(1)$ under f_θ, where $Z_n(\theta)$ is as in (2.74). For any $\eta > 0$ and any compact $H \subset \Theta$, we get, by a basic probability argument

$$P_\theta(|\log\{L_n(\theta^*)/L_n(\theta)\} - t_n^T Z_n(\theta) + \frac{1}{2}t_n^T I(\theta)t_n| > \eta)$$

$$\leq P_\theta\left(\sup_{t \in H}\left|\log\left\{\frac{L_n(\theta + tn^{-1/2})}{L_n(\theta)}\right\} - t^T Z_n(\theta) + \frac{1}{2}t^T I(\theta)t\right| > \eta\right)$$

$$+ P_\theta(t_n \in H^c). \tag{5.14}$$

The local asymptotic normality of the parametric family $\{f_\theta\}$ is implied by the given conditions (Ibragimov and Has'minskii, 1981, p. 123). Using this, and the tightness of the sequence $\{t_n\}$, one can make the right-hand side of the above equation smaller than any $\epsilon > 0$ by choosing H and n large enough; thus the desired result follows. $\square$

Lemma 5.2. [Simpson (1989b, Lemma 2)]. *Suppose that $\mathrm{HD}(g_n^*, f_\theta) \to 0$ under f_θ as $n \to \infty$. Suppose that conditions (i) to (vi) hold, θ is in the interior of Θ and θ^* is first-order efficient in the sense of (5.11). Then*

$$2n[\mathrm{HD}(g_n^*, f_\theta) - \mathrm{HD}(g_n^*, f_{\theta^*})] - Z_n^T(\theta)I^{-1}(\theta)Z_n(\theta) \to 0 \tag{5.15}$$

in probability under f_θ as $n \to \infty$.

Proof. Let t_n be as in Lemma 5.1. Under the given conditions we get, using relation (5.8),

$$2n[\mathrm{HD}(g_n^*, f_\theta) - \mathrm{HD}(g_n^*, f_{\theta^*})]$$

$$= -2\left[n^{1/2}t_n^T \nabla \mathrm{HD}(g_n^*, f_\theta) + \frac{1}{2}t_n^T \nabla_2 \mathrm{HD}(g_n^*, f_\theta)t_n\right] + o(\|t_n^2\|). \tag{5.16}$$

Then an application of (5.9) and (5.10) yields the result. $\square$

Theorem 5.3. [Simpson (1989b, Theorem 1)]. *For fixed $\theta_0 \in \Theta_0$ and $\tau \in \mathbb{R}^p$, let $\theta_n = \theta_0 + \tau n^{-1/2}$, $n = 1, 2, \ldots$. The conditions of Lemma 5.2 indicate that under f_{θ_n} and as $n \to \infty$,*

$$2n[\mathrm{HD}(g_n^*, f_{\hat{\theta}_0}) - \mathrm{HD}(g_n^*, f_{\hat{\theta}})]$$

$$- 2\left[\log\left(\prod_{i=1}^{n} f_{\hat{\theta}^{\mathrm{ML}}}(X_i)\right) - \log\left(\prod_{i=1}^{n} f_{\hat{\theta}_0^{\mathrm{ML}}}(X_i)\right)\right] \to 0 \tag{5.17}$$

in probability.

Proof. First consider the convergence under f_{θ_0}. If the null hypothesis is simple, the result follows directly from (5.13) and (5.15). For the composite null hypothesis, consider the reduced parameter space Γ, and from Lemmas 5.1 and 5.2

$$2n[\mathrm{HD}(g_n^*, f_{\theta_0}) - \mathrm{HD}(g_n^*, f_{b(\hat{\gamma})})] - 2\log\left\{\frac{L_n(b(\hat{\gamma}))}{L_n(\theta_0)}\right\} \to 0$$

under f_{θ_0}, where the above quantity represents the test statistic for testing $\{\theta_0\}$ versus $\theta \in \Theta \setminus \{\theta_0\}$. Note that the test statistics for $\{\theta_0\}$ versus $\Theta_0 \setminus \{\theta_0\}$ and Θ_0 versus $\Theta \setminus \Theta_0$ are additive, so that (5.17) holds under f_{θ_0}.

For the case under f_{θ_n}, an inspection of the log-likelihood ratio for f_{θ_n} versus f_{θ_0} shows that f_{θ_n} is contiguous to f_{θ_0}, so that convergence in probability under f_{θ_0} implies convergence in probability under f_{θ_n}.

More details about the proof are given in Simpson (1989b, Theorem 1). □

The above result shows that the asymptotic null distribution of the disparity difference test based on the Hellinger distance for the hypothesis in (5.1) is χ_r^2 where r is the number of independent restrictions under the null. In addition, the theorem implies that the test has the same asymptotic power as the likelihood ratio test for local parametric alternatives.

5.2   Disparity Difference Tests in Discrete Models

In order to extend Simpson's result to the general class of disparities in discrete models, we assume the setup and notation of Sections 2.3 and 5.1. We are interested in testing the null hypothesis in Equation (5.1) using a minimum distance approach based on an arbitrary disparity ρ_C satisfying the disparity conditions. The disparity difference test for the above hypothesis makes use of the unconstrained minimum distance estimator $\hat{\theta}$ and the value of the minimized distance. We have extensively discussed the limiting properties of this estimator in previous sections. In particular, it allows the expansion (5.11). In addition, we need a constrained estimator $\hat{\theta}_0$ under the null hypothesis if the latter is composite. In the notation of Section 5.1, the constrained minimum distance estimator is assumed to be of the form $\hat{\theta}_0 = b(\hat{\gamma})$, where $\hat{\gamma}$ is the first order efficient minimum distance estimator of γ based on the disparity ρ_C. Thus, $\hat{\gamma}$ allows a representation of the form (5.12). Our test statistic is given by

$$\mathrm{DDT}_{\rho_C}(d_n) = 2n[\rho_C(d_n, f_{\hat{\theta}_0}) - \rho_C(d_n, f_{\hat{\theta}})], \tag{5.18}$$

the asymptotic distribution of which is established in the following theorem. We first lay out the appropriate set of conditions.

(C1) The Assumptions (A1)–(A7) of Section 2.5 hold under the model conditions.

(C2) The unconstrained minimum distance estimator and the maximum likelihood estimator of θ are first-order efficient in the sense of (5.11).

(C3) The null hypothesis H_0 is either simple and $\Theta_0 = \{\theta_0\}$, where θ_0 is in the interior of Θ, or H_0 is composite and $\Theta_0 = \{b(\gamma) : \gamma \in \Gamma \subseteq \mathbb{R}^{p-r}\}$.

(C4) If H_0 is composite then the constrained estimators satisfy $\hat{\theta}_0 = b(\hat{\gamma})$ and $\hat{\theta}_0^{\mathrm{ML}} = b(\hat{\gamma}^{\mathrm{ML}})$ where $\hat{\gamma}$ and $\hat{\gamma}^{\mathrm{ML}}$ are first-order efficient in the sense of (5.12).

Theorem 5.4. *Suppose that Assumptions (C1)–(C4) hold. Under f_{θ_0}, $\theta_0 \in \Theta_0$, the limiting null distribution of the disparity difference test statistic in Equation (5.18) is χ_r^2, where r is the number of restrictions imposed by the null hypothesis in (5.1).*

Proof. A Taylor series expansion of (5.18) around $\hat{\theta}$ gives

$$\mathrm{DDT}_{\rho_C}(d_n) = 2n[\rho_C(d_n, f_{\hat{\theta}_0}) - \rho_C(d_n, f_{\hat{\theta}})]$$

$$= 2n\Big[\sum_j (\hat{\theta}_{0j} - \hat{\theta}_j)\nabla_j \rho_C(d_n, f_\theta)|_{\theta=\hat{\theta}}$$

$$+ \frac{1}{2}\sum_{j,k}(\hat{\theta}_{0j} - \hat{\theta}_j)(\hat{\theta}_{0k} - \hat{\theta}_k)\nabla_{jk}\rho_C(d_n, f_\theta)|_{\theta=\theta^*}\Big], \quad (5.19)$$

where the subscripts denote the indicated components of the vectors; also θ^* lies in the line segment joining $\hat{\theta}$ and $\hat{\theta}_0$. By definition, $\nabla_j \rho_C(d_n, f_\theta)|_{\theta=\hat{\theta}} = 0$, so that Equation (5.19) reduces to

$$\mathrm{DDT}_{\rho_C}(d_n) = n(\hat{\theta}_0 - \hat{\theta})^T [I(\theta_0)](\hat{\theta}_0 - \hat{\theta})$$

$$+ n(\hat{\theta}_0 - \hat{\theta})^T [\nabla_2 \rho_C(d_n, f_{\theta^*}) - I(\theta_0)](\hat{\theta}_0 - \hat{\theta}). \quad (5.20)$$

We will show that $\nabla_2 \rho_C(d_n, f_{\theta^*}) \to I(\theta_0)$ as $n \to \infty$. By another Taylor series expansion around the true value θ_0, we get, for some θ^{**} between θ_0 and θ^*,

$$\nabla_{jk}\rho_C(d_n, f_{\theta^*}) = \nabla_{jk}\rho_C(d_n, f_{\theta_0}) + (\theta_l^* - \theta_{0l})\nabla_{jkl}\rho_C(d_n, f_{\theta^{**}}). \quad (5.21)$$

From Theorem 2.19 and Remark 2.3 we have the convergences $\nabla_2 \rho_C(d_n, f_{\theta_0}) = I(\theta_0) + o_p(1)$, and $\nabla_{jkl}\rho_C(d_n, f_{\theta^{**}}) = O_p(1)$ for $\theta^{**} \in \omega$. Under the null hypothesis, the given conditions imply $(\hat{\theta} - \theta_0) = o_p(1)$ and $(\hat{\theta}_0 - \theta_0) = o_p(1)$. Thus, $(\theta^* - \theta_0) = o_p(1)$ and θ^{**} belongs to ω eventually. Also $n^{1/2}(\hat{\theta} - \hat{\theta}_0) = O_p(1)$ as shown in Equations (5.22) and (5.23). Hence Equation (5.21) reduces to $\nabla_2 \rho_C(d_n, f_{\theta^*}) = I(\theta_0) + o_p(1)$. Thus

$$\mathrm{DDT}_{\rho_C}(d_n) = n(\hat{\theta}_0 - \hat{\theta})^T I(\theta_0)(\hat{\theta}_0 - \hat{\theta}) + o_p(1).$$

From (5.11) and (5.12), the unrestricted and constrained minimum distance estimators have the same asymptotic distribution and the same convergence order as the corresponding maximum likelihood estimators. Thus

$$n^{1/2}(\hat{\theta} - \hat{\theta}_0) = n^{1/2}(\hat{\theta}^{\mathrm{ML}} - \hat{\theta}_0^{\mathrm{ML}}) + o_p(1), \quad (5.22)$$

where the involved quantities on the right-hand side are the unrestricted and constrained maximum likelihood estimators. From Serfling (1980), we have

$$n^{1/2}(\hat{\theta}^{\mathrm{ML}} - \hat{\theta}_0^{\mathrm{ML}}) = O_p(1), \tag{5.23}$$

and replacing these in Equation (5.20), we get

$$\mathrm{DDT}_{\rho C}(d_n) = n(\hat{\theta}_0^{\mathrm{ML}} - \hat{\theta}^{\mathrm{ML}})^T[I(\theta_0)](\hat{\theta}_0^{\mathrm{ML}} - \hat{\theta}^{\mathrm{ML}}) + o_p(1).$$

The limiting distribution of the quantity on the right-hand side of the above equation is a χ_r^2, as has been shown by Serfling (1980, Section 4.4.4). □

Remark 5.1. In statistical literature, two other kinds of tests, Wald tests and score (Rao) tests, are considered to be closely related to likelihood ratio tests. Here we briefly indicate why these test statistics also have the same asymptotic distribution under the null hypothesis and the conditions of the above theorem. In our case, the Wald statistics for testing the null hypothesis H_0 given in (5.1) has the form

$$W_{\rho C} = n(\hat{\theta} - \hat{\theta}_0)^T I(\hat{\theta})(\hat{\theta} - \hat{\theta}_0)$$

whose asymptotic null distribution comes out as an immediate byproduct of Theorem 5.4. Clearly $W_{\rho C} - \mathrm{DDT}_{\rho C} \to 0$ in probability under the conditions of this theorem for the general composite null hypothesis case.

For the score (Rao) statistic, define the quantity ϱ_n through the relation

$$\varrho_n(\theta) = -\nabla \rho_C(d_n, f_\theta) = n^{-1/2} Z_n(\theta) + o_p(n^{-1/2})$$

under f_θ where the last equality follows from Theorem 2.19. Since this implies $n^{1/2}(\hat{\theta} - \theta) = n^{1/2} I^{-1}(\theta)\varrho_n(\theta) + o_p(1)$, a score type test for the simple null hypothesis can be performed through the statistic

$$S_{\rho C} = n\varrho_n^T(\theta_0) I^{-1}(\theta_0) \varrho_n(\theta_0),$$

where θ_0 is the null value. It is obvious that the three test statistics $\mathrm{DDT}_{\rho C}$, $W_{\rho C}$ and $S_{\rho C}$ are asymptotically equivalent, and the null distribution of the three statistics are the same for the simple null hypothesis. Note that in the simple null case, the score statistic requires no parameter estimation.

For the composite null hypothesis, the equivalence of the statistics $\mathrm{DDT}_{\rho C}$ and $W_{\rho C}$ has been already indicated. Observe that the constrained maximum likelihood estimator $\hat{\theta}_0^{\mathrm{ML}}$ and the constrained minimum distance estimator $\hat{\theta}_0$ are asymptotically equivalent under the null hypothesis. Then the equivalence of the score statistic with the other two statistics under the null hypothesis follows along the lines of Rao (1973, Section 6e).

Remark 5.2. An inspection of Simpson's proof as described in Lemmas 5.1, 5.2 and Theorem 5.3 show that the actual forms of the distances are used only through the relations (5.8)–(5.12). When the above relations hold, under appropriate assumptions on the model, the distance and the nonparametric density estimator, one can mimic Simpson's proof to establish the asymptotic equivalence of the disparity difference test based on an arbitrary disparity and the likelihood ratio test. Such an approach works both for discrete and continuous models.

Example 5.1. Here we present a small example involving the likelihood ratio test and the disparity difference test based on the Hellinger distance. The tests are performed under the Poisson model, where the data generating distribution is a Poisson with mean $\theta = 5$, which is also the value specified by the null. We have taken different sample sizes, and plotted the likelihood ratio statistic against the disparity difference statistic based on the Hellinger distance. The scatter plots are presented in Figure 5.1 for sample sizes 25, 50, 100 and 200. Since the test statistics are asymptotically equivalent, the observations should settle on the diagonal line as the sample size increases to infinity. In this example we see that this does indeed happen, but the rate of this convergence is rather slow.

What is also observed on further exploration, but not presented here, is that when data are generated by a mixture of Poissons with Poisson(5) being the major component, the levels of the test for the same hypothesis based on the Hellinger distance are remarkably stable, but those for the likelihood ratio test based on the contaminated data quickly blow up. On the other hand, disparity difference tests based on distances with curvature parameter close to 0 (such as $\text{BWHD}_{1/3}$) show a much faster convergence to the chi-square limit, but turn out to be weak in terms of stability of inference under data contamination. This underscores the need for considering appropriate modifications to robust tests like that based on the Hellinger distance to improve their small sample properties without affecting their robustness. Such modifications are considered later in Chapter 6. ∥

5.2.1 Second-Order Effects in Testing

In this section we further explore the structure of the disparity difference tests with respect to its robustness. In particular, we consider the role of the estimation curvature in the stability of the limiting distribution of the disparity difference tests under contamination. Let $T(G)$ be our minimum distance functional corresponding to the disparity ρ_C, and let $\hat{\theta}_n = T(G_n)$ be the minimum disparity estimator of θ generated by the relative frequency vector d_n. For the unknown true distribution G, let

$$H_0 : T(G) = \theta_0 \tag{5.24}$$

be the null hypothesis of interest, where G may or may not be in the model. Let g be the true density. Our disparity difference test statistic for the above hypothesis is given by

$$\text{DDT}_{\rho_C}(d_n) = 2n[\rho_C(d_n, f_{\theta_0}) - \rho_C(d_n, f_{\hat{\theta}_n})], \tag{5.25}$$

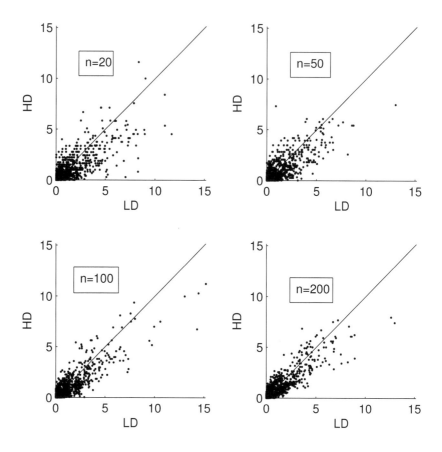

FIGURE 5.1
Scatter plots of the disparity difference statistic based on the Hellinger distance (HD) against the likelihood ratio statistic (LD).

where $\hat{\theta}_n = T(G_n)$, the minimum distance estimator of θ based on the disparity ρ_C. Under the conditions of Section 5.2, we have the following theorem (Lindsay, 1994) for the case where θ is a scalar parameter.

Theorem 5.5. [Lindsay (1994, Theorem 6)]. *Under the given conditions, let $g(x)$ be the true density, θ is a scalar parameter, and let the null hypothesis of interest be given by (5.24). Under the null hypothesis we have the convergence*

$$\mathrm{DDT}_{\rho_C}(d_n) \to c(g)\chi_1^2, \qquad (5.26)$$

where $c(g) = \mathrm{Var}_g[T'(X)] \cdot \nabla_2 \rho_C(g, f_\theta)|_{\theta=\theta_0}$, and $T'(X)$ is the influence function of the minimum distance estimator based on the disparity ρ_C.

Proof. We get, by a simple Taylor series expansion of the test statistic DDT_{ρ_C},

$$2n[\rho_C(d_n, f_{\theta_0}) - \rho_C(d_n, f_{\hat{\theta}_n})]$$

$$= 2n \left[\rho_C(d_n, f_{\hat{\theta}_n}) + (\theta_0 - \hat{\theta}_n)\nabla\rho_C(d_n, f_{\hat{\theta}_n}) + \frac{(\theta_0 - \hat{\theta}_n)^2}{2}\nabla_2\rho_C(d_n, f_{\hat{\theta}_n}) \right.$$

$$\left. + \frac{1}{6}(\theta_0 - \hat{\theta}_n)^3\nabla_3\rho_C(d_n, f_{\theta^*}) - \rho_C(d_n, f_{\hat{\theta}_n}) \right]$$

$$= 2n \left[\frac{1}{2}(\theta_0 - \hat{\theta}_n)^2\nabla_2\rho_C(d_n, f_{\hat{\theta}_n}) + \frac{1}{6}(\theta_0 - \hat{\theta}_n)^3\nabla_3\rho_C(d_n, f_{\theta^*}) \right],$$

where ∇_j indicates the gradient of the indicated order, and θ^* is on the line segment joint θ_0 and $\hat{\theta}_n$. The quantity $\rho_C(d_n, f_{\hat{\theta}})$ converges to $\rho_C(g, f_{\theta_0})$ by a simple extension of Theorem 2.19. Also $\nabla_3\rho_C(d_n, f_{\hat{\theta}}^*)$ is bounded in probability from Theorem 2.19. Thus

$$2n[\rho_C(d_n, f_{\theta_0}) - \rho_C(d_n, f_{\hat{\theta}_n})] = n(\hat{\theta}_n - \theta_0)^2\nabla_2\rho_C(g, f_{\theta_0}) + o_p(1). \quad (5.27)$$

But from Corollary 2.21,

$$n^{1/2}(T(G_n) - T(G)) = n^{1/2}(\hat{\theta}_n - \theta_0) = n^{-1/2}\sum_{i=1}^{n}T'(X_i) + o_p(1),$$

so that

$$\frac{n(\hat{\theta}_n - \theta_0)^2}{[\text{Var}_g(T'(X))]} \xrightarrow{\mathcal{D}} Y \sim \chi_1^2,$$

where χ_1^2 represents a chi-square random variable with one degree of freedom. The required result then follows from Equation (5.27). $\square$

We will refer to $c(g)$ as the chi-square inflation factor. The result of the previous theorem may be interpreted as follows. Suppose we use the relation (5.26) to determine a confidence interval for $T(G)$. The confidence level will be asymptotically correct if G is in the parametric model. But when G is outside the model, one would get confidence intervals which would be conservative or liberal depending on whether the quantity $c(g)$ is less than or greater than 1. A similar interpretation holds for the tests. Notice that for the disparity difference test based on the LD (i.e., the likelihood ratio test) the $c(g)$ values are given by the simple formula

$$c_{\text{LD}}(g) = 1 + \frac{E_g[u_{\theta_0}^2(X) + \nabla u_{\theta_0}(X)]}{E_g[-\nabla u_{\theta_0}(X)]},$$

where $\theta_0 = T_{\text{LD}}(G)$, the distribution corresponding to the density G. Clearly, $c(g) = 1$ when $g = f_{\theta_0}$.

TABLE 5.1
The values of the coefficient $c(f_\epsilon)$ at different contamination levels for several minimum distance estimators. The actual parameter estimates are given within the parentheses. The contaminating point is $y = 10$.

ϵ	$c_{\text{LD}}(f_\epsilon)$	$c_{\text{HD}}(f_\epsilon)$	$c_{\text{SCS}}(f_\epsilon)$
0.00005	1.00045 (0.50042)	1.00022 (0.50003)	1.00021 (0.50003)
0.0001	1.00090 (0.50004)	1.00044 (0.50006)	1.00041 (0.50005)
0.0005	1.00450 (0.50025)	1.00205 (0.50024)	1.00154 (0.50022)
0.001	1.00899 (0.5005)	1.00383 (0.50041)	1.00269 (0.50039)
0.005	1.04478 (0.5025)	1.01435 (0.50145)	1.01297 (0.50098)
0.01	1.08919 (0.505)	1.02431 (0.50234)	1.02154 (0.50119)
0.05	1.48571 (0.525)	1.08283 (0.50659)	1.05181 (0.50151)
0.1	1.81818 (0.550)	1.14787 (0.51029)	1.08033 (0.50162)

Example 5.2. The actual attained values of the quantity $c(g)$ give us some idea of the stability of the asymptotic distribution of the test statistic under small deviations. We go back to the model used in Example 4.2 to illustrate this. In Table 5.1 we compare the values of the chi-square inflation factor for the likelihood ratio test with those of the disparity difference tests based on the Hellinger distance (HD) and the symmetric chi-square (SCS).

The model is the binomial model, and f_θ represents the binomial $(10, \theta)$ density; the contamination puts a mass ϵ at $y = 10$, one of the extreme values. We define the contaminated density

$$g(x) = f_\epsilon(x) = (1 - \epsilon)f_\theta(x) + \epsilon\chi_y(x),$$

where χ_y is the indicator function defined in Section 1.1. In the table we have presented the $c(f_\epsilon)$ values for several contamination levels. For comparison we also present, within parentheses, the actual estimates $\hat{\theta}$ under a binomial model for each combination of method and contamination level. The results show that while increasing the contamination will quickly lead to overly liberal confidence intervals in case of the likelihood ratio test, the results are remarkably more stable for the robust distances. And between the robust distances, the SCS clearly outperforms the Hellinger distance in terms of stability except for very small values of ϵ.   ‖

To investigate the effect of such contaminations on the limiting distributions of the test statistics, and to further explore the role of the estimation curvature in this case, we choose the density g to be a point mass contamination f_ϵ of the model distribution and evaluate the derivative of $c(f_\epsilon)$ at $\epsilon = 0$. The lemma below (Lindsay, 1994), will show that the disparity difference tests whose chi-squared distributions are more stable under the null are the ones with negative values of A_2, with $A_2 = -1$ providing the optimal local stability.

To set up the notation for the lemma, we define

$$v(y) = u_{\theta_0}^2(y) + \nabla u_{\theta_0}(y), \quad \gamma = E[u_{\theta_0}(X)v(X)]/I(\theta_0).$$

Lemma 5.6. [Lindsay (1994, Proposition 7)]. *For the minimum distance estimator based on a disparity with curvature parameter A_2, we have*

$$\left.\frac{dc(f_\epsilon)}{d\epsilon}\right|_{\epsilon=0} = I^{-1}(\theta_0)[v(y) - \gamma u_{\theta_0}(y)]$$

$$+ A_2 I^{-1}(\theta_0)\left[u_{\theta_0}^2(y) - I(\theta_0) - \frac{u_{\theta_0}(y)E[u_{\theta_0}^3(X)]}{I(\theta_0)}\right].$$

For the one parameter exponential family case, the result reduces to

$$\left.\frac{dc(f_\epsilon)}{d\epsilon}\right|_{\epsilon=0} = I^{-1}(\mu)(1 + A_2)\left[(y - \mu) - I(\mu) - \gamma(y - \mu)\right],$$

where $I(\mu) = E(Y - \mu)^2$, so that $A_2 = -1$ provides the optimal stability for the limiting distribution.

Proof. From Theorem 5.5, we have $c(g) = \mathrm{Var}_g[T'(X)] \cdot \nabla_2\rho_C(g, f_\theta)\big|_{\theta=\theta_0}$. The form of the influence function, $T'(y)$ derived in Theorem 2.4, is

$$T'(y) = \frac{A'(\delta_0(y))u_{\theta_0}(y) - \sum_x A'(\delta_0(x))u_{\theta_0}(x)g(x)}{\sum_x A'(\delta_0(x))u_{\theta_0}^2(x)g(x) - \sum_x A(\delta_0(x))[\nabla u_{\theta_0}(x) + u_{\theta_0}^2(x)]f_{\theta_0}(x)},$$

$$\tag{5.28}$$

where $\theta_0 = T(G)$, and $\delta_0(x) = g(x)/f_{\theta_0}(x) - 1$. So

$$\mathrm{Var}_g[T'(X)]$$

$$= \frac{\sum_x [A'(\delta_0(x))u_{\theta_0}(x)]^2 g(x) - \left[\sum_x A'(\delta_0(x))u_{\theta_0}(x)g(x)\right]^2}{\left[\sum_x A'(\delta_0(x))u_{\theta_0}^2(x)g(x) - \sum_x A(\delta_0(x))\left[\nabla u_{\theta_0}(x) + u_{\theta_0}^2(x)\right]f_{\theta_0}(x)\right]^2}.$$

Now let $\delta(x) = g(x)/f_\theta(x) - 1$ for a generic θ. Then

$$\nabla_2\rho_C(g, f_\theta)|_{\theta=\theta_0}$$

$$= \nabla\left[-\sum A(\delta(x))\nabla f_\theta(x)\right]\Bigg|_{\theta=\theta_0}$$

$$= -\left\{\sum A'(\delta(x))\left[-\frac{g(x)}{f_\theta^2(x)}\nabla f_\theta(x)\right]\nabla f_\theta(x) + \sum A(\delta(x))\nabla_2 f_\theta(x)\right\}\Bigg|_{\theta=\theta_0}$$

$$= \sum_x A'(\delta_0(x))u_{\theta_0}^2(x)g(x) - \sum_x A(\delta_0(x))[\nabla u_{\theta_0}(x) + u_{\theta_0}^2(x)]f_{\theta_0}(x)$$

which is the same as the denominator in the expression of $T'(y)$ in (5.28). Thus, we have

$$\text{Var}_g[T'(X)].\nabla_2\rho_C(g, f_\theta)|_{\theta=\theta_0}$$

$$= \frac{\sum_x [A'(\delta_0(x))u_{\theta_0}(x)]^2\, g(x) - \left[\sum A'(\delta_0(x))u_{\theta_0}(x)g(x)\right]^2}{\sum_x A'(\delta_0(x))u_{\theta_0}^2(x)g(x) - \sum_x A(\delta_0(x))\left[\nabla u_{\theta_0}(x) + u_{\theta_0}^2(x)\right] f_{\theta_0}(x)}.$$

$$(5.29)$$

Let $f_\epsilon = (1-\epsilon)f_{\theta_0} + \epsilon\chi_y$, and let F_ϵ be the corresponding distribution. Also let $T(F_\epsilon) = \theta_\epsilon$. Let N_ϵ and D_ϵ be the numerator and denominator of (5.29) obtained by evaluating the same at $g = f_\epsilon$ and $\theta_0 = \theta_\epsilon$. When $\epsilon = 0$, $N_0 = D_0 = I(\theta)$; thus the ratio is 1, and there is no distortion in the asymptotic chi-square limit in this case. We get, through straightforward differentiation,

$$\frac{d}{d\epsilon}N_\epsilon|_{\epsilon=0} = 2A_2\left[u_{\theta_0}^2(y) - I(\theta_0) - T'(y)E_{\theta_0}[u_{\theta_0}^3(X)]\right]$$
$$+ 2T'(y)E[u_{\theta_0}(X)\nabla u_{\theta_0}(X)] + u_{\theta_0}^2(y) - I(\theta_0), \qquad (5.30)$$

and similarly,

$$\frac{d}{d\epsilon}D_\epsilon|_{\epsilon=0} = A_2\left[u_{\theta_0}^2(y) - I(\theta_0) - T'(y)E_{\theta_0}[u_{\theta_0}^3(X)]\right]$$
$$+ T'(y)E[u_{\theta_0}(X)\nabla u_{\theta_0}(X)] + u_{\theta_0}^2(y) - I(\theta_0)$$
$$- [\nabla u_{\theta_0}(y) + u_{\theta_0}^2(y)] + E_{\theta_0}[\nabla u_{\theta_0}(y) + u_{\theta_0}^2(y)]$$
$$+ T'(y)E[u_{\theta_0}(X)(\nabla u_{\theta_0}(X) + u_{\theta_0}^2(X))]. \qquad (5.31)$$

Using (5.30) and (5.31), the required result follows by a straightforward differentiation of N_ϵ/D_ϵ, with subsequent evaluation of the same at $\epsilon = 0$. $\qquad\square$

Notice that for Example 5.2, direct computation gives $C'_{\text{LD}}(f_\epsilon) = 9$ when evaluated at $\epsilon = 0$, while for the HD and SCS evaluation at $\epsilon = 0$ gives $C'_{\text{HD}}(f_\epsilon) = C'_{\text{SCS}}(f_\epsilon) = 4.5$. From Table 5.1 it may be seen that this derivative gives a fairly accurate first-order approximation for $C_{\text{LD}}(f_\epsilon)$ up to values of ϵ as high as 0.01, but is already quite inaccurate for the robust distances at a value of $\epsilon = 0.001$, again exhibiting the importance of the higher order terms for the robust distances.

5.3   Disparity Difference Tests: The Continuous Case

As mentioned in Chapter 3, there is no completely general approach for the minimum disparity estimation problem based on the entire class of disparities

TABLE 5.2
The t-test statistics (signed disparity statistics) with the corresponding p-values.

	LD	LD+D	HD	GKL$_{1/2}$	SCS
Test statistics	0.4524	3.201	2.8515	2.6989	2.5500
p-values	0.3292	0.0038	0.0068	0.0091	0.0121

under easily accessible general conditions as it is possible under the discrete model. As a consequence, a general class of hypothesis tests based on disparities is also not available in the continuous case, although specific distances, primarily those based on the Hellinger distance, have been considered by several authors. In order to provide a partially general solution, we consider the structure presented by Park and Basu (2004), and discuss the theory of parametric hypothesis testing under this setup. Given the results of Section 3.4, this extension is routine, and here we simply present the results without detailed proofs.

Assume the setup of Sections 3.4 and 5.1. Let $X_1, \ldots, X_n$ form a random sample of independently and identically distributed observations from a distribution with belongs to the parametric model family $\mathcal{F}$, where the component distributions have densities with respect to the Lebesgue measure. Notice that the results of Section 3.4 imply that

$$-n^{-1/2} \nabla \rho_C(g_n^*, f_\theta) = Z_n(\theta) + o_p(1)$$

and that the unrestricted minimum distance estimator satisfies Equation (5.11). Let $H_0 : \theta \in \Theta_0$ be the null hypothesis of interest, which is to be tested against $H_1 : \theta \in \Theta \setminus \Theta_0$. We consider the disparity difference statistic

$$\mathrm{DDT}_{\rho_C}(g_n^*) = 2n[\rho_C(g_n^*, f_{\hat\theta_0}) - \rho_C(g_n^*, f_{\hat\theta})]. \tag{5.32}$$

Theorem 5.7. *We assume that the conditions (a)–(h) of Section 3.4 hold, as do conditions (C2)–(C4) of Section 5.2. Then the statistic in (5.32) has the same asymptotic χ_r^2 distribution under the null hypothesis specified by (5.1), and the restrictions imposed by it. This statistic is asymptotically equivalent to the corresponding likelihood ratio test under the null hypothesis.*

Example 5.3. This example concerns an experiment to test a method of reducing faults on telephone lines (Welch, 1987). This data set was previously analyzed in Example 3.4. Once again we have used the Epanechnikov kernel with bandwidth $\hat{h}_n = 1.66 \, \mathrm{MAD} \, n^{-1/5}$ as in Example 3.4.

In Table 5.2, we have presented the signed disparity statistics with the corresponding p-values for the hypothesis test $H_0 : \mu = 0$ against $H_1 : \mu > 0$ under the normal model with σ unspecified.

In Table 5.2, the LD and LD+D column represents the t-test statistics and the corresponding p-values using the original data set and the data set after deleting the large outlier (-988). The likelihood ratio test (the one sided t-test in this case) produces p-values of 0.3292 (with the outlier) and 0.0038 (without the outlier) for the above data. The large outlier reverses the conclusion and forces the acceptance of H_0, while the tests based on the disparity difference statistics are not affected by the outlier.   ‖

5.3.1 The Smoothed Model Approach

Test statistics of the disparity difference type based on the smoothed model does not have asymptotic chi-square limits; instead such a test statistic has an asymptotic limit equivalent to the distribution of linear combinations of squared $N(0, 1)$ variables. We present the following theorem in this connection. Our results in this section follow those of Basu (1993); also see Agostinelli and Markatou (2001).

Theorem 5.8. *Suppose that $X_1, \ldots, X_n$ form a random sample from a distribution with density $f_\theta(x)$. Suppose that conditions (B1)-(B7) of Section 3.5 and conditions (C2)–(C4) of Section 5.2 hold. Consider the hypotheses*

$$H_0 : \theta \in \Theta_0 \quad \text{versus} \quad H_1 : \theta \in \Theta \setminus \Theta_0,$$

where Θ_0 is a proper subset of $\Theta \subseteq \mathbb{R}^p$. Let

$$\mathrm{DDT}^*_{\rho_C}(g^*_n) = 2n[\rho_C(g^*_n, f^*_{\hat\theta_0}) - \rho_C(g^*_n, f^*_{\hat\theta})]$$

represent the disparity difference statistic at g^*_n based on the smoothed model f^*_θ where the unrestricted and constrained estimates are expressed in their usual notation. Then the following results hold:

1. *Under f_{θ_0}, $\theta \in \Theta_0$,*

   $$\mathrm{DDT}^*_{\rho_C}(g^*_n) - n(\hat\theta_0 - \hat\theta)^T[J^*(\theta_0)](\hat\theta_0 - \hat\theta) \to 0$$

   in probability, where $J^*(\theta_0)$ is as defined in Definition 3.1.

2. *If the kernel K is a transparent kernel for the model,*

   $$\mathrm{DDT}^*_{\rho_C}(g^*_n) - n(\hat\theta_0 - \hat\theta)^T[AI(\theta_0)](\hat\theta_0 - \hat\theta) \to 0$$

   in probability, where A is the matrix defined in Equation (3.39).

Proof. From Theorem 3.19, $\nabla_2 \rho_C(g^*_n, f^*_\theta)|_{\theta=\theta^g}$ converges in probability to $J^{*g}(\theta^g)$, which reduces to $J^*(\theta_0)$ when $g = f_{\theta_0}$ belongs to the model. Under f_{θ_0}, where $\theta_0 \in \Theta_0$, we therefore get, as in the derivation of Theorem 5.4,

$$\mathrm{DDT}^*_{\rho_C}(g^*_n) = 2n[\rho_C(g^*_n, f^*_{\hat\theta_0}) - \rho_C(g^*_n, f^*_{\hat\theta})]$$
$$= n(\hat\theta_0 - \hat\theta)^T[J^*(\theta_0)](\hat\theta_0 - \hat\theta) + o_p(1).$$

In addition, when K is a transparent kernel, we get $J^*(\theta_0) = AI(\theta_0)$, so that the above result reduces to

$$\text{DDT}^*_{\rho_C}(g_n^*) - n(\hat{\theta}_0 - \hat{\theta})^T[AI(\theta_0)](\hat{\theta}_0 - \hat{\theta}). \tag{5.33}$$

Thus, the theorem is proved. □

The theorem has the following implications. The expression (5.33) shows that the asymptotic distribution of $\text{DDT}^*_{\rho_C}(g_n^*)$ in this case is not a χ^2 unless A is an identity matrix, which is true only if the smoothing parameter $h_n \to 0$. In the transparent kernel case the likelihood disparity constructed between the data and the smoothed model generates the ordinary maximum likelihood estimator for any fixed h as the transparent kernel adjusts the bias introduced into the data and the model appropriately. However, when one takes the difference between the disparities constructed between the smoothed model and the data, the effect of a fixed h actually shows up through the matrix A and does not vanish asymptotically.

Suppose the null hypothesis $H_0 : \theta = \theta_0$ is a simple one. Then the asymptotic null distribution of $\text{DDT}^*_{\rho_C}(g_n^*)$ is the same as the distribution of $\sum_{i=1}^p \lambda_i Z_i^2$, where Z_i are independent $N(0, 1)$ variables and λ_is are the eigen values of A. An analogous linear combination of squared $N(0, 1)$ variables continue to hold under the composite hypothesis.

Consider the $N(\mu, \sigma^2)$ model where $\theta = (\mu, \sigma^2)$ and both parameters are unknown. Assume that we have a normal kernel which is transparent for the above model. Consider the simple null hypothesis, which specifies both parameters. The coefficient matrix A defined in (3.39) is diagonal in this case (Example 3.1), and from the above theorem the asymptotic null distribution of the statistic is a linear combination of two independent χ_1^2 statistics, where the combination weights are completely determined by the null parameters.

Now consider the case where θ is a scalar, and let G be the true distribution not necessarily in the model. Consider the null hypothesis

$$H_0 : T(G) = \theta_0. \tag{5.34}$$

Then we have the following analog of Theorem 5.5.

Theorem 5.9. *Let $g(x)$ be the true distribution, θ be a scalar parameter, and the null hypothesis of interest be given by (5.34).*

1. *Under the null hypothesis we have the convergence*

$$\text{DDT}^*_{\rho_C}(g_n^*) \to c^*(g)\chi_1^2,$$

   where $c^*(g) = \text{Var}_g[T'(X)] \cdot \nabla_2 \rho_C(g^*, f_\theta^*)|_{\theta=\theta_0}$, and $T'(\cdot)$ is the influence function of the minimum disparity estimator.

2. *Suppose in addition the true distribution $G = F_{\theta_0}$ belongs to the model,*

and the kernel K is transparent for the model family. Then for any disparity difference test statistic $\mathrm{DDT}^*_{\rho_C}(g^*_n)$,

$$c^*(g) = c^*(f_{\theta_0}) = A,$$

where A is the quantity defined in Equation (3.39); it is a scalar in this case. Under a simple null, therefore, the limiting distribution of the disparity difference test statistic is completely known.

Proof. The proof of part (a) proceeds exactly as the proof of Theorem 5.5. For part (b) note that $\nabla_2 \rho_C(g^*_n, f^*_\theta)|_{\theta=\theta_0} = J^*(\theta_0)$ under $g = f_{\theta_0}$, which reduces to $AI(\theta_0)$ under a transparent kernel. Also at the model $T'(y) = I^{-1}(\theta_0)u_{\theta_0}(y)$, so that $\mathrm{Var}_{f_{\theta_0}}[T'(X)] = I^{-1}(\theta_0)$. Replacing these values in the expression of $c^*(f_{\theta_0})$, the result holds. $\square$

5.4   Power Breakdown of Disparity Difference Tests

Following Lindsay and Basu (2010), we present an application of the contamination envelope approach presented in Section 4.9 to the case of power breakdown of disparity difference tests. Suppose that we are interested in the amount of distortion which the test statistic may exhibit due to data contamination. We consider the disparity difference statistic as a functional on a space of distributions $\mathcal{G}$ which is assumed to be convex. As a function of the distribution point G (having density g), consider now the functional

$$\phi(G) = \rho_C(g, f_{\hat\theta_0(g)}) - \rho_C(g, f_{\hat\theta(g)}),$$

where

$$\hat\theta_0(g) = \arg\min_{\theta \in \Theta_0} \rho_C(g, f_\theta) \quad \text{and} \quad \hat\theta(g) = \arg\min_{\theta \in \Theta} \rho_C(g, f_\theta).$$

For a generic functional $t(G)$ of this type with domain $\mathcal{G}$ let

$$\epsilon_1(G; t) = \inf\{\epsilon : \inf_{V \in \mathcal{G}} t((1-\epsilon)G + \epsilon V) = t_{\min}\}.$$

Here V is a contaminating distribution, and $t_{\min} = \inf_{F \in \mathcal{G}} t(F)$. Thus, $\epsilon_1(G; t)$ is the smallest amount of contamination necessary to drive the p-value of the test to 1, leading to certain power-breakdown. For a contamination proportion below $\epsilon_1(G; t)$ the disparity difference test will be consistent in the sense that it will eventually reject a false hypothesis, so asymptotically there will be no power-breakdown. We will refer to $\epsilon_1(G; t)$ as the power-breakdown point. The level-breakdown point $\epsilon_0(G; t)$ can be similarly defined based on the reverse considerations. However, ϵ_1 is usually the more informative quantity, and here we will focus on ϵ_1 only.

For the true distribution G and the contaminating distribution V, let $h = (1 - \epsilon)g + \epsilon v$ represent the density of the contaminated distribution H, and the corresponding functional of interest then becomes

$$\phi(H) = \rho_C(h, f_{\hat{\theta}_0(h)}) - \rho_C(h, f_{\hat{\theta}(h)}).$$

Contamination at level ϵ can then cause power breakdown at G if for some $V \in \mathcal{G}$ one gets $\phi(H) = t_{\min}$. Notice that for disparity difference tests one has $t_{\min} = 0$, so that to show that power breakdown does not occur at a level of contamination ϵ one simply has to demonstrate that $\rho_C(h, f_{\hat{\theta}_0(h)})$ remains strictly greater than $\rho_C(h, f_{\hat{\theta}(h)})$, for all possible distributions V in $\mathcal{G}$. In this connection we will apply the contamination envelopes described in Section 4.9 (Lindsay and Basu 2010) to determine a lower bound for $\rho_C(h, f_{\hat{\theta}_0(h)})$ and an upper bound for $\rho_C(h, f_{\hat{\theta}(h)})$, for all possible distributions V in $\mathcal{G}$. As long as this lower bound is strictly above the upper bound, power breakdown cannot occur.

Simpson (1989b) derived a set of contamination bounds for the disparity difference test based on the Hellinger distance. In this case

$$\rho_C(g, f_\theta) = \mathrm{HD}(g, f_\theta) = 2 \int (g^{1/2} - f_\theta^{1/2})^2 = 4 - 4 \int g^{1/2} f_\theta^{1/2},$$

and

$$\phi_{\mathrm{HD}}(G) = 4 \left[\int g^{1/2} f_{\hat{\theta}(g)}^{1/2} - \int g^{1/2} f_{\hat{\theta}_0(g)}^{1/2} \right].$$

Simpson (1989b) obtained the upper and lower bounds as

$$\mathrm{HD}(h, f_{\hat{\theta}(h)}) \leq \mathrm{HD}(h, f_{\hat{\theta}(g)}) \leq 4 - 4(1 - \epsilon)^{1/2} \int g^{1/2} f_{\hat{\theta}(g)}^{1/2},$$

and

$$\mathrm{HD}(h, f_{\hat{\theta}_0(h)}) \geq 4 - 4(1 - \epsilon)^{1/2} \int g^{1/2} f_{\hat{\theta}_0(g)}^{1/2} - 4\epsilon^{1/2},$$

so that the restriction that the difference between the lower bound and the upper bound must remain positive produces the relation

$$4(1 - \epsilon)^{1/2} \left[\int g^{1/2} f_{\hat{\theta}(g)}^{1/2} - \int g^{1/2} f_{\hat{\theta}_0(g)}^{1/2} \right] - 4\epsilon^{1/2} > 0$$

which shows breakdown cannot occur for

$$\epsilon < \frac{\phi_{\mathrm{HD}}^2(G)}{(16 + \phi_{\mathrm{HD}}^2(G))}.$$

Breakdown bounds for the negative exponential disparity case have been considered by Bhandari, Basu and Sarkar (2006) and Dey and Basu (2007). Here we present a specific example in the context of the Hellinger distance.

Example 5.4. Consider the $N(\theta, 1)$ model. In order to compare the results of Lindsay and Basu (2010) with those of Simpson (1989b), we choose the null hypothesis $H_0 : \theta = 0$ versus $H_1 : \theta \neq 0$, and let the true distribution be $N(3,1)$ model density. In this case the true distribution belongs to the model, so that $c_1 = 0$ in Theorem 4.18, and for the Hellinger distance the Lindsay–Basu upper bound becomes $U_\epsilon = 4(1 - \bar{\epsilon}^{1/2})$, where $\bar{\epsilon} = (1 - \epsilon)$. This matches with Simpson's (1989b) upper bound. However, the Lindsay–Basu (2010) lower bound is substantially tighter than that of Simpson (1989b). The analytical bound for the Lindsay–Basu description in this case is given by

$$L_\epsilon^{LB} = 4 \left(1 - [K^*(\gamma_0)]^{1/2} \int_{S_{\gamma_0}^*} f(x) - \bar{\epsilon}^{1/2} e^{-9/8} \int_{S_{\gamma_0}^{*c}} \tilde{g}(x) \right),$$

while the corresponding bound by Simpson equals

$$L_\epsilon^S = 4 \left(1 - \epsilon^{1/2} - \bar{\epsilon}^{1/2} e^{-9/8} \right),$$

where f is the $N(0, 1)$ density and $\tilde{g}$ is the $N(1.5, 1)$ density, and γ_0, $K^*(\gamma_0)$ and $S_{\gamma_0}^*$ are specific to the levels of contamination as given in Theorem 4.19. The common upper bound, and both sets of lower bounds are plotted on Figure 5.2. In Simpson's case the curves intersect approximately at 0.313, showing that the maximum amount of contamination that the Hellinger distance test functional can withstand to preserve its power is no less than 0.313. For the Lindsay–Basu approach, the curves intersect approximately at 0.407, showing that the power breakdown of the disparity difference test based on the Hellinger distance is actually much higher than 0.313. ‖

5.5 Outlier Stability of Disparity Difference Tests

In this section we consider an outlier stability property of disparity difference tests in the spirit of the outlier stability analysis of the minimum distance estimators presented in Section 4.8. Consider a discrete parametric model, and let $\{f_\theta\}$ represent the model family of densities. Let d_n be the vector of relative frequencies based on a random sample of size n, and consider testing the simple null hypothesis $H_0 : \theta = \theta_0$ against the two sided alternative. Our disparity difference test for these hypotheses is given by the statistic

$$\mathrm{DDT}_{\rho_C}(d_n) = 2n[\rho_C(d_n, f_{\theta_0}) - \rho_C(d_n, f_{\hat{\theta}}], \tag{5.35}$$

where $\rho_C = \mathrm{LD}$ generates the likelihood ratio test. Here $\hat{\theta}$ is the minimum distance estimator of θ for the disparity ρ_C based on the data d_n. Let ξ_j be an

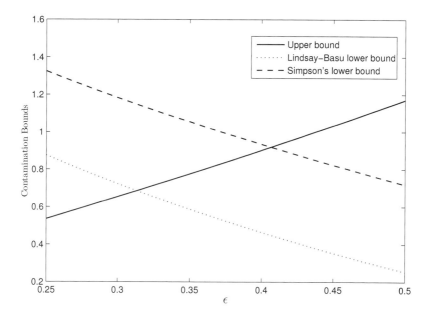

FIGURE 5.2
Contamination envelopes for the normal power breakdown example.

outlier sequence of elements of the sample space as defined in in Definition 4.2. As in Section 4.8, define

$$d_j = (1 - \epsilon)d_n + \epsilon \chi_{\xi_j}, \quad \text{and} \quad d_\epsilon^* = (1 - \epsilon)d_n.$$

We will say that the disparity difference test statistic defined in (5.35) is outlier stable if

$$\mathrm{DDT}_{\rho_C}(d_j) \to \mathrm{DDT}_{\rho_C}(d_\epsilon^*). \tag{5.36}$$

as $j \to \infty$.

Theorem 5.10. [Basu (2002, Theorem 1)]. *Under Assumptions 4.2, 4.3 and 4.4 of Chapter 4, the disparity difference test given by the statistic (5.35) is outlier stable.*

Proof. We have to prove the convergence in (5.36) holds for fixed n, as $j \to \infty$. Let T_j and θ^* be the minimizers of $\rho_C(d_j, f_\theta)$ and $\rho_C(d_\epsilon^*, f_\theta)$ respectively. Under the given conditions $T_j \to \theta^*$ as $j \to \infty$. By Scheffe's theorem (see, for example, Billingsley, 1986, pp. 218) $f_{T_j}(\xi_j) \to 0$ as $j \to \infty$. For a finite sample of size n, $d_n(\xi_j) = 0$ whenever $j > m$, for some positive integer depending on the sample. Note that for $j > m$,

$$\left| \text{DDT}_{\rho_C}(d_j) - \text{DDT}_{\rho_C}(d^*_\epsilon) \right|$$
$$\leq 2n\{|\rho_C(d_j, f_{\theta_0}) - \rho_C(d^*_\epsilon, f_{\theta_0})|\} + 2n\{|\rho_C(d_j, f_{T_j}) - \rho_C(d^*_\epsilon, f_{\theta^*})|\}.$$

Under the given conditions, $|\rho_C(d_j, f_{\theta_0}) - \rho_C(d^*_\epsilon, f_{\theta_0})| \to 0$ from relation (4.46). Rearranging terms as in Lemma 4.13

$$\left| \rho_C(d_j, f_{T_j}) - \rho_C(d^*_\epsilon, f_{\theta^*}) \right| \leq \left| \rho_C(d^*_\epsilon, f_{T_j}) - \rho_C(d^*_\epsilon, f_{\theta^*}) \right|$$
$$+ \left| C(-1) f_{T_j}(\xi_j) \right| + \left| C(\delta_j(\xi_j)) f_{T_j}(\xi_j) \right|$$

where $\delta_j(\xi_j) = d_j(\xi_j)/f_{T_j}(\xi_j) - 1$. The quantity

$$\left| \rho_C(d^*_\epsilon, f_{T_j}) - \rho_C(d^*_\epsilon, f_{\theta^*}) \right|$$

converges to zero by the given continuity assumptions, $|C(-1) f_{T_j}(\xi_j)|$ goes to zero by the convergence of $f_{T_j}(\xi_j)$ and the finiteness of $C(-1)$, and $|C(\delta_j(\xi_j)) f_{T_j}(\xi_j)|$ goes to zero as in Lemma 4.13. Thus, the desired convergence holds. $\qquad\square$

For the generalized Hellinger distance family we have

$$\text{DDT}_{\rho_C}(d^*_\epsilon) = (1 - \epsilon)^\alpha \text{DDT}_{\rho_C}(d_n).$$

Thus, for $\rho_C = \text{GHD}_\alpha$, the convergence result in (5.36) implies that the test statistic based on an outlier sequence $\{\xi_j\}$ converges to $(1 - \epsilon)^\alpha$ times the test statistic that one would have obtained with the original data d_n as $j \to \infty$. Thus, a single outlying value, however large, cannot arbitrarily perturb the test statistic.

Example 5.5. We now present a small example of this outlier stability property, given by Basu (2002), using the binomial $(12, \theta)$ model. A random sample of 50 observations was generated from the binomial $(12, 0.1)$ distribution. The hypothesis $H_0 : \theta = 0.1$ is to be tested against $H_1 : \theta \neq 0.1$. The likelihood ratio and the disparity difference test statistic based on the Hellinger distance for testing the above hypotheses are given by

$$\text{DDT}_{\text{LD}}(d_n) = 1.23545, \quad \text{and} \quad \text{DDT}_{\text{HD}}(d_n) = 1.61255.$$

A fixed contamination was then added to the observed sample at the extreme value $y = 12$ at a proportion of $\epsilon = 0.19$. The test statistics are recalculated at the contaminated data version

$$d_y(x) = 0.81 d_n(x) + 0.19 \chi_y(x)$$

with $y = 12$. The values of the test statistics are now

$$\text{DDT}_{\text{LD}}(d_y) = 124.5748 \quad \text{and} \quad \text{DDT}_{\text{HD}}(d_y) = 1.45119.$$

TABLE 5.3
Observed test statistics for the contaminated binomial distribution. Model is binomial $(12, \theta)$, sample size $n = 50$.

			α			
	0.1	0.2	0.3	0.4	0.5	LRT
Without contamination	3.78409	2.44005	1.99613	1.76313	1.61255	1.2355
$y = 8$	3.69676	2.32694	1.84874	1.56294	1.31025	50.2715
$y = 9$	3.70459	2.33815	1.87053	1.61021	1.41639	66.2377
$y = 10$	3.70516	2.33927	1.87352	1.61915	1.44437	84.0018
$y = 11$	3.70518	2.33935	1.87382	1.62045	1.45024	103.4733
$y = 12$	3.70518	2.33935	1.87384	1.62060	1.45119	124.5748
$\vdots$	$\vdots$	$\vdots$	$\vdots$	$\vdots$	$\vdots$	$\vdots$
$(1 - \epsilon)^\alpha \times$ Uncontaminated Statistic	3.70519	2.33935	1.87385	1.62061	1.45129	

The presence of the outlying component clearly blows up the likelihood ratio test, but the $\mathrm{DDT_{HD}}$ statistic remains stable. It may be noticed that the value of $\mathrm{DDT_{HD}}(d_y)$ is practically equivalent to $(1 - \epsilon)^{1/2}\mathrm{DDT_{HD}}(d_n) = 0.9 \times \mathrm{DDT_{HD}}(d_n) = 1.45129$, as one would expect. This is true for other disparity difference tests based on the generalized Hellinger distance as well, and this phenomenon is highlighted in Table 5.3. The table presents the values of the statistics $\mathrm{DDT_{GHD_\alpha}}(d_y)$ for $\alpha = 0.1, 0.2, \ldots, 0.5$ and y is allowed to slide to its maximum value 12 starting from 8. The numbers show that as y tends to its most extreme values, the test statistics all tend toward $(1 - \epsilon)^\alpha \times$ [the uncontaminated statistic]. $\quad\|$

5.5.1   The GHD and the Chi-Square Inflation Factor

Empirical investigations involving the generalized Hellinger distance family under point mass contaminations at extreme values suggest that the chi-square inflation factor has a remarkably close approximation based on the tuning parameter α and the contamination proportion ϵ in these cases. The results presented in this section show that this is no accident and indeed this is what one would expect in such situations.

Let $\{f_\theta\}$ be a discrete parametric model, $g(x)$ be the true density, and θ be a scalar parameter. Under the notation of Section 5.2, consider testing the null hypothesis $H_0 : T(G) = \theta_0$, where G is the distribution corresponding to the density g. For our purpose, we define the density

$$g(x) = (1 - \epsilon) f_{\theta_0}(x) + \epsilon \chi_\xi(x). \tag{5.37}$$

To determine the approximate value of the chi-square inflation factor, we first present two useful lemmas.

Lemma 5.11. [Basu (2002, Proposition 2)]. *Given $\theta_0 \in \Theta$, let g be the density defined in Equation (5.37). For fixed $\alpha \in (0,1)$, let ξ be an element of the sample space such that $u_{\theta_0}^2(\xi) f_{\theta_0}^{1-\alpha}(\xi)$ and $\nabla u_{\theta_0}(\xi) f_{\theta_0}^{1-\alpha}(\xi)$ are approximately zero. Then $\nabla_2 \mathrm{GHD}_\alpha(g, f_\theta)|_{\theta=\theta_0}$ is approximately equal to $(1 - \epsilon)^\alpha I(\theta_0)$.*

Proof. Direct differentiation gives

$$\begin{aligned}
\nabla_2 \mathrm{GHD}_\alpha(g, f_\theta) &= \nabla \left[-\sum g^\alpha(x) f_\theta^{1-\alpha}(x) u_\theta(x) \right] / \alpha \\
&= -(1 - \alpha) \sum g^\alpha(x) f_\theta^{1-\alpha}(x) u_\theta^2(x) / \alpha \\
&\quad - \sum g^\alpha(x) f_\theta^{1-\alpha}(x) \nabla u_\theta(x) / \alpha.
\end{aligned}$$

Replacing the value of $g(x)$, and rearranging terms, we get

$$\begin{aligned}
&\nabla_2 \mathrm{GHD}_\alpha(g, f_\theta)|_{\theta=\theta_0} \\
&= -(1 - \epsilon)^\alpha (1 - \alpha) \sum u_{\theta_0}^2(x) f_{\theta_0}(x) / \alpha \\
&\quad -(1 - \alpha) \left[(1 - \epsilon) f_{\theta_0}(\xi) + \epsilon \right]^\alpha f_{\theta_0}^{1-\alpha}(\xi) u_{\theta_0}^2(\xi) / \alpha \\
&\quad +(1 - \alpha)(1 - \epsilon)^\alpha f_{\theta_0}(\xi) u_{\theta_0}^2(\xi) / \alpha - (1 - \epsilon)^\alpha \sum \nabla u_{\theta_0}(x) f_{\theta_0}(x) / \alpha \\
&\quad -\left[(1 - \epsilon) f_{\theta_0}(\xi) + \epsilon \right]^\alpha f_{\theta_0}^{1-\alpha}(\xi) \nabla u_{\theta_0}(\xi) / \alpha + (1 - \epsilon)^\alpha f_{\theta_0}(\xi) \nabla u_{\theta_0}(\xi) / \alpha \\
&= (1 - \epsilon)^\alpha [I(\theta_0) - (1 - \alpha) I(\theta_0)] / \alpha \\
&\quad -(1 - \alpha) \left[(1 - \epsilon) f_{\theta_0}(\xi) + \epsilon \right]^\alpha f_{\theta_0}^{1-\alpha}(\xi) u_{\theta_0}^2(\xi) / \alpha \\
&\quad +(1 - \alpha)(1 - \epsilon)^\alpha f_{\theta_0}(\xi) u_{\theta_0}^2(\xi) / \alpha \\
&\quad -\left[(1 - \epsilon) f_{\theta_0}(\xi) + \epsilon \right]^\alpha f_{\theta_0}^{1-\alpha}(\xi) \nabla u_{\theta_0}(\xi) / \alpha + (1 - \epsilon)^\alpha f_{\theta_0}(\xi) \nabla u_{\theta_0}(\xi) / \alpha \\
&\approx (1 - \epsilon)^\alpha I(\theta_0)
\end{aligned}$$

under the stated conditions. □

Lemma 5.12. [Basu (2002, Proposition 3)]. *Let g be as defined in Equation (5.37), and ξ and α, belonging to their respective spaces, be such that the conditions of Lemma 5.11 are satisfied. Then $\mathrm{Var}_g[T'(X)]$ is approximately equal to $[(1 - \epsilon) I(\theta_0)]^{-1}$.*

Proof. The influence function $T'_\alpha(y)$ of the minimum GHD_α functional is given by $D_\alpha^{-1} N_\alpha$, where D_α and N_α are as given in Theorem 4.1, evaluated at the best fitting value of the functional. As θ is a scalar parameter, writing N_α explicitly as a function of the random variable, we get

$$\mathrm{Var}_g[T'_\alpha(X)] = \mathrm{Var}_g[N_\alpha(X)] / D_\alpha^2,$$

where the relevant quantities are evaluated at $T(G)$. From the results of Section 4.8 it follows, using the properties of the generalized Hellinger distance, that for a sufficiently extreme value ξ of the sample space, and for g as defined in Equation (5.37), the minimizer of $\text{GHD}_\alpha(g, f_\theta)$ is approximately equal to θ_0. Thus, for a sufficiently extreme ξ, the null hypothesis $T(G) = \theta_0$ is approximately true.

Evaluating D_α at $\theta = \theta_0$, and using manipulations similar to Lemma 5.11 we get,

$$D_\alpha = D_\alpha(\theta_0) \approx \alpha(1 - \epsilon)^\alpha I(\theta_0). \tag{5.38}$$

Similar manipulations and evaluation under $\theta = \theta_0$ give, $E_g(N_\alpha(X)) \approx 0$ and

$$\text{Var}_g[N_\alpha(X)] = E_g[N_\alpha(X)]^2 = \alpha^2(1 - \epsilon)^{2\alpha-1}I(\theta_0). \tag{5.39}$$

The final result is then a combination of the relations (5.38) and (5.39). $\qquad\square$

Combining the above two Lemmas, we get the chi-square inflation factor (as defined in Theorem 5.5) to be approximately equal to

$$(1 - \epsilon)^\alpha I(\theta_0)/[(1 - \epsilon)I(\theta_0)] = (1 - \epsilon)^{\alpha-1}.$$

As an illustration of the above, we consider the binomial $(20, \theta)$ model, and let f_θ be the binomial $(20, \theta)$ density. Let $g(x) = (1 - \epsilon)f_{0.1}(x) + \epsilon\xi_{20}(x)$, where $\epsilon = 0.19$. Consider testing the hypothesis $H_0 : \theta = \theta_0$ where θ_0 is the minimizer of $\rho_C(g, f_\theta)$. Direct calculation of the inflation factor via (5.26) gives $c(g) = 1.11111$, which is equal, at least up to five decimal places after the decimal point, to $(1 - \epsilon)^{\alpha-1} = (0.9)^{-1}$ for $\alpha = 0.5$.

5.6   The Two-Sample Problem

In this chapter we have discussed the parametric hypothesis testing problem based on an independently and identically distributed random sample drawn from a given population. Sometimes it is of interest to test the equality or other relationships between the parameters (or subsets of parameters) of two or more different populations. In this section we will briefly discuss the problem of parametric hypothesis testing based on disparities when independent random samples from two discrete populations are available. Suppose that random samples of size n_i are available from the i-th population, $i = 1, 2$. Suppose also that f_{θ_1} and f_{θ_2} be the parametric densities representing the two populations. A possible hypothesis of interest would be $H_0 : \theta_1 = \theta_2$, to be tested against an appropriate alternative. More generally, one could test the equality of the subset of the vectors θ_1 and θ_2, or consider some sort of dependent structure linking the two.

Basu, Mandal and Pardo (2009) have considered several possible test statistics for testing the null hypothesis of the equality of two scalar parameters representing two different populations, and have derived the asymptotic null distributions of these statistics. Let $\rho_C(d_i, f_{\theta_i})$ be the distance based on the disparity ρ_C constructed for the i-th population, $i = 1, 2$. Two possible tests (among many others) for testing the hypothesis of equality of the parameters θ_1 and θ_2 are given by the statistics

$$T_n = 2\frac{n_1 n_2}{n_1 + n_2}\rho_C(f_{\hat{\theta}_1}, f_{\hat{\theta}_2}),$$

where $\hat{\theta}_1$ and $\hat{\theta}_2$ are the unconstrained minimum distance estimators of θ_1 and θ_2, and

$$S_n = 2n[\min_{\theta_1=\theta_2} \rho_O(d_1, d_2, f_{\theta_1}, f_{\theta_2}) - \min_{\theta_1, \theta_2} \rho_O(d_1, d_2, f_{\theta_1}, f_{\theta_2})],$$

where

$$\rho_O(d_1, d_2, f_{\theta_1}, f_{\theta_2}) = \frac{n_1}{n_1 + n_2}\rho_C(d_1, f_{\theta_1}) + \frac{n_2}{n_1 + n_2}\rho_C(d_2, f_{\theta_2})$$

is an overall measure of disparity encompassing the two populations, and $n = n_1 + n_2$. Both the statistics T and S have asymptotic χ_1^2 null distributions; see Basu, Mandal and Pardo (2009).

The test statistic T_n as defined above is specific to the two-sample problem. However the statistic S_n can be be extended, in principle, for testing hypothesis relating to several populations.

Example 5.6. We apply the disparity difference tests on a two-sample problem with two sets of chemical mutagenicity data (Table 5.4) involving Drosophila, also analyzed by Simpson (1989b, p.112). The response variable in the two samples represent the numbers of daughters with a recessive lethal mutations among a certain variety of flies (Drosophila) exposed to chemicals and subject to control conditions respectively. The responses are modeled as random samples from Poisson (θ_1) for the control group and Poisson (θ_2) for the exposed group. The hypotheses of interest are

$$H_0 : \theta_1 \geq \theta_2 \quad \text{versus} \quad H_1 : \theta_1 < \theta_2. \tag{5.40}$$

Let n_i, d_i and f_{θ_i} be as defined earlier in this section. A test for the hypothesis of the equality of the means will be given by the disparity difference statistic

$$S_n = 2n\left[\min_{\theta_1=\theta_2} \rho_O(d_1, d_2, f_{\theta_1}, f_{\theta_2}) - \min_{\theta_1, \theta_2} \rho_O(d_1, d_2, f_{\theta_1}, f_{\theta_2})\right]. \tag{5.41}$$

For testing the null hypothesis of the equality of the two means against the not equal to alternative, the test statistic in (5.41) is directly applicable. However, for testing the hypothesis in (5.40), a signed disparity difference test is appropriate. In our case, this signed disparity difference test is given by

TABLE 5.4

Frequencies of the number of recessive lethal daughters for the control and exposed groups.

	Number of recessive lethal daughters							
	0	1	2	3	4	5	6	7
Control	159	15	3	0	0	0	0	0
Exposed	110	11	5	0	0	0	1	1

TABLE 5.5

The performance of disparity difference tests for the two-sample Drosophila example.

Distance	Full Data		Reduced Data	
	statistic	p-value	statistic	p-value
LD	2.9587	0.0015	1.0987	0.1359
HD	0.6982	0.2425	0.6940	0.2438
NED	0.4619	0.3221	0.4388	0.3304
PCS	56.1299	0.0000	1.8067	0.0354
$PD_{-0.9}$	0.4165	0.3385	0.4147	0.3392
SCS	0.4537	0.3250	0.4446	0.3283
$BWCS_{0.2}$	1.2099	0.1132	1.2012	0.1148
$BWHD_{1/3}$	1.0952	0.1367	1.0800	0.1401
$GKL_{1/3}$	0.7248	0.2342	0.7197	0.2358
$RLD_{1/3}$	0.1602	0.4364	0.1533	0.4391

$s_n^{1/2} \mathrm{sign}(\hat{\theta}_1 - \hat{\theta}_2)$, where s_n is the observed value of the statistic in (5.41), and $\hat{\theta}_1$ and $\hat{\theta}_2$ are the unrestricted minimum distance estimates of the two parameters. Under the null hypothesis, this statistic is asymptotically equivalent to the signed likelihood ratio test. For the full and reduced (after removing the two large counts for the exposed group) data, the signed disparity statistics and the associated p-values are given in Table 5.5.

The results show that the presence of the two large counts lead to a false positive for the likelihood ratio test. With the exclusion of these two values, the p-value of the likelihood ratio test jumps up to 0.1359 from 0.0015. Apart from the Pearson's chi-square, which is completely overwhelmed by these two large counts, all our robust tests perform adequately in the presence of these outliers. The p-values of these robust tests hardly show any change when the outliers are removed from the data. This even applies to the disparity difference tests based on the $BWHD_{1/3}$ and $BWCS_{0.2}$ distances. The $BWHD_{1/3}$ statistic has a value that is very close to the value of the likelihood ratio test under the reduced data, but the re-introduction of the contaminating values produces an insignificant shift in this statistic. Any of the statistics considered here, apart from the likelihood ratio and the Pearson's chi-square, will conclude that there is insufficient evidence against the null, whether with or without the outliers, and hence prevent a false declaration of significance in this example.   ‖

6

Techniques for Inlier Modification

6.1 Minimum Distance Estimation: Inlier Correction in Small Samples

As we have seen earlier, many members within the class of minimum distance estimators based on disparities provide excellent theoretical alternatives to the classical estimators both in terms of efficiency and robustness. Mathematically, these estimators retain their first-order efficiency at the model while having strong robustness features. They have easy interpretability as they are generated naturally through minimum distance procedures and can be computed using simple algorithms.

Yet, there is a great deal of variation in the behavior of the minimum distance estimators based on disparities, and there are times when the theoretical advantages do not translate readily into practical advantages with real data, particularly when the sample sizes are small. This is very prominently observed among the more robust members within the class of minimum distance estimators based on disparities which have large negative values for the estimation curvature. For such estimators it is usually observed that the small sample performance of the estimator under pure data is often substantially poorer than that of the maximum likelihood estimator. This is quite unfortunate, since their strong robustness properties make these estimators otherwise desirable.

We have introduced the notion of "inliers" and provided a short preliminary discussion of their role in parametric inference in Section 2.3.5. These are observations that have fewer data than what is predicted by the model. The role of inliers in case of the minimum distance estimation problem has so far received less attention in the literature compared to outliers. Lindsay (1994) observed that while the minimum Hellinger distance estimator performs very satisfactorily in dealing with outlying observations, its performance is actually worse than that of the maximum likelihood estimator under the presence of inliers. There is overwhelming empirical evidence which shows that the small sample performances of the minimum Hellinger distance estimator and several other minimum distance estimators within the class of disparities are significantly deficient compared to the performance of the maximum likelihood estimator. This appears to be at least partially due to the nature of the treatment of the inliers by these estimators. In particular, the weight attached to the empty cells (cells that correspond to $\delta = -1$, where δ is the Pearson residual) seem to have a major impact on the performance of the corresponding estimator. Although it shrinks the effect of large outliers, the Hellinger distance also magnifies the effect of the inliers. The residual adjustment function of the Hellinger distance shows a sharp dip in the left tail as δ approaches -1. This dip is more acute for, say, the disparities within the Cressie–Read family with tuning parameter $\lambda < -0.5$ (see Figure 2.1). In particular, $A(-1)$ is equal to $-\infty$ for all $\lambda \leq -1$. This leads to an infinite weight for the empty cells, and as a result such a distance is not defined when empty cells are present. Note that empty cells represent the most extreme cases of inliers.

The perception that proper control of inliers is necessary for good small sample efficiency is further enhanced by the fact that estimators which provide a reasonable treatment of inliers do not appear to have this deficient performance in small samples. This includes, for example, the class of minimum distance estimators within the Cressie–Read family with positive values of λ; this also includes the minimum negative exponential disparity estimator (Lindsay, 1994; Basu, Sarkar and Vidyashankar, 1997), or the family of minimum generalized negative exponential disparity estimators (Bhandari, Basu and Sarkar, 2006) which have been shown to perform well in dealing with inliers. Yet, within the class of minimum distance estimators based on disparities, the minimum Hellinger distance estimator still represents the most popular standard; the performance of the other estimators is often judged in comparison with the minimum Hellinger distance estimator. In this chapter, we will consider several modifications of many of our robust distances including the Hellinger distance with the aim of altering their inlier treatments appropriately while keeping their outlier treatments intact, so that the small sample efficiency of these estimators could be improved without significantly affecting their robustness properties.

The deficient behavior of the robust minimum distance estimators in small samples has also been observed in parametric hypothesis testing problems. The test statistics resulting from many of these robust disparities are very poorly approximated by their chi-square limits except for very large sample sizes (e.g., Simpson, 1989b, Table 3). It is difficult to attach much importance to its observed power if a test cannot hold its own level. We will show later in this section that the modifications considered herein appear to bring the tests closer in performance to the likelihood ratio test under the null when the model is correct, without any appreciable change in their robustness properties when the data are obtained from contaminated sources.

In this chapter, we will discuss five different modifications of disparity based distances. All of them are motivated by the need to moderate the effect of inliers and essentially shrink the magnitude of the inliers in one way or another. We will describe the development of each of these methods, discuss their theoretical properties, and provide numerical calculations which demonstrate the improved performance of these five methods in small samples.

Of the five inlier correction techniques discussed in this chapter, the first one – the method based on penalized distances – is specific to discrete models. All of the other four are applicable in both discrete and continuous models. In extensive simulations it does appear that the benefits derived from these inlier correction methods are more pronounced for the discrete model, although they are useful in continuous models as well. For illustration we will focus primarily on discrete models.

Throughout this chapter we will assume that the regular disparities involved in the relevant inlier correction analysis are standardized so that the summand in the summation expression of the disparities is nonnegative. As we have already observed, the condition described in Equation (2.16) together with the disparity conditions is sufficient to guarantee this.

6.2   Penalized Distances

Among the five different methods considered in this chapter, this is the only one which is limited to discrete distributions as it involves modifying the weight of a particular cell. In this method, we consider a modification of the distance which imposes a different, artificial weight on the empty cells. We will refer to this as the "empty cell penalty." Like many of the other minimum distance methods based on disparities, the application of this empty cell penalty was first developed in the context of the Hellinger distance. In the historical spirit we will first provide a description of minimum penalized Hellinger distance estimation, before considering the case of penalized distances more generally.

The penalized Hellinger distance evolved, rather accidentally, through an effort by Harris and Basu (1994) to pose the Hellinger distance as a log likelihood (possibly for a different model). While doing this, Harris and Basu observed that the performance of the minimum Hellinger distance estimator can be significantly improved if the weight of the empty cells was replaced by half of their original weight. Incidentally, this choice makes the new weight of the empty cells equal to the natural weight of the empty cells for the likelihood disparity. Since then there has been a fair amount of application of the penalized Hellinger distance and other penalized disparities in the literature. See, among others, Basu, Harris and Basu (1996), Basu and Basu (1998), Sarkar, Song and Jeong (1998), Yan (2001), Pardo and Pardo (2003), Alin (2007, 2008), Alin and Kurt (2008), and Mandal, Basu and Pardo (2010).

6.2.1   The Penalized Hellinger Distance

As pointed out earlier, empirical evidence strongly suggests that the relatively poor behavior of the minimum Hellinger distance estimator in small samples is at least partially due to the large weight that the Hellinger distance puts on the inliers (values with less data than expected under the model). Empty cells, in particular, constitute part of the inlier problem of the Hellinger distance. To understand the nature of the difference in their treatment of the empty cells, let us contrast the structure of the likelihood disparity with that of the Hellinger distance. Suppose that a random sample of n independently and identically distributed observations is obtained from the true distribution. We will assume a discrete parametric model, and the setup and notation of Section 2.3. Note that one can write the likelihood disparity $\mathrm{LD}(d_n, f_\theta)$ in the form

$$\sum_x \left[d_n(x) \log(d_n(x)/f_\theta(x)) + (f_\theta(x) - d_n(x)) \right]$$

$$= \sum_{x:d_n(x)>0} \left[d_n(x) \log \left(d_n(x)/f_\theta(x) \right) + \left(f_\theta(x) - d_n(x) \right) \right]$$

$$+ \sum_{x:d_n(x)=0} f_\theta(x). \tag{6.1}$$

When written in this form, each term in the summand in the left-hand side of the above equation is itself nonnegative. Comparing the corresponding decomposition of the Hellinger distance $\mathrm{HD}(d_n, f_\theta)$ given by

$$2 \sum_x \left(d_n^{1/2}(x) - f_\theta^{1/2}(x) \right)^2$$

$$= 2 \sum_{x:d_n(x)>0} \left(d_n^{1/2}(x) - f_\theta^{1/2}(x) \right)^2 + 2 \sum_{x:d_n(x)=0} f_\theta(x) \tag{6.2}$$

with the form in (6.1), one can see that the contribution of a cell with $d_n(x) = 0$ to the Hellinger distance is $2f_\theta(x)$, but equals just half of that in case of the likelihood disparity. For a given value of θ, one can modify the Hellinger distance in (6.2) by considering the distance

$$2 \sum_{x:d_n(x)>0} (d_n^{1/2}(x) - f_\theta^{1/2}(x))^2 + \sum_{x:d_n(x)=0} f_\theta(x), \tag{6.3}$$

which provides an identical treatment of the empty cells compared to the likelihood disparity in (6.1). Notice that the distance in (6.3) satisfies the basic properties of a statistical distance; it is nonnegative, and equals zero only if d_n and f_θ are identically equal. Harris and Basu (1994) considered a generalization of this and defined the penalized Hellinger distance (PHD) family as

$$\mathrm{PHD}_h(d_n, f_\theta) = 2 \sum_{x:d_n(x)>0} (d_n^{1/2}(x) - f_\theta^{1/2}(x))^2 + h \sum_{x:d_n(x)=0} f_\theta(x), \quad h \geq 0, \tag{6.4}$$

where $h = 2$ generates the ordinary Hellinger distance in (6.2), and $h = 1$ generates the distance in (6.3). As described in Equation (2.33), one gets the minimum penalized Hellinger distance estimator of θ corresponding to tuning parameter h by minimizing the distance in (6.4) over Θ.

Harris and Basu (1994) have studied the effect of modifying the weight of the empty cells in parametric estimation; their results show that often the estimator obtained by minimizing (6.3) can have substantially smaller mean square errors than the ordinary minimum Hellinger distance estimator (MHDE) – the minimizer of (6.2) – in small samples. Note that since the difference between the ordinary Hellinger distance and the other members of the penalized Hellinger distance family is only in the empty cells, the outlier resistant properties of the MHDE – which essentially relate to the behavior of the residual adjustment function and the estimator for large positive values along the δ axis – are shared by the minimum penalized Hellinger distance estimators for all values of the tuning parameter h. Unless otherwise stated, our default value of h will be equal to 1 in all applications involving the penalized Hellinger distance or other penalized distances.

Notice that the disparity generating function of the penalized Hellinger distance corresponding to the tuning parameter h is given by

$$C_{\mathrm{PHD}}(\delta) = 2[(\delta + 1)^{1/2} - 1]^2 \text{ for } \delta > -1, \text{ and } C_{\mathrm{PHD}}(-1) = h. \qquad (6.5)$$

By taking a derivative of the penalized Hellinger distance with respect to the parameter and by equating it to zero we get an estimating equation of the form (2.39), where the following residual adjustment function is obtained:

$$A_{\mathrm{PHD}}(\delta) = 2[(\delta + 1)^{1/2} - 1] \text{ for } \delta > -1, \text{ and } A_{\mathrm{PHD}}(-1) = -h. \qquad (6.6)$$

In particular when $h = 1$ this distance matches the likelihood disparity in its treatment of the empty cells. Note that the forms of the C and A functions of the penalized distance in (6.5) and (6.6) match those of the ordinary Hellinger distance at all points except $\delta = -1$. Unfortunately, however, except in case of the natural weight $h = 2$, the penalized Hellinger distance no longer retains the basic structure of a disparity. In particular, the disparity generating function C_{PHD} and residual adjustment function A_{PHD} are no longer continuous at $\delta = -1$. Thus, the asymptotic distribution of the minimum penalized Hellinger distance estimator (or other minimum penalized distance estimators in general) no longer follow directly from the asymptotic properties of the minimum distance estimators within the class of disparities as obtained in Chapter 2. As a result, one has to derive the asymptotics freshly in this case. We present this later in Section 6.2.3.

Next we turn our attention to the hypothesis testing problem. Suppose that the parametric hypothesis

$$H_0 : \theta \in \Theta_0 \text{ versus } H_1 : \theta \in \Theta \setminus \Theta_0, \qquad (6.7)$$

is of interest where Θ_0 is a proper subset of Θ. Given a general distance measure ρ_C satisfying the disparity conditions, the disparity difference test based on the disparity ρ_C is given by the test statistic

$$\mathrm{DDT}_{\rho_C}(d_n) = 2n\left[\rho_C(d_n, f_{\hat{\theta}_0}) - \rho_C(d_n, f_{\hat{\theta}})\right], \tag{6.8}$$

where the quantities $\hat{\theta}_0$ and $\hat{\theta}$ represent the minimum distance estimators corresponding to ρ_C over the null space Θ_0 and the unrestricted parameter space Θ respectively. The null distribution of the test statistic in (6.8) has been discussed extensively in Chapter 5. Under the null hypothesis, the above statistic has an asymptotic χ_r^2 distribution, where r represents the number of independent restrictions imposed by H_0. In particular, $\rho_C = $ HD generates the disparity difference test based on the Hellinger distance. Analogously, the disparity difference test based on the penalized Hellinger distance may be performed with the test statistic

$$\mathrm{DDT}_{\mathrm{PHD}}(d_n) = 2n\left[\mathrm{PHD}_h(d_n, f_{\hat{\theta}_0}) - \mathrm{PHD}_h(d_n, f_{\hat{\theta}})\right], \tag{6.9}$$

where the parameter estimates are now the indicated minimizers of $\mathrm{PHD}_h(d_n, f_\theta)$. However, as in the case of the estimation problem, the asymptotic null distribution of the test statistic in (6.9) does not directly follow from the results of Chapter 5 and has to be derived afresh, which we present in Section 6.2.4.

In the following, we will refer to the disparity generating function of the penalized disparity as C_h, to make the dependence on h explicit.

6.2.2   Minimum Penalized Distance Estimators

Our discussion so far has centered around the Hellinger distance and its penalized versions. The familiarity of the Hellinger distance and the simplicity of its mathematical structure makes it one of the most popular methods within the class of disparities, but there is nothing special about the Hellinger distance when the empty cell penalties are concerned; the Hellinger distance is just one of many disparities with a large negative value of the estimation curvature parameter. In general, given the distance

$$\rho_C(d_n, f_\theta) = \sum_x C(\delta(x)) f_\theta(x)$$

$$= \sum_{x:d_n(x)>0} C(\delta(x)) f_\theta(x) + C(-1) \sum_{x:d_n(x)=0} f_\theta(x),$$

one can consider the corresponding penalized distance

$$\rho_{C_h}(d_n, f_\theta) = \sum_{x:d_n(x)>0} C(\delta(x)) f_\theta(x) + h \sum_{x:d_n(x)=0} f_\theta(x), \tag{6.10}$$

where $\delta(x) = d_n(x)/f_\theta(x) - 1$. Here $C(-1)$ is the natural weight applied to the empty cells by the distance, and h is the artificial weight replacing it.

Once again our default value of h will be equal to 1, and the term penalized distance estimator will, unless otherwise mentioned, refer to this case. The larger the value of $C(-1)$, the greater is the expected improvement in the performance of the penalized distance estimator. Note that the modification in Equation (6.10) does not violate the nonnegativity condition of the distance.

The estimating equation of the penalized distance in (6.10) can be expressed as

$$-\nabla \rho_{C_h}(d_n, f_\theta) = \sum_x A_{C_h}(\delta(x))\nabla f_\theta(x) = 0,$$

where the residual adjustment function A_{C_h} of the penalized disparity is linked to the residual adjustment function A_C of the ordinary distance as

$$A_{C_h}(\delta) = \begin{cases} A_C(\delta) & \text{for } \delta > -1 \\ -h & \text{for } \delta = -1. \end{cases} \tag{6.11}$$

We will use the notation A_C and A_{C_h} in the following discussion to distinguish between the residual adjustment function of the ordinary and penalized distances.

6.2.3 Asymptotic Distribution of the Minimum Penalized Distance Estimator

We are now ready to state and prove the consistency and asymptotic normality of the minimum penalized distance estimator. Suppose $X_1, \ldots, X_n$ are n independently and identically distributed observations from a discrete distribution modeled by $\mathcal{F} = \{F_\theta : \theta \in \Theta \subseteq \mathbb{R}^p\}$ and $\mathcal{X} = \{0, 1, \ldots, \}$. To keep a clear focus here, we look at the case where the true distribution belongs to the model so that $g = f_{\theta_0}$ for some $\theta_0 \in \Theta$. Consider a disparity $\rho_C(d_n, f_\theta)$, where C is the disparity generating function. Let A_C represent the associated residual adjustment function; let ρ_{C_h} be the penalized disparity as defined in (6.10) and let A_{C_h} be its associated residual adjustment function as defined in (6.11).

As discussed in Remark 2.3, we will, essentially, prove the analogs of the convergence of the linear, quadratic and cubic terms of the derivatives of the disparity. The results will be proved under Assumptions (A1)–(A7) in Section 2.5, with suitable simplifications to reflect that $g = f_{\theta_0}$, i.e., the true distribution now belongs to the model. In particular, this implies $\theta^g = \theta_0$ and $J_g = I(\theta_0)$.

Lemma 6.1. *Assume that the true distribution $G = F_{\theta_0}$ belongs to the model. Under Assumptions (A1)–(A7) of Section 2.5, the linear term of the derivative of the disparity in the estimating equation of the minimum penalized distance estimator allows the following approximation:*

$$\nabla_j \rho_{C_h}(d_n, f_\theta)\Big|_{\theta=\theta_0} = -\sum_x d_n(x)u_{j\theta_0}(x) + o_p(n^{-1/2}). \tag{6.12}$$

Proof. Consider the difference between the ordinary distance and the penalized distance. We have

$$
\begin{aligned}
\rho_C(d_n, f_\theta) - \rho_{C_h}(d_n, f_\theta) &= (C(-1) - h) \sum_{x:d_n(x)=0} f_\theta(x) \\
&= (C(-1) - h) \sum_x f_\theta(x) \varsigma(d_n(x)) \\
&= R_n(\theta), \text{ (say)}, \tag{6.13}
\end{aligned}
$$

where $\varsigma(y) = 1$ if $y = 0$ and 0 otherwise. Taking a partial derivative with respect to θ_j, we get

$$
\begin{aligned}
\nabla_j R_n(\theta) &= (C(-1) - h) \sum_x \nabla_j f_\theta(x) \varsigma(d_n(x)) \\
&= (C(-1) - h) \sum_x f_\theta(x) u_{j\theta}(x) \varsigma(d_n(x)), \tag{6.14}
\end{aligned}
$$

where $u_{j\theta}(x)$ is as defined in Section 2.5. Thus, we get

$$
\begin{aligned}
&E\left[n^{1/2}|\nabla_j R_n(\theta)|\right] \\
&= n^{1/2}|C(-1) - h| \sum_x |f_\theta(x) u_{j\theta}(x)| E\left[\varsigma(d_n(x))\right] \\
&= n^{1/2}|C(-1) - h| \sum_x |f_\theta(x) u_{j\theta}(x)|\{1 - f_\theta(x)\}^n \\
&= |C(-1) - h| \sum_x \left|f_\theta^{1/2}(x) u_{j\theta}(x)\right| \left[n^{1/2} f_\theta^{1/2}(x)\{1 - f_\theta(x)\}^n\right].
\end{aligned}
$$

Suppose $\varphi_n(x) = n^{1/2} x^{1/2}(1 - x)^n$, where $0 < x < 1$. Note $\varphi_n(x) \to 0$ for all $0 < x < 1$ as $n \to \infty$. Now

$$
\max_{0<x<1} \varphi_n(x) = \frac{1}{\sqrt{2}} \left\{ \frac{2n}{2n+1} \right\}^{\frac{2n+1}{2}} \leq \frac{1}{\sqrt{2}}.
$$

Therefore, using Assumption (A5), it follows from an application of the dominated convergence theorem that

$$
E\left[n^{1/2}|\nabla_j R_n(\theta)|\right] \to 0 \text{ as } n \to \infty,
$$

and hence using Markov's inequality we get

$$
\nabla_j R_n(\theta) = o_p\left(n^{-1/2}\right).
$$

for any $\theta \in \omega$, where ω is the neighborhood defined in Assumption (A3). For such θ, Equation (6.13) therefore yields

$$
\nabla_j \rho_C(d_n, f_\theta) - \nabla_j \rho_{C_h}(d_n, f_\theta) = \nabla_j R_n(\theta) = o_p\left(n^{-1/2}\right),
$$

and by substituting θ_0 for θ in the last equation we get

$$\nabla_j \rho_{C_h}(d_n, f_\theta)\Big|_{\theta=\theta_0} = \nabla_j \rho_C(d_n, f_\theta)\Big|_{\theta=\theta_0} + o_p\left(n^{-1/2}\right). \tag{6.15}$$

The result

$$\nabla_j \rho_C(d_n, f_\theta)\Big|_{\theta=\theta_0} = -\sum_x d_n(x) u_{j\theta_0}(x) + o_p(n^{-1/2}) \tag{6.16}$$

holds at the model and has been established and encountered several times in Chapter 2; see, e.g., Equation (2.73). Combining (6.15) and (6.16), the lemma is proved. □

Lemma 6.2. *Assume that the true distribution $G = F_{\theta_0}$ belongs to the model. Under Assumptions (A1)–(A7) of Section 2.5, the quadratic term in the derivative of the disparity in the estimating equation of the minimum penalized distance estimator allows the following approximation:*

$$\nabla_{jk} \rho_{C_h}(d_n, f_\theta)\Big|_{\theta=\theta_0} = \sum_x f_{\theta_0}(x) u_{j\theta_0}(x) u_{k\theta_0}(x) + o_p(1).$$

Proof. Taking the second-order partial derivative of $R_n(\theta)$, the difference defined in (6.13), we get

$$\nabla_{jk} R_n(\theta) = (C(-1) - h) \sum_x \nabla_{jk} f_\theta(x) \varsigma(d_n(x)),$$

where $\varsigma(y) = 1$ if $y = 0$ and 0 otherwise. Taking expectations, we get

$$E\Big[\big|\nabla_{jk} R_n(\theta)\big|\Big]$$

$$= \big|C(-1) - h\big| \sum_x \big|\nabla_{jk} f_\theta(x)\big| E\left[\varsigma(d_n(x))\right]$$

$$= \big|C(-1) - h\big| \sum_x \big|\nabla_{jk} f_\theta(x)\big| \{1 - f_\theta(x)\}^n$$

$$= \big|C(-1) - h\big| \sum_x \big|f_\theta(x) u_{j\theta}(x) u_{k\theta}(x) + f_\theta(x) u_{jk\theta}(x)\big| \{1 - f_\theta(x)\}^n.$$

Now $\{1 - f_\theta(x)\}^n \to 0$ as $n \to \infty$ and $(1 - f_\theta(x)) < 1$. As in the proof of Lemma 6.1, it follows from Assumption (A5) and an application of the dominated convergence theorem that

$$E\Big[\big|\nabla_{jk} R_{jkn}(\theta)\big|\Big] \to 0 \text{ as } n \to \infty,$$

so that Markov's inequality now yields

$$\nabla_{jk} R_n(\theta) = o_p(1),$$

for any $\theta \in \omega$, where ω is the neighborhood defined in Assumption (A3). Using the above equation, and taking the second-order partial derivatives of (6.13), we get

$$\nabla_{jk}\rho_C(d_n, f_\theta) - \nabla_{jk}\ \rho_{C_h}(d_n, f_\theta) = \nabla_{jk}R_n(\theta) = o_p(1).$$

By substituting θ_0 for θ in the above equation it follows that

$$\nabla_{jk}\rho_{C_h}(d_n, f_\theta)\Big|_{\theta=\theta_0} = \nabla_{jk}\rho_C(d_n, f_\theta)\Big|_{\theta=\theta_0} + o_p(1). \tag{6.17}$$

In Theorem 2.19 we have shown that under the Assumptions (A1)–(A7),

$$\nabla_{jk}\rho_C(d_n, f_\theta)\Big|_{\theta=\theta^g} \to J_g^{jk} \quad \text{as } n \to \infty.$$

But at the model $\theta^g = \theta_0$ and $J_g = I(\theta_0)$, so that the above yields

$$\nabla_{jk}\rho_C(d_n, f_\theta)\Big|_{\theta=\theta_0} = \sum_x f_{\theta_0}(x)u_{j\theta_0}(x)u_{k\theta_0}(x) + o_p(1). \tag{6.18}$$

Combining (6.17) and (6.18), the lemma is proved. $\square$

Lemma 6.3. *Assume that the true distribution $G = F_{\theta_0}$ belongs to the model. Under Assumptions (A1)–(A7) of Section 2.5, there exists a finite number γ such that*

$$|\nabla_{jkl}\rho_{C_h}(d_n, f_\theta)| < \gamma, \tag{6.19}$$

with probability tending to 1 for all j, k and l and for all $\theta \in \omega$, where ω is the neighborhood defined in (A3).

Proof. A third-order partial derivative of the penalized disparity gives

$$\nabla_{jkl}\rho_{C_h}(d_n, f_\theta)$$
$$= -\sum_{x:d_n(x)>0} \Big[A_C''(\delta_n)(1+\delta_n)^2 u_{j\theta}u_{k\theta}u_{l\theta}f_\theta$$
$$- A_C'(\delta_n)(1+\delta_n)\{u_j u_{kl\theta} + u_{k\theta}u_{jl\theta} + u_{l\theta}u_{jk\theta} + u_{j\theta}u_{k\theta}u_{l\theta}\}f_\theta$$
$$+ A_C(\delta_n)\{u_{j\theta}u_{kl\theta} + u_{k\theta}u_{jl\theta} + u_{l\theta}u_{jk\theta} + u_{j\theta}u_{k\theta}u_{l\theta} + u_{jkl\theta}\}f_\theta \Big]$$
$$+ h\sum_{x:d_n(x)=0} [u_{jkl\theta} + u_{j\theta}u_{kl\theta} + u_{k\theta}u_{jl\theta} + u_{l\theta}u_{jk\theta} + u_{j\theta}u_{k\theta}u_{l\theta}]f_\theta,$$

where $\delta_n(x) = d_n(x)/f_\theta(x) - 1$. Using Assumption (A7), and the fact that

$|A_C(\delta)| \leq K_1|\delta|$ we get

$|\nabla_{jkl}\rho_{C_h}(d_n, f_\theta)|$

$\leq \sum_{x:d_n(x)>0} \Big[K_2(1+\delta_n)|u_{j\theta}||u_{k\theta}||u_{l\theta}|f_\theta$

$\quad - K_1(1+\delta_n)\{|u_{j\theta}||u_{kl\theta}| + |u_{k\theta}||u_{jl\theta}| + |u_{l\theta}||u_{jk\theta}| + |u_{j\theta}||u_{k\theta}||u_{l\theta}|\}f_\theta$

$\quad + K_1|\delta_n|\{|u_{j\theta}||u_{kl\theta}| + |u_{k\theta}||u_{jl\theta}| + |u_{l\theta}||u_{jk\theta}| + |u_{j\theta}||u_{k\theta}||u_{l\theta}| + |u_{jkl\theta}|\}f_\theta\Big]$

$\quad + h\sum_{x:d_n(x)=0} [|u_{jkl\theta}| + |u_{j\theta}||u_{kl\theta}| + |u_{k\theta}||u_{jl\theta}| + |u_{l\theta}||u_{jk\theta}| + |u_{j\theta}||u_{k\theta}||u_{l\theta}|]f_\theta.$

Routine simplification of the above gives

$|\nabla_{jkl}\rho_{C_h}(d_n, f_\theta)|$

$\leq \sum_{x:d_n(x)>0} \Big[K_2|u_{j\theta}||u_{k\theta}||u_{l\theta}|d_n$

$\quad - K_1\{|u_{j\theta}||u_{kl\theta}| + |u_{k\theta}||u_{jl\theta}| + |u_{l\theta}||u_{jk\theta}| + |u_{j\theta}||u_{k\theta}||u_{l\theta}|\}d_n$

$\quad + K_1\{|u_{j\theta}||u_{kl\theta}| + |u_{k\theta}||u_{jl\theta}| + |u_{l\theta}||u_{jk\theta}| + |u_{j\theta}||u_{k\theta}||u_{l\theta}|$

$\qquad + |u_{jkl\theta}|\}(d_n + f_\theta)\Big]$

$\quad + h\sum_{x:d_n(x)=0} [|u_{jkl\theta}| + |u_{j\theta}||u_{kl\theta}| + |u_{k\theta}||u_{jl\theta}| + |u_{l\theta}||u_{jk\theta}| + |u_{j\theta}||u_{k\theta}||u_{l\theta}|]f_\theta,$

and thus there exists a positive constant M^* such that

$|\nabla_{jkl}\rho_{C_h}(d_n, f_\theta)|$

$\leq M^*\sum_x (d_n + f_\theta)\Big[|u_{j\theta}||u_{k\theta}||u_{l\theta}|$

$\quad + |u_{j\theta}||u_{kl\theta}| + |u_{k\theta}||u_{jl\theta}| + |u_{l\theta}||u_{jk\theta}| + |u_{j\theta}||u_{k\theta}||u_{l\theta}|$

$\quad + |u_{j\theta}||u_{kl\theta}| + |u_{k\theta}||u_{jl\theta}| + |u_{l\theta}||u_{jk\theta}| + |u_{j\theta}||u_{k\theta}||u_{l\theta}| + |u_{jkl\theta}|\Big].$

It follows from Assumption (A4) and an application of the central limit theorem that the right-hand side of the above equation is bounded with probability tending to 1 for $\theta \in \omega$. This completes the proof. $\qquad\square$

The consistency and the asymptotic normality of the minimum penalized disparity estimator is then stated in the following theorem. The proof is a simple consequence of the above three lemmas.

Theorem 6.4. *Suppose that the true distribution* $G = F_{\theta_0}$ *belongs to the model. Under the Assumptions (A1)–(A7) of Section 2.5, the minimum penalized disparity estimating equation*

$$\nabla\rho_{C_h}(d_n, f_\theta) = 0 \tag{6.20}$$

has a consistent sequence of roots θ_n. The sequence θ_n has an asymptotic multivariate normal distribution given by

$$n^{1/2}(\theta_n - \theta_0) \xrightarrow{\mathcal{D}} Z^* \sim N_p(0, I^{-1}(\theta_0)), \qquad (6.21)$$

where $I(\theta_0)$ is the Fisher information matrix at f_{θ_0}.

The three lemmas proved above actually do more than simply establish the asymptotic distribution of the minimum penalized distance estimator. A quick comparison of Theorem 6.4 with Theorem 2.19 shows that the minimum penalized distance estimator is asymptotically equivalent to any other first-order efficient estimator characterized by Equation (2.84). This is expressed through the following corollary.

Corollary 6.5. *Let θ_n be the minimum penalized distance estimator of θ, and $\tilde{\theta}_n$ be any other first-order efficient estimator characterized by Equation (2.84). Then these estimators are asymptotically equivalent in the sense*

$$n^{1/2}(\theta_n - \tilde{\theta}_n) \to 0$$

in probability as $n \to \infty$.

6.2.4 Penalized Disparity Difference Tests: Asymptotic Results

We now consider the hypothesis testing problem. We are interested in testing the null hypothesis

$$H_0 : \theta \in \Theta_0 \quad \text{versus} \quad H_1 : \theta \in \Theta \setminus \Theta_0. \qquad (6.22)$$

We assume the setup of Sections 2.3 and 5.1. Let us suppose that H_0 is defined by a set of r ($\leq p$) restrictions on Θ defined by $R_i(\theta) = 0$, $i = 1, \ldots, r$, and $\theta = b(\gamma)$, $\gamma \in \Gamma \subseteq \mathbb{R}^{p-r}$ describes the parameter space under the null. The function $b : \mathbb{R}^{p-r} \to \mathbb{R}^p$ has continuous derivatives $\dot{b}(\gamma)$ of order $p \times (p - r)$ with rank $p - r$. The test statistic for testing the null hypothesis in (6.22) is given by

$$\text{DDT}_{\rho_{C_h}}(d_n) = 2n[\rho_{C_h}(d_n, f_{\hat{\theta}_0}) - \rho_{C_h}(d_n, f_{\hat{\theta}})] \qquad (6.23)$$

where $\hat{\theta}$ and $\hat{\theta}_0$ are the unrestricted and constrained minimum penalized distance estimators of θ respectively. Under the notation of Section 5.1, the constrained minimum distance estimator is assumed to be of the form $\hat{\theta}_0 = b(\hat{\gamma})$, where $\hat{\gamma}$ is the first-order efficient minimum penalized distance estimator of γ based on the penalized disparity ρ_{C_h}. We now state the theorem which formalizes the asymptotic properties of the disparity difference tests based on penalized distances. The theorem is proved by retracing the steps of Theorem 5.4, taken together with the fact that the essential convergence of the terms of the three orders are now proved in Lemmas 6.1–6.3.

Theorem 6.6. *Under Assumptions (B1)–(B4) of Section 5.2, the limiting null distribution of the disparity difference test statistic in Equation (6.23) is χ_r^2, where r is the number of restrictions imposed by the null hypothesis in (6.22).*

6.2.5 The Power Divergence Family versus the Blended Weight Hellinger Distance Family

As we have discussed, the power divergence family of Cressie and Read (1984) represents a very rich subclass of density-based distances. It has wide familiarity relative to other subclasses of distances, and also contains most of the well known density-based distances within the class of disparities. Many members of this class provide sturdy treatment of outliers leading to minimum distance estimators with good robustness and stability properties. In many ways, this class is the representative family within the collection of disparities, primarily because of the unifying work of Cressie and Read (1984) and Read and Cressie (1988), although the main focus of these works was on goodness-of-fit testing, rather than on minimum distance estimation.

The inlier problem is a hindrance to the practical applicability of this family, however, which has been described in the previous sections. In fact the members of the power divergence family are not even defined for $\lambda \leq -1$ in case there is a single empty cell. In this section, we briefly discuss the connection of the power divergence family with another subclass of disparities – the blended weight Hellinger distance BWHD_α – which is less known than the Cressie–Read family, although some of its properties have been discussed in limited setups (e.g., Lindsay 1994; Basu and Sarkar, 1994b; Shin, Basu and Sarkar, 1995; Basu and Basu, 1998). This family has been introduced in Chapter 2 of this book and is defined by Equation (2.18). Since each summand in the expression of this distance is a squared term, $\text{BWHD}_\alpha(d_n, f_\theta)$ is a valid statistical distance for any $\alpha \in (-\infty, \infty)$ in the sense described in Chapter 2. However, the interpretation of the distance becomes more complicated when α lies outside the range $[0, 1]$. The BWHD_α family generates the Pearson's chi-square, the Hellinger distance and the Neyman's chi-square for $\alpha = 0, 1/2$ and 1 respectively. The family does not have the empty cell limitation of the Cressie–Read family (except for the case $\alpha = 1$).

Here we show that for each member of the Cressie–Read family, there is a corresponding member within the family of blended weight Hellinger distances BWHD_α which is very close to the former distance, so that even when a distance within the Cressie–Read family is incomputable, another distance with very similar properties is available. Our discussion of this item follows that of Basu and Basu (1998).

To describe the equivalence between the power-divergence family of Cressie and Read and the blended weight Hellinger distances, we briefly introduce the setup of multinomial goodness-of-fit testing which will be discussed in greater detail in Chapter 8. This is done primarily because the expressions through which the equivalence is studied were originally proposed in connection with the use of the distances based on disparities in the context of multinomial goodness-of-fit tests.

Consider the multinomial setup with k cells, having cell probabilities $\boldsymbol{\pi} = (\pi_1, \ldots, \pi_k)$ with a sample of size n. To keep a clear focus in our presentation, let us assume that the null hypothesis is a simple one, given by

$$H_0 : \pi_i = \pi_{0i}, \quad \pi_{0i} > 0, i = 1, 2, \ldots, k, \quad \sum_{i=1}^{k} \pi_{0i} = 1.$$

Let $(x_1, \ldots, x_k)$ denote the vector of observed frequencies for the k categories, $\sum_{i=1}^{k} x_i = n$ and let $\boldsymbol{p} = (p_1, \ldots, p_k) = (x_1/n, \ldots, x_k/n)$ be the vector of observed proportions. As discussed in Cressie and Read (1984) and in Chapter 8 of this book, the disparity test statistics $2n\mathrm{PD}_\lambda(\boldsymbol{p}, \boldsymbol{\pi}_0)$ and $2n\mathrm{BWHD}_\alpha(\boldsymbol{p}, \boldsymbol{\pi}_0)$ are all eligible candidates for performing the relevant goodness-of-fit test. Let $W_i = n^{1/2}(p_i - \pi_{0i})$. Given a null vector $\boldsymbol{\pi}_0 = (\pi_{01}, \ldots, \pi_{0k})$ we get, expanding $2n\mathrm{PD}_\lambda(\boldsymbol{p}, \boldsymbol{\pi}_0)$ in a Taylor series

$$2n\mathrm{PD}_\lambda(\boldsymbol{p}, \boldsymbol{\pi}_0) = \sum_{i=1}^{k} \frac{W_i^2}{\pi_{0i}} + \frac{\lambda - 1}{3n^{1/2}} \sum_{i=1}^{k} \frac{W_i^3}{\pi_{0i}^2}$$

$$+ \frac{(\lambda - 1)(\lambda - 2)}{12n} \sum_{i=1}^{k} \frac{W_i^4}{\pi_{0i}^3} + O_p(n^{-3/2}). \quad (6.24)$$

The derivation is straightforward, and is available in Read and Cressie (1988, page 176). A similar expansion for the BWHD_α, derived by Basu and Sarkar (1994b) has the form

$$2n\mathrm{BWHD}_\alpha(\boldsymbol{p}, \boldsymbol{\pi}_0) = \sum_{i=1}^{k} \frac{W_i^2}{\pi_{0i}} - \frac{\alpha}{n^{1/2}} \sum_{i=1}^{k} \frac{W_i^3}{\pi_{0i}^2}$$

$$+ \frac{3\alpha^2 + \alpha}{4n} \sum_{i=1}^{k} \frac{W_i^4}{\pi_{0i}^3} + O_p(n^{-3/2}). \quad (6.25)$$

Comparing (6.24) and (6.25), we see that for any given $\boldsymbol{p}$ and $\boldsymbol{\pi}_0$, the Cressie–Read disparity corresponding to the parameter λ is exactly equivalent (excluding terms of the order of $O_p(n^{-3/2})$) to the member of the blended weight Hellinger distance family with $\alpha = (1 - \lambda)/3$. (Also see Shin, Basu and Sarkar, 1995).

In the context of multivariate goodness-of-fit tests, Cressie and Read have derived the first three moments of the test statistic $2n\mathrm{PD}_\lambda(\boldsymbol{p}, \boldsymbol{\pi}_0)$ under the simple null hypothesis; the moment calculations are based on Equation (6.24), and are done excluding the $O_p(n^{-3/2})$ terms. See Read and Cressie (1988, Appendix A11, pp. 175–181). However, the moments are derived only for the cases $\lambda > -1$, since the moments are not defined otherwise. Since the expansions

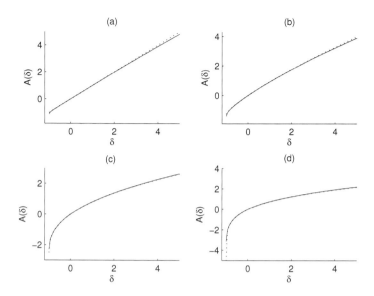

FIGURE 6.1
Residual adjustment functions of the PD_λ family (dashed line) and the $BWHD_\alpha$ family (solid line) for several combinations of (λ, α).

(6.24) and (6.25) are identical with the exclusion of the $O_p(n^{-3/2})$ terms, a similar calculation, based on Equation (6.25), will give exactly the same moment expressions for the corresponding $2n\text{BWHD}_{(1-\lambda)/3}(\boldsymbol{p}, \boldsymbol{\pi}_0)$ statistic. But unlike the $2n\text{PD}_\lambda(\boldsymbol{p}, \boldsymbol{\pi}_0)$ statistic, empty cells do not pose any difficulty in the definition of the blended weight Hellinger distance or in the calculation of the moments of the $2n\text{BWHD}_\alpha(\boldsymbol{p}, \boldsymbol{\pi}_0)$ statistic except in the case that $\alpha = 1$ (the Neyman's chi-square). Thus, for example, while the power divergence test statistic $2n\text{PD}_\lambda(\boldsymbol{p}, \boldsymbol{\pi}_0)$ may be undefined for $\lambda = -1$, there is no difficulty in the construction of the corresponding statistic based on the $BWHD_\alpha$ family for $\alpha = (1 - (-1))/3 = 2/3$.

A visual inspection gives a better illustration of the equivalence between the two families. For this purpose we have plotted the residual adjustment function of several members of the Cressie–Read family of distances (with parameter λ) with the equivalent member of the blended weight Hellinger distance family (with parameter $\alpha = (1 - \lambda)/3$) on the same axes in Figures 6.1 (a)–(d). Four combinations, $(0, 1/3)$, $(-0.2, 0.4)$, $(-0.6, 8/15)$ and $(-0.8, 0.6)$, of (λ, α) are selected. In each case, the correspondence between the functions can be seen to be extremely close; the only hint of a slight discrepancy is observed in the left-hand tail of the fourth combination, where the residual adjustment function of the Cressie–Read distance appears to dip further down compared to the blended weight Hellinger distance.

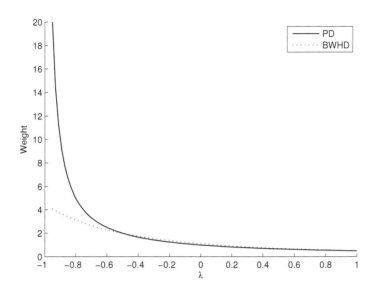

FIGURE 6.2
Weight functions for the empty cells in the PD and BWHD families.

We follow this up with a comparison of the empty cell behavior of the members of the Cressie–Read family of distances and blended weight Hellinger distance family for different values of the tuning parameters λ and α. A direct comparison of the expressions for the Cressie Read distance PD_λ and the blended weight Hellinger distance $BWHD_\alpha$ shows that the natural weights for the empty cells for the two families are $(1+\lambda)^{-1}$ and $[2(1-\alpha)]^{-1}$ where the equivalent distances correspond to $\alpha = (1-\lambda)/3$. In Figure 6.2 we have plotted the weights of the empty cells for the Cressie–Read family in the Y-axis against the values of $\lambda \in (-1,1]$ in the X-axis; in the same axes we have also plotted the weights of the empty cells for the corresponding member $BWHD_{(1-\lambda)/3}$ of the blended weight Hellinger distance family. The graph reveals that the weights of the empty cells match each other very closely in the range $[-0.6,1]$; the weights for the Cressie–Read family blow up very fast in the range $\lambda \in (-1,-0.6]$, but remain relatively stable for the blended weight Hellinger distance family. This is what is observed in the left tail of Figure 6.1 (d). This also shows that in spite of the equivalence demonstrated through equations (6.24) and (6.25), the residual adjustment function for the two families can be somewhat different as one moves away from the central part of the curve, particularly for the more robust distances.

It is easy to see that the penalizing of the empty cells of the distances with a fixed, finite, penalty weight eliminates the possibility of the distance being undefined when any particular cell is empty. Thus, while the Cressie–Read distances may not be defined for $\lambda \leq -1$, their penalized versions remain perfect candidates that may be used in minimum distance inference or in disparity goodness-of-fit tests without any such technical difficulty. The discussion of the previous paragraphs also shows that the penalized power divergence families and the penalized blended weight Hellinger distance families are likely to lead to similar inference, whether the original distances are defined or not.

Example 6.1. This example is reproduced from Basu and Basu (1998). We consider a numerical study illustrating the closeness in performance of the minimum power divergence estimators (wherever defined) and the minimum blended weight Hellinger distance estimators, as well as their penalized versions. For this we choose a multinomial distribution supported on four cells, with cell probabilities $\pi_1 = \theta$, $\pi_2 = \theta(1 - \theta)$, $\pi_3 = \theta(1 - \theta)^2$ and $\pi_4 = (1 - \sum_{i=1}^{3} \pi_i)$, $\theta \in (0, 1)$. Ten observations are generated from this distribution and our aim is to estimate the value of θ. The mean square errors of the estimators are obtained through complete enumeration of the sample space (rather than through any Monte-Carlo type method), which generates the exact distribution of the estimators. Thus, we enumerate all possible sample combinations in the space D given by

$$D = \{(x_1, x_2, x_3, x_4) : \sum_{i=1}^{4} x_i = 10, \text{ each } x_i \text{ is an integer between 0 and 10}\},$$

and the probability of each such sample combination can be obtained using the multinomial probability function. Since each sample combination generates a particular value of the parameter estimate, the entire probability distribution of the estimator becomes known and the mean square error of the estimator around the true value of θ, in this case chosen to be $\theta = 0.1$, can be exactly computed. The four minimum distance estimators considered in this example are denoted by $\hat{\theta}_I$, $\hat{\theta}_{PI}$, $\hat{\theta}_B$, and $\hat{\theta}_{PB}$, and correspond to the minimum PD$_\lambda$ estimator, the minimum penalized PD$_\lambda$ estimator, the minimum BWHD$_\alpha$ estimator and the minimum penalized BWHD$_\alpha$ estimator respectively. In either case the penalty weight used is $h = 1$. For each value of the tuning parameter λ in the power divergence family, the corresponding tuning parameter in the blended weight Hellinger distance is chosen as $\alpha = (1 - \lambda)/3$.

Notice that the estimators are quite similar in performance for $\lambda = 0$ ($\alpha = 1/3$), $\lambda = -0.5$ ($\alpha = 0.5$) and $\lambda = -0.6$ ($\alpha = 8/15$). However, as λ tends toward 1 (and α tends toward 2/3), the minimum PD$_\lambda$ estimator exhibits a substantial underperformance compared to the minimum BWHD$_\alpha$ estimator. This is quite expected from Figure 6.2. The improvement in the performance of each set of estimators because of the application of the penalty is evident in Table 6.1. The ordinary estimator in the power divergence family does not exist for $\lambda \leq -1$, but the corresponding estimator in the blended weight Hellinger distance family has no such problem. Also, the difference in the performance of the ordinary estimators is almost completely eliminated by the application of the penalty. ‖

TABLE 6.1
Comparison of exact mean square errors of the minimum power divergence and minimum blended weight Hellinger distance estimators, as well as their penalized versions.

$\lambda(\alpha)$	MSE $(\hat{\theta}_I)$	MSE $(\hat{\theta}_{PI})$	MSE $(\hat{\theta}_B)$	MSE $(\hat{\theta}_{PB})$
0 (1/3)	0.003631	0.003631	0.003636	0.003631
−0.5(0.5)	0.004306	0.003630	0.004306	0.003630
−0.6(8/15)	0.004883	0.003652	0.004720	0.003653
−0.7(17/30)	0.005762	0.003695	0.005158	0.003695
−0.8(0.6)	0.007438	0.003745	0.005709	0.003740
−0.9(19/30)	0.010111	0.003807	0.006558	0.003792
−1.1(0.7)	.	0.003974	0.009025	0.003928
−1.3(23/30)	.	0.004256	0.010238	0.004117
−1.5(5/6)	.	0.004583	0.010759	0.004421
−1.7(0.9)	.	0.004786	0.010814	0.004666
−1.9(29/30)	.	0.004879	0.015892	0.004873

For systematic reference we will refer to the penalized distance within the power divergence family by $\text{PPD}_{(h,\lambda)}$, where h denotes the penalty weight for the empty cells, and λ is the tuning parameter of the distance within the PD family. Similar notation will be used for other disparities and other families of disparities.

6.3   Combined Distances

Another natural way to control the inlier problem is to consider "combined distances." In Figures 6.3 and 6.4 we have presented the graphs of the disparity generating functions of several members of the power divergence family of Cressie and Read and several members of the blended weight Hellinger distance family respectively. In either figure, notice that the disparity generating functions which are reasonably flat on the right-hand side of the δ axis – being associated with robust disparities – rise sharply on the left-hand side of the δ axis, making them sensitive to inliers. However, disparity generating functions which are sharply rising on the right-hand side do have a significantly dampened response on the left-hand side, so that inlying observations are more sensibly treated by these disparities. None of the natural disparities within these families appear to give a satisfactory treatment of inliers and outliers simultaneously.

This observation is further strengthened by taking a look at the residual adjustment functions of these disparities. In Figures 6.5 and 6.6, we have plotted the residual adjustment functions of different disparities of these two

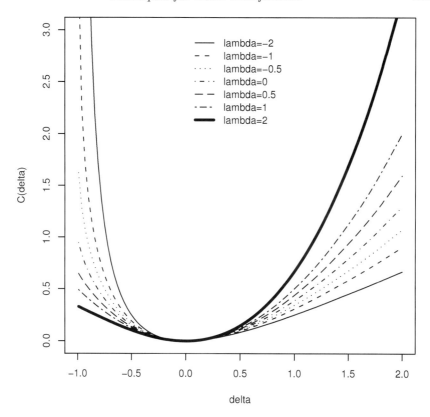

FIGURE 6.3
Disparity generating functions for several members of the power divergence family.

families. The RAFs of the more robust disparities dip sharply in the left-hand tail, magnifying the effect of inliers. However, residual adjustment functions which fail to control the effect of large outliers on the positive side of the axis, do provide a more solid treatment of the inliers. This is not just a feature of the power divergence and the blended weight Hellinger distance families, this appears to be the norm for most families of density-based distances.

The method of penalized distances tries to control this behavior by appropriately modifying the effect of the empty cells. While the method of penalty deals with the most serious inliers, a more comprehensive remedy to this problem may perhaps be achieved through the combined distance approach. Instead of adjusting for the empty cells alone, here we propose to combine two different disparities – the one at the left providing appropriate treatment of inliers, and the one on the right providing suitable downweighting of the outliers – at the origin of the δ axis. We hope to control the outlier and inlier problems simultaneously through this construction.

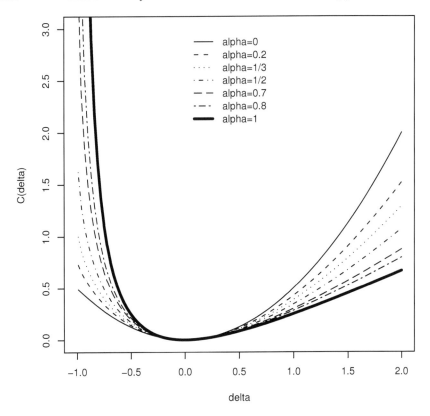

FIGURE 6.4
Disparity generating functions for several members of the blended weight
Hellinger distance family.

Formally, this combination process may be described as follows. Suppose
that we have two distances generated by the functions C_1 and C_2 satisfying
the disparity conditions. Then the combined distance ρ_{C_m} is defined by

$$\rho_{C_m}(d_n, f_\theta) = \sum_x C_m(\delta(x)) f_\theta(x), \qquad (6.26)$$

where

$$C_m(\delta) = \begin{cases} C_1(\delta) & \text{for } \delta \leq 0 \\ C_2(\delta) & \text{for } \delta > 0. \end{cases}$$

If A_1 and A_2 are the associated residual adjustment functions, the residual
adjustment function A_m of the combined distance turns out to be

$$A_m(\delta) = \begin{cases} A_1(\delta) & \text{for } \delta \leq 0 \\ A_2(\delta) & \text{for } \delta > 0. \end{cases}$$

Under the combined distance approach, our aim will be to combine a robust

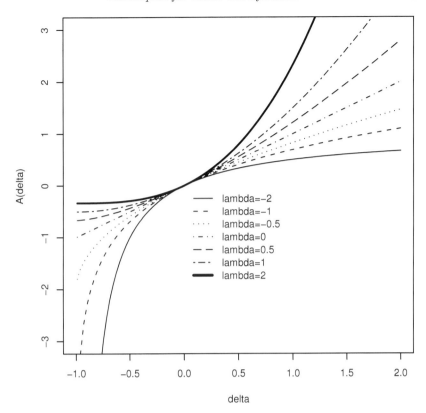

FIGURE 6.5
Residual adjustment functions for several members of the power divergence family.

RAF on the positive side of the axis with an RAF which provides a controlled treatment of inliers on the negative side of the axis. For example, a natural choice may be to combine the Hellinger distance on the positive side with the likelihood disparity on the negative side of the δ axis. In Figures 6.7 and 6.8 we have presented the residual adjustment function of the disparity resulting from this combination, and its disparity generating function respectively. The solid line represents the Hellinger distance and the dashed line stands for the likelihood disparity. The modified RAF obtained by combining the solid RAF on the right with the dashed RAF on the left generates the combined RAF in Figure 6.7. Similarly the combined disparity generating function is constituted by the dashed line on the left and the solid line on the right in Figure 6.8.

For a systematic reference, we will refer to the disparity obtained by combining the inlier component of PD_{λ_1} with the outlier component of PD_{λ_2} as $CPD_{(\lambda_1,\lambda_2)}$. Also $CBWHD_{(\alpha_1,\alpha_2)}$ will represent the indicated combined dis-

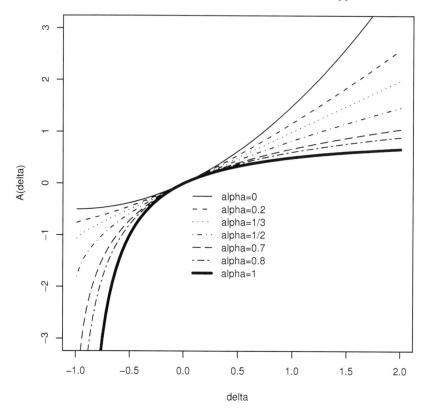

FIGURE 6.6
Residual adjustment functions for several members of the blended weight
Hellinger distance family.

parity based on the blended weight Hellinger distance family. Similar notation
will be used for other disparities and families of disparities.

6.3.1 Asymptotic Distribution of the Minimum Combined Distance Estimators

Let $X_1, \ldots, X_n$ be n independently and identically distributed observations
from a discrete distribution modeled by the family $\mathcal{F} = \{F_\theta : \theta \in \Theta \subseteq \mathbb{R}^p\}$.
Without loss of generality assume that the sample space $\mathcal{X} = \{0, 1, \ldots, \}$. We
assume that the true distribution belongs to the model and the true value
of the parameter is θ_0. Consider two distances within the class of disparities
generated by functions C_1 and C_2 satisfying the disparity conditions, and let
$\rho_{C_m}(d_n, f_\theta)$ be the corresponding combined distance as defined in (6.26). Let
the minimum combined distance estimator be the minimizer of $\rho_{C_m}(d_n, f_\theta)$ in
the spirit of Equation (2.33), provided such a minimizer exists. In this section,

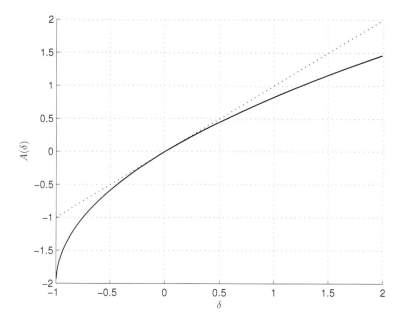

FIGURE 6.7
The combined residual adjustment function where the inlier component is the likelihood disparity (dashed line) and the outlier component is the Hellinger distance (solid line).

we will consider the asymptotic properties of the minimum combined distance estimator and the corresponding test of hypothesis.

We first prove three lemmas to establish the convergence of the linear, quadratic and cubic terms under the combined distance scenario.

Lemma 6.7. *Let ρ_{C_m} be a combined disparity generated by the functions C_1 and C_2. Assume that the true distribution $G = F_{\theta_0}$ belongs to the model, and Assumptions (A1)–(A7) of Section 2.5 hold, with the quantities K_1 and K_2 in Assumption (A7) now representing the maximum of the bounds over the two disparities. Then the linear term of the derivative of the disparity in the estimating equation of the minimum combined distance estimator allows the following approximation:*

$$\nabla_j \rho_{C_m}(d_n, f_\theta)\Big|_{\theta=\theta_0} = -\sum_x d_n(x) u_{j\theta_0}(x) + o_p(n^{-1/2}).$$

Proof. For any fixed $x \in \mathcal{X}$, define

$$Y_n(x) = n^{1/2} \left[\left\{ \frac{d_n(x)}{f_{\theta_0}(x)} \right\}^{1/2} - 1 \right]^2, \qquad (6.27)$$

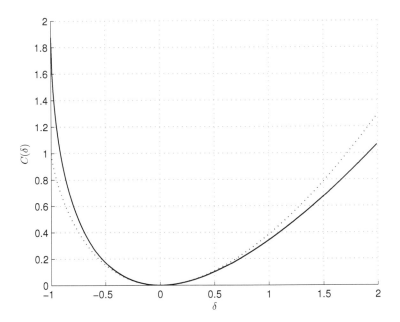

FIGURE 6.8
The combined disparity generating function where the inlier component is the likelihood disparity (dashed line) and the outlier component is the Hellinger distance (solid line).

and let
$$\delta_{n\theta_0}(x) = d_n(x)/f_{\theta_0}(x) - 1. \tag{6.28}$$

Using $k = 1$ and $g = f_{\theta_0}$, we get, from Lemmas 2.13 and 2.14,

$$\lim_{n\to\infty} E_{\theta_0}[Y_n(x)] = 0, \ \forall x \in \mathcal{X}. \tag{6.29}$$

Now using a Taylor series expansion of $A_i(r^2 - 1) - (r^2 - 1)$, $i = 1, 2$, about $r = 1$, we get

$$A_i(r^2 - 1) - (r^2 - 1) = A_i(0) + (r - 1)\{2A_i'(0) - 2\}$$
$$+ \frac{1}{2}(r - 1)^2\{4r_*^2 A_i''(r_*^2 - 1) + 2A_i'(r_*^2 - 1) - 2\}$$
$$= (r - 1)^2\{2r_*^2 A_i''(r_*^2 - 1) + A_i'(r_*^2 - 1) - 1\}, \tag{6.30}$$

where $r_* \in (r, 1)$. So we get, from Assumption (A7)

$$\left| A_i(r^2 - 1) - (r^2 - 1) \right| \le (r - 1)^2 (K_1 + 2K_2 - 1). \tag{6.31}$$

Notice that in the expansion (6.30), r_* and r lie on the same side of 1 irrespective of the value of i, so that all the arguments of A, A' and A'' in (6.30) lie on the same side of $\delta = 0$. Thus, the expansion (6.30) holds – for every nonnegative r – not only just for $A_i, i = 1, 2$, but for A_m as well. Therefore for every nonnegative r one gets

$$\left| A_m(r^2 - 1) - (r^2 - 1) \right| \leq (r - 1)^2 (K_1 + 2K_2 - 1). \tag{6.32}$$

Now we will show

$$n^{1/2} \sum_x \left| \left[A_m(\delta_{n\theta_0}(x)) - \delta_{n\theta_0}(x) \right] \nabla_j f_{\theta_0}(x) \right| = o_p(1). \tag{6.33}$$

The left-hand side of (6.33) is bounded in expectation by

$$n^{1/2} \sum_x E_{\theta_0} \left| A_m(\delta_{n\theta_0}(x)) - \delta_{n\theta_0}(x) \right| \left| \nabla_j f_{\theta_0}(x) \right|. \tag{6.34}$$

Putting $r = \left\{ \frac{d_n(x)}{f_{\theta_0}(x)} \right\}^{1/2}$, and using the bound in (6.32), the expression in (6.34) is bounded by

$$(K_1 + 2K_2 - 1) \sum_x E_{\theta_0}(Y_n(x)) \left| \nabla_j f_{\theta_0}(x) \right|. \tag{6.35}$$

As in Equation (2.77) in Theorem 2.19, the integrand in (6.35) goes to zero pointwise from Equation (6.29). This integrand in (6.35) goes to zero pointwise from Equation (6.29). Thus, by the Dominated Convergence Theorem, the expectation of the left-hand side of (6.33) goes to zero as $n \to \infty$. Hence by Markov's inequality Equation (6.33) holds. Thus

$$\nabla_j \rho_{C_m}(d_n, f_\theta) \Big|_{\theta=\theta_0} = -\sum A_m(\delta_{n\theta_0}(x)) \nabla_j f_{\theta_0}(x)$$
$$= -\left[\sum \delta_{n\theta_0}(x) \nabla_j f_{\theta_0}(x) + o_p(n^{-1/2}) \right]$$
$$= -\sum_x d_n(x) u_{j\theta_0}(x) + o_p(n^{-1/2}),$$

which completes the proof. □

Lemma 6.8. *Let ρ_{C_m} be a combined disparity generated by the functions C_1 and C_2. Assume that the true distribution $G = F_{\theta_0}$ belongs to the model, and Assumptions (A1)–(A7) of Section 2.5 hold, with the quantities K_1 and K_2 in (A7) now representing the maximum of the bounds over the two disparities. Then the quadratic term of the derivative of the disparity in the estimating equation of the minimum combined disparity estimator allows the following approximation:*

$$\nabla_{jk} \rho_{C_m}(d_n, f_\theta) \Big|_{\theta=\theta_0} = \sum_x f_{\theta_0}(x) u_{j\theta_0}(x) u_{k\theta_0}(x) + o_p(1). \tag{6.36}$$

Proof. For notational simplicity, we have dropped the argument "(x)" in the expressions below. Let $\delta_{n\theta_0}$ be as in Equation (6.28). Direct differentiation gives

$$\nabla_{jk}\rho_{C_m}(d_n, f_\theta)\Big|_{\theta=\theta_0}$$
$$= \sum A'_m(\delta_{n\theta_0})(1+\delta_{n\theta_0})u_{j\theta_0}u_{k\theta_0}f_{\theta_0} - \sum A_m(\delta_{n\theta_0})\nabla_{jk}f_{\theta_0}. \quad (6.37)$$

We will first show that

$$\left|\sum A'_m(\delta_{n\theta_0})(1+\delta_{n\theta_0})u_{j\theta_0}u_{k\theta_0}f_{\theta_0} - \sum u_{j\theta_0}u_{k\theta_0}f_{\theta_0}\right| \to 0 \quad (6.38)$$

in probability as $n \to \infty$. Doing a one term Taylor series of the difference

$$|A'_m(\delta_{n\theta_0})(1+\delta_{n\theta_0}) - 1|$$

around $\delta_{n\theta_0} = 0$, we get, by a manipulation similar to that in Theorem 2.19,

$$\left|\sum_x [A'_m(\delta_{n\theta_0})(1+\delta_{n\theta_0}) - 1]u_{j\theta_0}u_{k\theta_0}f_{\theta^0}\right|$$
$$\leq (K_1+K_2)\sum_x \left|\delta_{n\theta^0}u_{j\theta^0}u_{k\theta^0}m_{\theta^0}\right|. \quad (6.39)$$

For each $i = 1, 2$, we have

$$A'_i(\delta_{n\theta_0})(1+\delta_{n\theta_0}) - 1 = \delta_{n\theta_0}[A''_i(\delta^*)(1+\delta^*) + A'_i(\delta^*)], \quad (6.40)$$

and thus from the given conditions we have the bound

$$|A'_i(\delta_{n\theta_0})(1+\delta_{n\theta_0}) - 1| \leq (K_1+K_2)|\delta_{n\theta_0}|.$$

In Equation (6.40), δ^* lies in $(0, \delta_{n\theta^0})$, so that $\delta_{n\theta^0}$ and δ^* are on the same side of zero. Thus, the expansion in (6.40) is valid for each A_i, and for A_m as well. Now use Lemma 2.13 (i), to get the convergence to zero in expectation of the residual difference in the right-hand side of (6.39). Also using Lemma 2.13 (ii), the expectation on the right-hand side of (6.39) is bounded by

$$2(K_1+K_2)\sum f_{\theta_0}^{1/2}(x)|u_{j\theta_0}||u_{k\theta_0}|$$

which is finite by Assumption (A5). Thus, by the Dominated Convergence Theorem and Markov's inequality, the right-hand side of Equation (6.39) goes to zero in probability. Thus, the convergence in (6.38) holds.

Next we will show that

$$\sum A_m(\delta_{n\theta_0})\nabla_{jk}f_{\theta_0} \to 0 \quad (6.41)$$

in probability. Now, as in Theorem 2.19, $|A_m(\delta)| \leq K_1|\delta|$, and we get

$$\left|\sum A_m(\delta_{n\theta_0})\nabla_{jk}f_{\theta_0}\right| \leq \sum_x |A_m(\delta_{n\theta_0})||\nabla_{jk}f_{\theta^0}| \leq K_1\sum_x |\delta_{n\theta_0}||\nabla_{jk}f_{\theta^0}|. \quad (6.42)$$

Noting that $\nabla_{jk} f_\theta(x)/f_\theta(x) = u_{j\theta} u_{k\theta}(x) f_\theta(x) + u_{jk\theta}(x) f_\theta(x)$, the expression in (6.42) goes to zero by a similar argument. Thus, the convergence in (6.41) holds. This proves the lemma. □

Lemma 6.9. *Let ρ_{C_m} be a combined disparity generated by the functions C_1 and C_2. Assume that the true distribution $G = F_{\theta_0}$ belongs to the model, and Assumptions (A1)–(A7) of Section 2.5, with the quantities K_1 and K_2 in (A7) now representing the maximum of the bounds over the two disparities. Then, with probability tending to 1 there exists a finite constant γ such that*

$$|\nabla_{jkl} \rho_{C_m}(d_n, f_\theta)| < \gamma$$

for all j, k and l and all $\theta \in \omega$, where ω is defined in (A3).

Proof. We drop the argument x below to simplify the notation. Notice that

$$\nabla_{jkl} \rho_{C_m}(d_n, f_\theta)$$
$$= -\nabla_{kl} \sum_x A(\delta_{n\theta}) \nabla_j f_\theta$$
$$= -\nabla_{kl} \sum_{\delta_{n\theta} \neq 0} A(\delta_{n\theta}) \nabla_j f_\theta$$
$$= - \sum_{\delta_{n\theta} \neq 0} \big[A_m''(\delta_{n\theta})(1 + \delta_{n\theta})^2 u_{j\theta} u_{k\theta} u_{l\theta} f_\theta$$
$$\qquad - A_m'(\delta_{n\theta})(1 + \delta_{n\theta})\{u_{j\theta} u_{kl\theta} + u_{k\theta} u_{jl\theta} + u_{l\theta} u_{jk\theta} + u_{j\theta} u_{k\theta} u_{l\theta}\} f_\theta$$
$$\qquad + A_m(\delta_{n\theta})\{u_{j\theta} u_{kl\theta} + u_{k\theta} u_{jl\theta} + u_{l\theta} u_{jk\theta} + u_{j\theta} u_{k\theta} u_{l\theta} + u_{jkl\theta}\} f_\theta \big].$$

Now from Assumption (A7) it follows

$$|\nabla_{jkl} \rho_{C_m}(d_n, f_\theta)|$$
$$\leq K_2 \sum_{\delta_{n\theta} \neq 0} |(1 + \delta_{n\theta}) u_{j\theta} u_{k\theta} u_{l\theta} f_\theta|$$
$$+ K_1 \sum_{\delta_{n\theta} \neq 0} |(1 + \delta_{n\theta})\{u_{j\theta} u_{kl\theta} + u_{k\theta} u_{jl\theta} + u_{l\theta} u_{jk\theta} + u_{j\theta} u_{k\theta} u_{l\theta}\} f_\theta|$$
$$+ K_1 \sum_{\delta_{n\theta} \neq 0} |\delta_{n\theta}\{u_{j\theta} u_{kl\theta} + u_{k\theta} u_{jl\theta} + u_{l\theta} u_{jk\theta} + u_{j\theta} u_{k\theta} u_{l\theta} + u_{jkl\theta}\} f_\theta|$$
$$\leq K_2 \sum_x |(1 + \delta_{n\theta}) u_{j\theta} u_{k\theta} u_{l\theta} f_\theta|$$
$$+ K_1 \sum_x |(1 + \delta_{n\theta})\{u_{j\theta} u_{kl\theta} + u_{k\theta} u_{jl\theta} + u_{l\theta} u_{jk\theta} + u_{j\theta} u_{k\theta} u_{l\theta}\} f_\theta|$$
$$+ K_1 \sum_x |\delta_{n\theta}\{u_{j\theta} u_{kl\theta} + u_{k\theta} u_{jl\theta} + u_{l\theta} u_{jk\theta} + u_{j\theta} u_{k\theta} u_{l\theta} + u_{jkl\theta}\} f_\theta|.$$

The right-hand side of the above equation is bounded with probability tending to 1 as $n \to \infty$. □

With this background we are now equipped to state the final results in connection with minimum combined distance estimators and the corresponding tests of hypotheses. As stated before, given the three convergence results of Lemmas 6.7–6.9, the final outcome is routinely established, and here we simply state the appropriate results.

Theorem 6.10. *Suppose that the true distribution $G = F_{\theta_0}$ belongs to the model. Consider a combined disparity $\rho_{C_m}(d_n, f_\theta)$, where the component disparities C_1 and C_2 satisfy the disparity conditions. Suppose that Assumptions (A1)–(A7) of Section 2.5 hold, where the quantities K_1 and K_2 now represent the bounds over the maximum of the two disparities involved in the combination. Let*

$$\nabla \rho_{C_m}(d_n, f_\theta) = 0$$

represent the minimum combined distance estimating equation. Then

(a) The minimum disparity estimating equation has a consistent sequence of roots θ_n.

(b) The sequence θ_n has an asymptotic multivariate normal distribution given by

$$n^{1/2}(\theta_n - \theta_0) \xrightarrow{\mathcal{D}} Z^* \sim N_p(0, I^{-1}(\theta_0)).$$

(c) The estimator θ_n is asymptotically equivalent to any other first-order efficient estimator $\tilde{\theta}_n$ characterized by Equation (2.84) in the sense

$$n^{1/2}(\theta_n - \tilde{\theta}_n) \to 0$$

in probability as $n \to \infty$.

The analog of Theorem 6.6 dealing with the asymptotic null distribution of the disparity difference test based on combined disparities can be similarly established.

6.4 ϵ-Combined Distances

In this scheme also we combine two different residual adjustment functions of two different disparities on either side of the origin of the δ axis as in the case of combined disparities. However, instead of extending both the RAFs up to the origin and combining them by brute force (and thereby sacrificing the smoothness of the first derivative of the combined RAF at the origin), we use a third function which is defined in a small band $[-\epsilon, \epsilon]$ around the origin and smoothly combines the two different RAFs so that the resulting ϵ-combined RAF corresponds to a regular distance satisfying all the disparity conditions.

It turns out that the function on this central band – which we will refer to as

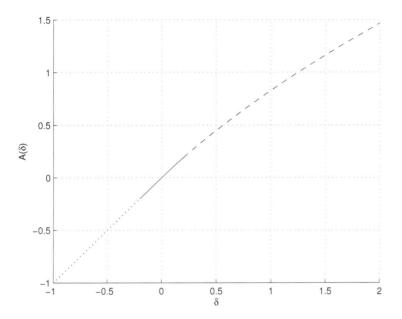

FIGURE 6.9
The ϵ-combined residual adjustment function with the likelihood disparity in the inlier component (dotted line), Hellinger distance in the outlier component (dashed line), and the appropriate seven degree polynomial in the $[-0.2, 0.2]$ band in the centre (solid line).

the central function – must be a seven degree polynomial on $\delta \in [-\epsilon, \epsilon]$ so that all the smoothness conditions are satisfied. If the RAFs on the positive side and the negative side of the axis are denoted by $A_1(\delta)$ and $A_2(\delta)$ respectively, the complete ϵ-combined RAF has the form

$$A_\epsilon(\delta) = \begin{cases} A_1(\delta), & \text{if } \delta > \epsilon, \\ \sum_{i=0}^{7} k_i \delta^i, & \text{if } -\epsilon \leq \delta \leq \epsilon, \\ A_2(\delta), & \text{if } \delta < -\epsilon. \end{cases}$$

The central function satisfies eight constraints, which leads to a seven degree polynomial. These eight constraints are listed below.

1. $A_\epsilon(0) = 0$ and $A'_\epsilon(0) = 1$.

2. $A_\epsilon(\delta)$ is a continuous function for $-1 \leq \delta < \infty$. So $A_\epsilon(\epsilon) = A_1(\epsilon)$ and $A_\epsilon(-\epsilon) = A_2(-\epsilon)$.

3. The first derivative of $A_\epsilon(\delta)$ exists for all values of δ in the interval $(-1, \infty)$. So $A'_\epsilon(\epsilon) = A'_1(\epsilon)$ and $A'_\epsilon(-\epsilon) = A'_2(-\epsilon)$.

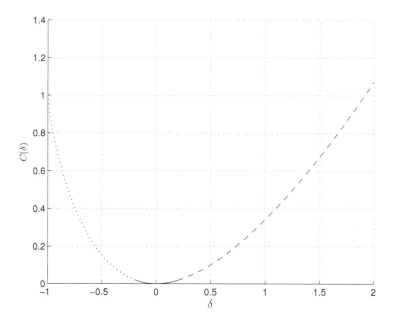

FIGURE 6.10
The disparity generating function corresponding to the residual adjustment function of Figure 6.9, with the same coding for the line types of the different components.

4. The second derivative of $A_\epsilon(\delta)$ exists for all values of δ in the interval $(-1, \infty)$. So $A_\epsilon''(\epsilon) = A_1''(\epsilon)$ and $A_\epsilon''(-\epsilon) = A_2''(-\epsilon)$.

In Figure 6.9 the RAF of the ϵ-combined distance consisting of the Hellinger distance on the right-hand side (dashed line), the likelihood disparity on the left-hand side (dotted line) and the corresponding 7 degree polynomial (solid line) on the band $[-0.2, 0.2]$ is plotted. For $\delta < -0.2$, the RAF of the ϵ-combined distance is the same as that of the likelihood disparity, for $\delta > 0.2$ it is the same as that of the Hellinger distance, and in the central band it is the seven degree polynomial defined by the above set of conditions. Similarly in Figure 6.10 the disparity generating function of the ϵ-combined distance is plotted with the same coding for the line types.

The ϵ-combined distance constructed using the disparity generating function $C_\epsilon(\delta)$ recovered from the residual adjustment function A_ϵ by solving the differential Equation (2.38) satisfies all the disparity conditions if A_ϵ is increasing (in which case C_ϵ is convex). Checking for the convexity of C_ϵ, or the increasing nature of A_ϵ through a general analytical formula is difficult, and usually has to be checked on a case by case basis. However, it is usually true for reasonable tuning parameters (as it is in the case of the example in Figure 6.9). In such a case, all the general results about minimum distance estimators based on disparities, including the asymptotic distribution of the corresponding minimum distance estimator, apply to the minimum ϵ-combined distance estimator. The ϵ-combination approach, apart from making the disparity generating function appropriately smooth in this case, also suggests other benefits in case of other distances where smoothness properties may have been lost due to similar or other modifications. For example, the rough modifications of the robustified likelihood disparity can be made smooth by putting similar bands at the roughness spots. When the band is over a small region, the impact of this smoothing on parameter estimation is expected to be small. By the same token, the combined disparity approach and the ϵ-combined disparity approach are likely to lead to similar inference when ϵ is small.

6.5 Coupled Distances

Consider a robust distance ρ_C having corresponding residual adjustment function $A(\delta)$. In the coupled distance scheme, one constructs a "coupled RAF" A_{CP} to deal with inliers using the following strategy. Let the chosen RAF $A(\delta)$ represent $A_{\mathrm{CP}}(\delta)$ on the positive side of the δ axis. On the negative side of the axis, one chooses a function $A_{\mathrm{L}}(\delta)$ which conforms to the necessary smoothness conditions at $\delta = 0$ when coupled with the chosen RAF $A(\delta)$ at the origin, but at the same time it also ensures that $A_{\mathrm{CP}}(-1) = k_0$ where $k_0(< 0)$ is suitably chosen to arrest any sharp dip in the left tail of the residual adjustment function.

The necessary conditions constrain the required function on the left-hand side of the δ axis to be a third degree polynomial. The coupled RAF is given by

$$A_{\mathrm{CP}}(\delta) = \begin{cases} A(\delta), & \text{if } \delta \geq 0 \\ A_{\mathrm{L}}(\delta), & \text{if } \delta < 0 \end{cases} \tag{6.43}$$

subject to the following set of conditions.

1. The coupled RAF $A_{\mathrm{CP}}(\delta)$ is a continuous function for $-1 \leq \delta < \infty$. So $A_{\mathrm{L}}(0) = 0$.

2. First two derivatives of $A_{\mathrm{L}}(\delta)$ at $\delta = 0$ match with the original residual adjustment function $A(\delta)$, i.e., $A_{\mathrm{L}}'(0) = 1$ and $A_{\mathrm{L}}''(0) = A''(0)$.

3. $A_{CP}(\delta)$ gives a desired weight to the empty cells (and correspondingly downweights the other inliers). Thus, $A_{CP}(-1) = A_{\mathrm{L}}(-1) = k_0$ where $k_0 < 0$ is a suitable tuning parameter. A reasonable choice may be $k_0 = -1$, in which case $A_{\mathrm{CP}}(\delta)$ matches $A_{\mathrm{LD}}(\delta)$ at $\delta = -1$, where LD represents the likelihood disparity.

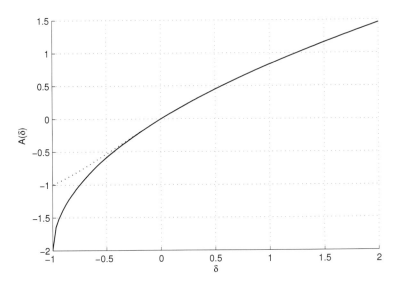

FIGURE 6.11
The residual adjustment function of the coupled disparity obtained by coupling the Hellinger distance in the outlier component with the inlier component generated by $k_0 = -1$.

As the above set of conditions generates four independent constraints, one needs a third degree polynomial to specify the function $A_L(\delta)$.

The disparity generating function C_{CP} of the coupled distance can then be constructed by solving the differential equation in (2.38). When the recovered $C_{CP}(\cdot)$ functions are convex, all the necessary disparity conditions are satisfied with respect to the corresponding coupled disparity. Thus, all the general results about minimum distance estimators based on disparities, including the asymptotic distribution of the corresponding minimum distance estimator, apply to the minimum coupled distance estimator.

Finding exact conditions for the convexity of the C_{CP} function may not be easy. However, Mandal and Basu (2010a) provide the following upper bound of possible values of k_0 (in relation to the original distance) under which the convexity of C_{CP} is preserved.

Lemma 6.11. *Let A_2 be the curvature parameter of the original distance. For choices of k_0 satisfying $k_0 < -\frac{1}{6}[A_2 + 4]$, the disparity generating function of the resulting coupled distance is strictly convex.*

Once the convexity of the function is established, the properties of the corresponding minimum distance estimators and disparity difference test statistics follow from the standard theory established in earlier chapters.

As an example, we choose the coupled disparity obtained by using the

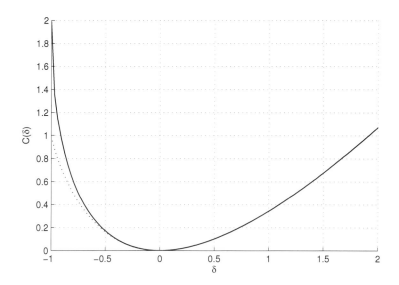

FIGURE 6.12
The reconstructed disparity generating function corresponding to the RAF in
Figure 6.11 (Hellinger distance coupled with the inlier component generated
by $k_0 = -1$).

Hellinger distance for the outlier component and the tuning parameter $k_0 =
-1$ for the inlier component. The conditions of the above lemma are satisfied
in this case. The graphs of the corresponding residual adjustment function
and the disparity generating function are presented in Figures 6.11 and 6.12.
How this inlier modification allows a more moderate treatment of the inliers
in obvious. The solid line in either case represents the ordinary Hellinger
distance, and the dashed line represents the coupling component.

Again, for a systematic reference, we will denote by $\text{CpPD}_{(k_0,\lambda)}$ the disparity obtained by coupling the PD_λ statistic with the inlier component generated
by k_0. Also $\text{CpBWHD}_{(k_0,\alpha)}$ will denote the indicated coupled disparity based
on the BWHD family. Similar notation will be used for the coupled version of
other disparities and families of disparities.

6.6 The Inlier-Shrunk Distances

In this section, we present yet another strategy to deal with the inlier problem.
Here the aim is to shrink the disparity generating function of a robust disparity

toward zero in the inlier part by a factor which converges to 1 as $\delta \to 0$, while keeping the outlier part intact. Under the proper choice of parameters for this shrinkage factor, this may lead to a modified function with the right properties (the resulting A function will be strictly increasing and continuously differentiable up to the second order over the entire applicable range). We formally describe this proposal below. The discussion of this function follows the development in Patra, Mandal and Basu (2008).

Let $C(\cdot)$ be a function satisfying the disparity conditions of Definition 2.1. Define the corresponding inlier-shrunk class of disparity generating functions indexed by the tuning parameter γ through the relation

$$
C_\gamma(\delta) = \begin{cases} \dfrac{C(\delta)}{(1+\delta^2)^\gamma} & \delta \leq 0, \gamma \geq 0 \\ C(\delta) & \delta > 0 \end{cases}, \tag{6.44}
$$

which can then be used to generate the associated family of inlier-shrunk distances based on the function C using Equation (2.7). Notice that this strategy again keeps the $C(\cdot)$ function intact on the outlier side but shrinks it closer to zero on the inlier side. Clearly there is no shrinkage for the case $\gamma = 0$; on the other hand, as the tuning parameter increases, the inlier component is subjected to greater shrinkage. It can be easily verified that $C_\gamma'''(\delta)$, the third derivative of the function $C_\gamma(\delta)$, exists and is continuous at $\delta = 0$; the same is true for the corresponding second derivative of the residual adjustment function. Thus, one would only need to verify that C_γ is a convex function to establish that the corresponding inlier-shrunk distance satisfies the disparity conditions.

For illustration, let $C(\delta)$ represent the disparity generating function of the Hellinger distance, and denote the associated family of inlier-shrunk distances as the inlier-shrunk Hellinger distance (ISHD_γ) family. For the corresponding disparity measure we get $C_\gamma(-1) = 2^{1-\gamma}$. In particular for $\gamma = 1$, the disparity generating function of ISHD_1 matches the disparity generating function of the likelihood disparity at $\delta = -1$. However, the $\text{ISHD}_{0.5}$ measure has a somewhat better correspondence with the disparity generating function of the likelihood disparity over a larger range of the inlier part, although it has a sharper dip near the left boundary. The $\text{ISHD}_{0.5}$ is also intuitive in that its denominator $(1+\delta^2)^{1/2}$ over the inlier part is of the order $O(\delta)$, which is of the same order as the numerator, and hence has the same scale factor. However, the value of the residual adjustment function at $\delta = -1$ is -1.414 for the $\text{ISHD}_{0.5}$ distance, a number which is still substantially larger than 1 in absolute magnitude. In practice, a tuning parameter γ roughly midway between 0.5 and 1 may perform well with real data.

The differentiability conditions of the disparity generating function and the residual adjustment function of the inlier-shrunk distances are satisfied in more general settings where the denominator of the inlier part of the function defined in (6.44) is replaced by $(1 + \delta^2 P(\delta))^\gamma$, where $P(\delta)$ is a polynomial in

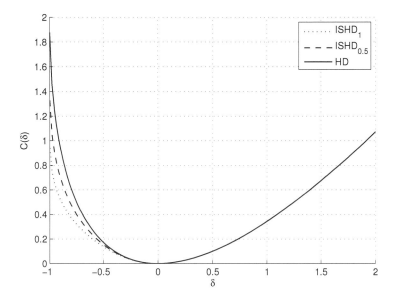

FIGURE 6.13
The disparity generating functions of the Hellinger distance, the $ISHD_{0.5}$ and $ISHD_1$.

δ. However, this would not be the case if the denominator of the inlier part were replaced by $(1 + \delta P(\gamma))^\gamma$ instead.

It is not easy to analytically verify that the disparity generating functions of the $ISHD_\gamma$ is convex on $[-1, \infty)$. However, direct evaluation shows that this is indeed the case, at least for $\gamma \in [0, 1]$. In Figures 6.13 and 6.14 we have presented the plots of the disparity generating functions and the residual adjustment functions of members of the $ISHD_\gamma$ family for $\gamma = 0, 0.5$ and 1. Since $HD = ISHD_0$, the above graphs give an indication of how the weights of the inliers are reduced in these applications with increasing γ. Based on our discussion in this chapter, it would be fair to expect that the minimum distance estimators based on $ISHD_{0.5}$ and $ISHD_1$ will exhibit much better small sample performance compared to the minimum Hellinger distance estimator, with little or no change in the robustness features.

Of course there is nothing particularly special about the Hellinger distance when choosing a basic disparity while constructing a family of inlier-shrunk distances. Any disparity with a reasonable treatment of outliers will be a suitable candidate for the generating a useful class of inlier-shrunk distances.

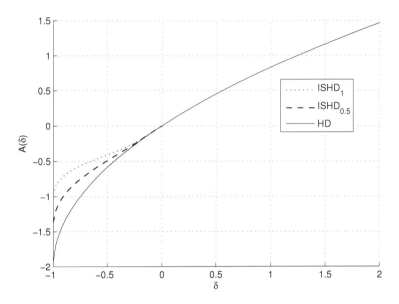

FIGURE 6.14
The residual adjustment functions of the Hellinger distance, the $ISHD_{0.5}$ and $ISHD_1$.

6.7   Numerical Simulations and Examples

Example 6.2. This example illustrates the performance of our inlier modified statistics in a simulation study. Samples of size 50 are drawn from a geometric distribution with true parameter (success probability) $\theta = 0.5$. A sample size of 50 cannot be considered to be very small in this context, but even at this sample size the estimated mean square of the minimum Hellinger distance estimator is nearly twice that of the maximum likelihood estimator. For the minimum $PD_{-0.9}$ estimator, the MSE is about 9 times larger than that of the maximum likelihood estimator.

We have provided several inlier modified estimators for both the above statistics in Table 6.2. It is clear how these estimators improve the mean square performance of the estimator. The performance of the same estimators are given under a mixture contamination in Table 6.3. Notice that in this case the improved small sample performance of the inlier modified statistics in Table 6.2 have not been at the cost of compromising its robustness in Table 6.3.

Determining the exact tuning parameters that will lead to the best overall performance is a difficult problem. However, considering the Hellinger distance

TABLE 6.2
Performance of the inlier modified estimators in the geometric model. True parameter $\theta = 0.5$, sample size $n = 50$, number of replications is 10000.

Distance	Bias	Variance	MSE
LD	0.0053	0.0026	0.0026
HD	0.0462	0.0027	0.0049
$PPD_{(1,-0.5)}$	0.0206	0.0028	0.0032
$PPD_{(0.5,-0.5)}$	0.0025	0.0029	0.0029
$CPD_{(0,-0.5)}$	0.0190	0.0027	0.0031
$CpPD_{(-1,-0.5)}$	0.0194	0.0027	0.0031
$PD_{-0.9}$	0.1381	0.0037	0.0228
$PPD_{(1,-0.9)}$	0.0309	0.0030	0.0040
$CPD_{(0,-0.9)}$	0.0275	0.0029	0.0036
$CpPD_{(-1,-0.5)}$	0.0281	0.0028	0.0036
$ISHD_{0.7}$	0.0253	0.0027	0.0033
SCS	0.0270	0.0030	0.0037
NED	0.0114	0.0030	0.0031

to be the standard, the distances $PPD_{(1,-0.5)}$, $CPD_{(0,-0.5)}$ and $CpPD_{(-1,-0.5)}$ appear to give reasonable (and similar) results. A penalty weight of $h = 0.5$ seems to be slightly better than $h = 1$ in terms of improving the small sample mean square error of estimator (mainly by reducing the bias), but appears to be significantly poorer on the robustness front.

We have not presented the number corresponding to the ϵ-combined disparities, since their performance seems to be very similar to the combined distance when the band $(-\epsilon, \epsilon)$ is small. ‖

Example 6.3. In Chapter 5 we had considered the scatter plots of the likelihood ratio test statistics against disparity difference tests based on the Hellinger distance. In this example, we take this further and demonstrate that our inlier modified techniques can make the tests much more improved in terms of their attained levels and closeness to the likelihood ratio test. Between Figures 6.15 and 6.16, six pairs of scatter plots are provided where different test statistics are plotted against the likelihood ratio test. The tests are performed under the Poisson model, where the data generating distribution is a Poisson with mean $\theta = 5$, which is also the null parameter. For each of the six sets, the figure on the left presents the scatter plot of the statistics for sample size 25, while the plot on the right presents the scatter plot for sample size 200. The six sets, in the order in which they appear, present the scatter plots of the likelihood ratio test against the disparity difference tests based on (i) the ordinary Hellinger distance; (ii) the penalized Hellinger distance with $h = 1$;

TABLE 6.3
Performance of some inlier modified estimators under contaminated geometric data. The data are generated from the 0.9Geometric(0.5)+0.1Geometric(0.2) distribution and sample size is $n = 50$, number of replications is 10000. The mean square error is computed around the target value of $\theta = 0.5$

Distance	Bias	Variance	MSE
LD	−0.0584	0.0032	0.0066
HD	0.0094	0.0030	0.0031
$PPD_{(1,-0.5)}$	−0.0192	0.0030	0.0034
$PPD_{(0.5,-0.5)}$	0.0398	0.0032	0.0048
$CPD_{(0,-0.5)}$	0.0214	0.0030	0.0034
$CpPD_{(-1,-0.5)}$	−0.0208	0.0030	0.0034
$PD_{-0.9}$	0.1114	0.0044	0.0168
$PPD_{(1,-0.9)}$	0.0027	0.0033	0.0033
$CPD_{(0,-0.9)}$	0.0070	0.0031	0.0032
$CpPD_{(-1,-0.9)}$	−0.0051	0.0032	0.0033
$ISHD_{0.7}$	−0.01433	0.0029	0.0031
SCS	−0.0062	0.0032	0.0033
NED	−0.0228	0.0032	0.0038

(iii) the coupled distance which combines the Hellinger distance with an inlier component given by $k_0 = -1$; (iv) the combined distance which combines the likelihood disparity (inlier part) and the Hellinger distance (outlier part) (v) the inlier shrunk Hellinger distance with $\gamma = 0.7$; and (vi) the ϵ-combined distance which combines the likelihood disparity (inlier side) and the Hellinger distance (outlier side) with a central band on $[-0.1, 0.1]$;

The improvement due to inlier modification for any of the other five tests, compared to the test based on the ordinary Hellinger distance is immediately apparent. All the observations of the scatter plot are already quite close to the diagonal line at $n = 25$ for each of the inlier modified statistics. At $n = 200$, the scatter plots of all the other five tests have practically settled on the diagonal line, whereas the Hellinger distance case still shows a lot of random fluctuation.   ‖

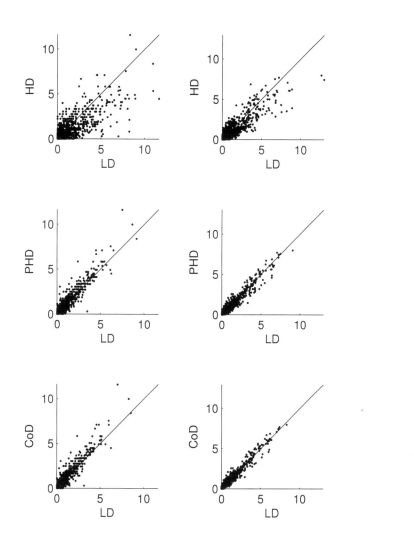

FIGURE 6.15

Scatter plots of the likelihood ratio statistics versus other disparity difference statistics for testing hypothesis about the Poisson mean. The disparities in the three rows are the ordinary Hellinger distance, the penalized Hellinger distance ($h = 1$) and the coupled distance $\text{CpPD}_{(-1,0.5)}$. The left column represents a sample size of $n = 25$, while the right column corresponds to $n = 200$.

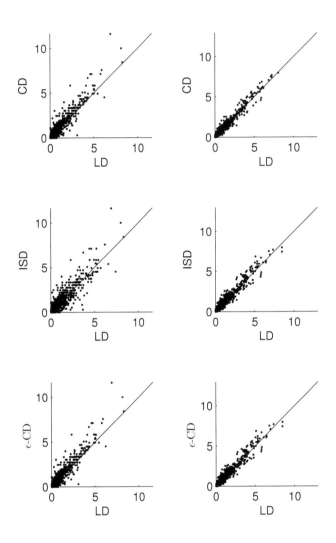

FIGURE 6.16
Scatter plots of the likelihood ratio statistics versus other disparity difference statistics for testing hypothesis about the Poisson mean. The disparities in the three rows are the combined distance $\mathrm{CPD}_{(0,-0.5)}$ (denoted by CD), the Inlier shrunk Hellinger distance with $\gamma = 0.7$, and the ϵ combined distance combining the likelihood disparity and the Hellinger distance to a central band. The left column represents a sample size of $n = 25$, while the right column corresponds to $n = 200$.

7

Weighted Likelihood Estimation

In Chapters 2 and 3 we have emphasized the fact that the minimum distance estimation method based on disparities generates a rich and general class of estimators which includes the maximum likelihood estimator as a special case. Under differentiability of the model, the estimators are obtained by solving estimating equations of the form (2.39), or their analogs for the continuous case. In this chapter we will further explore the estimating functions under the weighted likelihood estimating equation approach to this problem. We will show that weighted likelihood estimating functions based on disparities also generate a very broad class of estimation processes containing the maximum likelihood score function as a special case. Under standard regularity conditions the above approach produces a class of asymptotically efficient procedures; in addition, many members of this class have very strong robustness features.

A very important work done by Green (1984), published in a Royal Statistical Society, Series B, discussion paper, extensively discussed the theory and use of iteratively reweighted least squares for maximum likelihood, and suggested the replacement of the usual maximum likelihood score equations with weighted score equations. In particular, Green (1984) considered the weight function

$$\Delta_\theta(x) = -2\{\log f_\theta(x) - \sup_t \log f_t(x)\}.$$

Lenth and Green (1987) use the weights

$$w\{\Delta_\theta^{1/2}(x)\} = \frac{\psi_c\left(\Delta_\theta^{1/2}(x)\right)}{\Delta_\theta^{1/2}(x)},$$

where ψ_c is the Huber's ψ function. Field and Smith (1994) used weights based on the empirical cumulative distribution function.

In this section another method of constructing weighted likelihood score equations and the corresponding estimators are discussed. We will show that the proposed estimation process arises naturally from the minimum distance estimating equations studied in Chapters 2 and 3, and the estimating equations asymptotically behave like the ordinary maximum likelihood score equations.

The material presented in this chapter primarily follows Markatou, Basu and Lindsay (1997, 1998) and Agostinelli and Markatou (2001), as well as several other works by these authors and their associates. Also see Basu (1991) and Basu and Lindsay (2004) for a description of the iteratively reweighted estimating equation algorithm. This algorithm is also described in Section 4.10 of this book and forms the basis for this approach.

7.1 The Discrete Case

Consider a parametric model $\mathcal{F}$ as defined in Section 1.1, and let $X_1, \ldots, X_n$ be a random sample of independently and identically distributed observations. We assume the notation and setup of Section 2.3. We also assume that the model follows appropriate regularity and differentiability conditions so that the maximum likelihood estimator of the parameter θ is obtained as a solution of the estimating equation

$$\sum_{i=1}^{n} u_\theta(X_i) = 0,$$

where $u_\theta(x) = \nabla \log f_\theta(x) = \nabla f_\theta(x)/f_\theta(x)$ is the maximum likelihood score function. Given any point x in the sample space, the approach taken here is to construct a nonnegative weight function which depends on the value x, the model distribution F_θ, and the empirical cumulative distribution function G_n. Denoting the weight function by $w(x) = w(x, F_\theta, G_n)$, our interest will be in the class of solutions of all weighted likelihood estimating equations of the type

$$\sum_{i=1}^{n} w(X_i, F_\theta, G_n) u_\theta(X_i) = 0. \tag{7.1}$$

The guiding principle for selecting the weight function is to choose it in a manner so that it has a value close to 1 at a data point X_i when the empirical distribution function provides little or no evidence of model violation at that point. On the other hand, the weight at X_i should be close to zero if the empirical cumulative distribution indicates a lack-of-fit at that point. Such an overall strategy ensures that the method does not automatically down-weight a proportion of the data, but only does so for the points which are inconsistent with the model; this allows the accommodation of optimal model efficiency simultaneously with strong robustness features. When the model is true, the construction should ensure that all the weights asymptotically tend to 1 as $n \to \infty$. This will guarantee that the procedure does not automatically downweight a proportion of the data even when the model is true.

The weighted likelihood estimation procedure presented here is not just a parameter estimation procedure. The technique has an equally important use for diagnostic purposes. Once a parametric model has been fitted to a set of data, the final fitted weights provide a most natural measure of the relevance of particular data points to the overall data set in relation to the parametric model under consideration. The fitted weights are extremely useful quantities for future data scrutiny; they indicate which of the data points were downweighted, relative to the maximum likelihood estimator, in the final solution. In addition, the final solution(s) of the weighted likelihood procedure – the root(s) of the estimating equation – provide a great deal of information about the structure of the data. Multiple roots are, in fact, further indicators of the peculiarity of the structure of the data in relation to the model, and show that the data merit further scrutiny. Example 7.5, presented later in this chapter, discusses such a case.

Another advantage of the method is its computational simplicity. Equation (7.1) provides an important motivation for the weighted likelihood approach, as it immediately suggests a natural algorithm based on iterative reweighting. One can start with an appropriate initial value of the parameter, calculate the weights, solve Equation (7.1) treating the weights to be fixed constants, and repeat the exercise with the new solution. The procedure is continued until convergence is achieved.

7.1.1 The Disparity Weights

Let $X_1, \ldots, X_n$ represent a random sample of independently and identically distributed observations from a discrete distribution with density g, and let the parametric model $\mathcal{F}$ be as defined in Section 1.1. The weighting proposal considered here downweights the outliers or surprising observations manifested through large Pearson residuals. The weight of an observation is chosen as a function of the corresponding Pearson residual and has a value 1 when the Pearson residual is zero. In particular, the weight may be chosen as

$$w(x) = w(x, F_\theta, G_n) = w(\delta(x)) = \frac{A(\delta(x)) + 1}{\delta(x) + 1}, \qquad (7.2)$$

where $\delta(x) = d_n(x)/f_\theta(x) - 1$, and $A(\cdot)$ is a strictly increasing, twice differentiable function on $[-1, \infty)$, which satisfies $A(0) = 0$ and $A'(0) = 1$. This is, in particular, satisfied by the natural or appropriately modified residual adjustment functions for our class of disparities. A little algebra shows that the estimating Equation (2.39) can be written as

$$\sum_{x:d_n(x)\neq 0} \frac{A(\delta(x)) + 1}{\delta(x) + 1} d_n(x) u_\theta(x) + \sum_{x:d_n(x)=0} [A(-1) + 1]\nabla f_\theta(x) = 0.$$

For disparities such as the likelihood disparity which satisfy $A(-1) = -1$, we can rewrite the above equation as

$$\sum_{x:d_n(x)\neq 0} w(x)d_n(x)u_\theta(x) = \frac{1}{n}\sum_{i=1}^{n} w(X_i)u_\theta(X_i) = 0, \qquad (7.3)$$

where $w(x) = w(\delta(x))$ is as in Equation (7.2). The equation in (7.3) can be viewed as a weighted likelihood score equation in the spirit of Equation (7.1). The solution to this equation will be referred to as the weighted likelihood estimator. If $w(x) \equiv 1$ identically in x, Equation (7.3) reduces to the ordinary maximum likelihood score equation. For fixed weights w, a closed form solution of Equation (7.3) is usually available. Hence, Equation (7.3) can be solved via a fixed point iteration algorithm similar to iteratively reweighted least squares (see the iteratively reweighted estimating equation algorithm described in Section 4.10). One can directly construct weighted likelihood estimating equations as in Equation (7.3) by constructing the weight functions in Equation (7.2) starting from the residual adjustment function of any appropriate disparity. We will refer to the weights in Equation (7.2) as disparity weights.

Suppose that the true data generating distribution g belongs to the model family, and let θ_0 be the true value of the parameter. The asymptotic distribution of the weighted likelihood estimator is then presented in the following theorem.

Theorem 7.1. *Suppose $A(\cdot)$ represents the residual adjustment function of a distance satisfying the disparity conditions as stated in Definition 2.1. Consider the corresponding weighted likelihood estimating equation as represented by (7.3), where the weights are as in Equation (7.2). Suppose that conditions (A1)–(A7) of Section 2.5 are satisfied. Then there exists a consistent sequence θ_n of roots to Equation (7.3) such that $n^{1/2}(\theta_n - \theta_0)$ has an asymptotic p-dimensional multivariate normal distribution with mean vector 0 and covariance matrix $I^{-1}(\theta_0)$.*

Proof. When the residual adjustment function satisfies $A(-1) = -1$, the estimating equation in (7.3) is simply the estimating equation of a minimum distance estimator based on a distance which satisfies the disparity conditions. Thus, the consistency and asymptotic distribution of the corresponding estimator follows from Theorem 2.19 and Corollary 2.20.

When $A(-1) \neq -1$ for the chosen distance, the weighted likelihood equation in (7.3) represents the estimating equation of the corresponding penalized distance with penalty weight h equal to 1. Thus, the consistency and asymptotic normality of the corresponding estimator follows from Theorem 6.4. □

The above theorem shows that the weighted likelihood estimating equation corresponds to the estimating equation of a minimum disparity problem; therefore the corresponding disparity may be used for the selection of the correct, robust root, in case the weighted likelihood estimating equation allows multiple roots.

Because of the robustness issue, our primary interest – when constructing weighted likelihood estimating equations based on disparity weights – is in disparities for which the corresponding residual adjustment function curves down sharply on the positive side of the δ axis. Notice that as long as the residual adjustment function of the disparity remains below that of the likelihood disparity on the positive side of the δ axis, the generated disparity weights will always be smaller than or equal to 1 for $\delta > 0$.

We will express the desired structure of our weight function through the following definition.

Definition 7.1. The weight function will be called regular if $w(\delta)$ is a nonnegative function on $[0, 1]$, is unimodal, and satisfies $w(0) = 1$, $w'(0) = 0$, and $w''(0) < 0$.

For disparity weights, the conditions $w(0) = 1$ and $w'(0) = 0$ are automatically satisfied; in addition $w''(0)$ is the estimation curvature A_2 of the disparity, so that the requirement $w''(0) < 0$ essentially restricts the weight function to disparities having a local robustness property. In order to restrict the weight function to $[0, 1]$ and to achieve the necessary downweighting, we will adopt a more restricted sense for our robust disparities compared to Definition 4.1; the disparities that satisfy, beyond the conditions of Definition 4.1, $A(\delta) < \delta$ for all $\delta > 0$, will be treated as robust disparities in this context.

While Definition 7.1 represents a very convenient format for our discussion, minor variations of this structure are not critical for the functioning of the procedure. However, some care is needed in developing regular weight functions starting from the disparity weights in (7.2). Notice that for the Hellinger distance, Equation (7.2) will lead to negative weights whenever $A_{\mathrm{HD}}(\delta) < -1$. On the other hand, for a distance like the negative exponential disparity, all the weights will be greater than 1 for $\delta < 0$. One simple remedy for this is to force the function $w(\delta)$ to have weights equal to 1, identically, for $\delta \leq 0$. Note that such a modification essentially reduces the weighted likelihood estimating equation to the estimating equation of the combined disparity ρ_{C_m}, which combines a robust disparity on the right with the likelihood disparity on the left. Such a modification does not significantly affect the robustness properties of the corresponding weighted likelihood estimator, nor does it affect the asymptotic optimality of the estimator at the model.

With the restriction $w(\delta) = 1$ for $\delta \leq 0$, our condition $w''(0) < 0$ needs to be restated and will now relate to the downweighting properties of the weight function for large positive values of δ. In this case we will require that the right-hand derivative of $w'(\delta)$ at $\delta = 0$ is negative so that our robustness requirement – which really relates to the positive side of the δ axis anyway – is satisfied.

A weight function restricted to $[0, 1]$ which satisfies $w(\delta) = 1$ at $\delta \leq 0$, $w'(\delta) = 0$ at $\delta = 0$, and the right-hand derivative of $w'(\delta)$ is negative at $\delta = 0$, may be generated by considering the weights

$$w(\delta) = \begin{cases} 1 & \text{for } \delta \leq 0, \\ \dfrac{A(\delta) + 1}{\delta + 1} & \text{for } \delta > 0. \end{cases} \tag{7.4}$$

in connection with any of our robust disparities. Disparities such as the negative exponential disparity, for which $w''(\delta) = 0$ at $\delta = 0$ may be accommodated

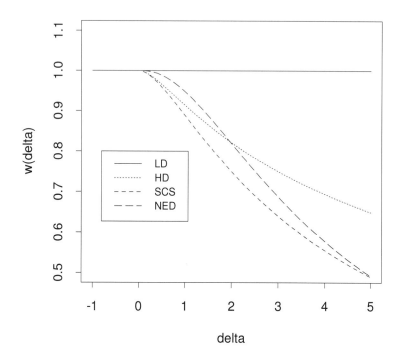

FIGURE 7.1
Modified Weights Functions for LD, PCS, HD and NED.

by requiring that the right-hand derivative of $w''(\delta)$ at $\delta = 0$ is negative. The weighted likelihood estimating equation based on the negative exponential disparity possesses a remarkable degree of robustness, and although in the immediate vicinity of $\delta = 0$ the corresponding disparity weights are less sensitive than those of the Hellinger distance, it is competitive or better than the Hellinger distance for larger values of δ.

Suppose that the true distribution belongs to the model and θ_0 represents the true parameter.

Theorem 7.2. *Consider a robust disparity ρ_C satisfying the disparity conditions in Definition 2.1, and let $A(\delta)$ be the associated residual adjustment function. Consider the weighted likelihood estimating equation given in Equation (7.3), where the weights are as given in Equation (7.4). Then there exists a consistent sequence θ_n to the weighted likelihood estimating Equation (7.3)* *such that $n^{1/2}(\theta_n - \theta_0)$ has an asymptotic p-dimensional multivariate normal distribution with mean vector 0 and covariance matrix $I^{-1}(\theta_0)$.*

Proof. Note that the weighted likelihood estimating equation as constructed in the statement of the theorem is simply the estimating equation of the minimum combined distance estimator obtained by combining the likelihood disparity in the negative side of the δ axis with the distance ρ_C in question on the positive side of the δ axis. Thus, the asymptotic properties of the corresponding estimator follow from Theorem 6.10. □

The choice of the disparity weights represented by Equation (7.4) are in conformity with the main goals of the downweighting philosophy discussed in Section 7.1. These weights allow smooth downweighting of observations inconsistent with the model. For suitably chosen disparity weights, the degree of downweighting increases as the observations become more and more aberrant. When $A(\delta)$ represents the RAF of an appropriate robust disparity, wildly discrepant observations will get weights practically equal to zero. On the other hand, the Pearson residual $\delta(x)$ converges to zero for each fixed x under the model, and from the property of residual adjustment functions, the weights will all converge to 1 in that case; thus the weighted likelihood estimating equation asymptotically behaves like the maximum likelihood score equation. In addition, since the weighted likelihood estimating equation corresponds to the estimating equation of a minimum disparity problem, the corresponding disparity may be used for the selection of the correct, robust root.

A higher degree of robustness can, in fact, be achieved, if we make the weights stronger and more sensitive. Since the weight function is restricted to $[0, 1]$, this may be accomplished by choosing a higher power of the weights. To include such cases within the scope of our weighted likelihood estimation technique, we define the general weighted likelihood estimation method through the estimating equation

$$\sum_{i=1}^{n}[w(X_i, F_\theta, G_n)]^c u_\theta(X_i) = 0, \quad c \geq 0, \tag{7.5}$$

where the weight function w is obtained through disparity weights or other suitable techniques, and satisfies appropriate conditions discussed earlier in this section. The choice $c = 0$ will recover the ordinary maximum likelihood score equation, while $c = 1$ will reproduce the weighted likelihood estimating equation in (7.1). For values of c larger than 1, the equation provides stronger downweighting for observations inconsistent with the model. Yet, it is easy to see that the weights in Equation (7.5) still converge to 1 under the model for any finite c.

Comparing Equation (7.5) with the equivalent version of Equation (2.39), we can see the weight function in (7.5) corresponds to an RAF given by

$$A(\delta) = -1 + (\delta + 1)[w(\delta)]^c. \tag{7.6}$$

As long as this $A(\delta)$ represents an increasing function, one can recover a convex $C(\delta)$ which generates a disparity measure ρ_C as in (2.7), such that the estimating equation of the corresponding disparity matches Equation (7.5). Direct application of Equation (2.44) on $A'(\delta) = (1 + \delta)c[w(\delta)]^{c-1}w'(\delta) + [w(\delta)]^c$ gives

$$C(\delta) = \int_0^\delta ([w(t)]^c - 1)dt + \int_0^\delta \int_0^t \frac{[w(s)]^c}{(1+s)}dsdt. \qquad (7.7)$$

Thus, all the appropriate results about the robustness of the minimum distance estimator based on disparities carry over to the case of the weighted likelihood estimator.

If $A(\delta)$ in (7.6) is not an increasing function, the above procedure will produce a criterion function through (2.7), although it will no longer be a formal disparity measure. However, the criterion will still have a local minimum at the true parameter, and can be used to select the robust root.

For the rest of the chapter, unless otherwise mentioned, the term "weight function" will refer to disparity weights as given by Equation (7.4).

7.1.2   Influence Function and Standard Error

To determine the influence function of the weighted likelihood estimator, we adopt the notation of Section 2.3.7. We consider a general distribution G not necessarily in the model. Let $T(G) = \theta^g$ represent the weighted likelihood functional which solves

$$\int w(\delta(x))u_\theta(x)dG(x) = 0$$

over $\theta \in \Theta$, where g represents the density function of G and $\delta(x) = g(x)/f_\theta(x) - 1$. We set

$$G_\epsilon(x) = (1 - \epsilon)G(x) + \epsilon \wedge_y (x),$$

where $0 < \epsilon < 1$, and $\wedge_y(x)$ is as defined in Section 1.1. Thus, the corresponding density function may be represented as $g_\epsilon(x) = (1 - \epsilon)g(x) + \epsilon\chi_y(x)$. Let θ_ϵ denote the weighted likelihood functional $T(G_\epsilon)$.

Theorem 7.3. [Markatou, Basu and Lindsay (1997, Proposition 1)]. *The influence function of the weighted likelihood estimator is given by*

$$T'(y) = \frac{\partial}{\partial\epsilon}\theta_\epsilon|_{\epsilon=0} = D^{-1}N, \qquad (7.8)$$

where

$$D = \int w'(\delta(x))u_{\theta^g}(x)u_{\theta^g}^T(x)(\delta(x) + 1)dG(x) + \int w(\delta(x))(-\nabla u_{\theta^g}(x))dG(x),$$

$$N = w(\delta(y))u_{\theta^g}(y) + w'(\delta(y))u_{\theta^g}(y)(\delta(y) + 1) - \int w'(\delta(x))g(x)\frac{u_{\theta^g}(x)}{f_{\theta^g}(x)}dG(x),$$

$\theta^g = T(G)$ and $\delta(x) = g(x)/f_{\theta^g}(x) - 1$. When the true distribution G belongs to the model, so that $G = F_\theta$ for some $\theta \in \Theta$, the influence function has the simple form

$$T'(y) = \left[\int u_\theta(x)u_\theta^T(x)dF_\theta(x)\right]^{-1} u_\theta(y) = I^{-1}(\theta)u_\theta(y)$$

which is the same as the influence function of the maximum likelihood estimator as well as the class of minimum distance estimators based on disparities at the model.

Proof. Let $\delta_\epsilon(x) = g_\epsilon(x)/f_{\theta_\epsilon}(x) - 1$. Implicit differentiation of the weighted likelihood estimating equation

$$\int w(\delta_\epsilon(x))u_{\theta_\epsilon}(x)dG_\epsilon(x) = 0$$

leads to the equation

$$\int w'(\delta_\epsilon(x))\delta_\epsilon'(x)u_{\theta_\epsilon}(x)dG_\epsilon(x) + \theta_\epsilon'\left[\int w(\delta_\epsilon(x))\nabla u_{\theta_\epsilon}(x)dG_\epsilon(x)\right]$$
$$+ \int w(\delta_\epsilon(x))u_{\theta_\epsilon}(x)d(\wedge_y - G)(x) = 0. \quad (7.9)$$

Note that

$$\delta_\epsilon'(x)|_{\epsilon=0} = \frac{\partial}{\partial \epsilon}\left[\frac{g_\epsilon}{f_{\theta_\epsilon}} - 1\right]\Big|_{\epsilon=0}$$
$$= [\chi_y(x) - g(x) - u_{\theta^g}^T(x)T'(y)g(x)]/f_{\theta^g}(x). \quad (7.10)$$

When Equation (7.10) is substituted in Equation (7.9) one gets the form described in Equation (7.8).

When $G = F_\theta$ is the true distribution function (and hence $g = f_\theta$ is the true density), we get $\theta^g = \theta$ and $\delta(x) = 0$, so that the influence function now has the simple form

$$T'(y) = \left[\int u_\theta(x)u_\theta^T(x)dF_\theta(x)\right]^{-1} u_\theta(y) = I^{-1}(\theta)u_\theta(y),$$

which is the same as the influence function of the maximum likelihood estimator at the model. □

The quantity D is a function of the underlying distribution G, and N is a function of G and the contaminating point y, so that we denote the quantities as $D(G)$ and $N(y, G)$ wherever necessary. From the previous theorem, the asymptotic variance of $n^{1/2}$ times the estimator can be estimated consistently in the "sandwich" fashion by the quantity

$$\hat{\Sigma} = D^{-1}(G_n)\left[\frac{1}{n-1}\sum_{i=1}^{n} N(X_i, G_n)N(X_i, G_n)^T\right]D^{-1}(G_n).$$

Consider a scalar parameter θ for the model family of densities $\{f_\theta\}$, and let $I(\theta)$ be the Fisher information. From Theorem 7.3, the influence function of the weighted likelihood estimator is exactly same as that of the maximum likelihood estimator at the model. To find the corresponding second-order term

$$T''(y) = \frac{\partial^2}{\partial \epsilon^2} \theta_\epsilon |_{\epsilon=0},$$

we take a second derivative of the weighted likelihood estimating equation under contamination and evaluate at $\epsilon = 0$. A straightforward computation gives

$$T''(y) = T'(y)I^{-1}(\theta)[m_1(y) + A_2 m_2(y)],$$

where the quantities $m_1(y)$ and $m_2(y)$ are exactly the same as those in Theorem 4.2. Thus, from this theorem and Corollary 4.3, all the results involving the second-order prediction of bias for minimum distance estimators and the adequacy of the first-order approximation in relation to the second carry over to the case of weighted likelihood estimators with disparity weights.

7.1.3   The Mean Downweighting Parameter

One of the vital points in the development of the theory of weighted likelihood equations is the issue of the calibration of the weights. That is, how do we select an appropriate weight function and how can we measure its impact on our procedure. Some key insights into this question can be obtained by examining the behavior and variability of the weights when the model is correct. Let

$$\tilde{w} = n^{-1} \sum w(X_i, F_\theta, G_n). \tag{7.11}$$

The next, very important result shows that, in the multinomial case, the final sum of the fitted weights provides a chi-square goodness-of-fit test for the model against the general multinomial alternative. Let A_2 be the curvature parameter of the disparity generating the weights.

Theorem 7.4. [Markatou, Basu and Lindsay (1997, Proposition 2)]. *Let* $W_2 = -A''(0) = -A_2$, *and assume that* $W_2 \neq 0$. *Assume the k-cell multinomial model. Under standard regularity conditions, when the model is correctly specified,*

$$2n(1 - \tilde{w})W_2^{-1} - P^2 \to 0$$

in probability, where $P^2 = 2n\mathrm{PCS}$ *is the Pearson's chi-square.*

Proof. Do a Taylor series expansion of $w(\delta(x))$ around $\delta(x) = 0$, and obtain

$$n(1 - \tilde{w}) = \frac{W_2}{2} \left[n \sum d_n(x)\delta^2(x) \right] + o_p(1)$$

where the bracketed quantity on the right is asymptotically equivalent to the Pearson's chi-square goodness-of-fit statistic, as the parameter estimate of θ is first-order efficient. $\qquad\square$

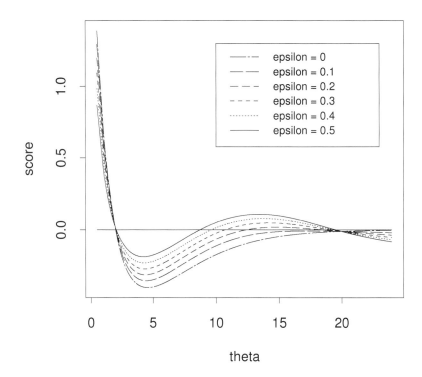

FIGURE 7.2
The weighted likelihood score function and the attained roots for the $(1 - \epsilon)f_2 + \epsilon f_{20}$ mixture.

The relevance of the result to the estimation process is that we can expect the sum of weights, when the model is correct, to be roughly equal in magnitude to $(n - n^*)$, wth $n^* = \frac{1}{2}W_2(k - 1 - p)$, where p represents the dimension of θ. Thus, in an intuitive sense n^* reflects the loss of sample size necessary to achieve the improved robustness properties.

7.1.4 Examples

In this section we apply the weighted likelihood technique to some discrete models, to demonstrate the usefulness of the method in robust parametric estimation, in generating diagnostics, and in performing further data scrutiny. In order to illustrate several aspects of this procedure in a simple setting involving verifications at the true model, we have considered mixture distributions of the form

$$F_m(x) = \alpha F_1(x) + (1 - \alpha)F_2(x),$$

where $\alpha \in [0, 1]$, and the distributions F_1 and F_2 belong to or are in the neighborhood of the assumed parametric model. F_m is also a function of α, but we keep that dependence implicit. Denoting the density function of F_m by f_m, our aim is to solve weighted likelihood estimating equations of the form

$$\int w(\delta(x))u_\theta(x)dF_m(x) = 0,$$

where $\delta(x) = f_m(x)/f_\theta(x) - 1$.

We hope to show, when α is equal to or in the neighborhood of 0.5, that the weighted likelihood estimating equation allows the existence of a root near each of the components. Normally this is accompanied by another crossing of the θ-axis (X-axis) by the weighted likelihood score function, leading to the existence of another "MLE-like" root. When the contamination proportion leads to a substantial overweighting of one of the components, we expect to observe the existence of a root near the component with the larger mass. Such findings are, however, expected to be related to the downweighting strength of the weight function, in our case, determined primarily by the power of the weight c, as well as the separation of the components.

We choose the Poisson model for illustration, and present the following three examples (7.1, 7.2, 7.3) with this aim in this context. Let $\{f_\theta\}$ represent the Poisson model density with parameter θ. See Markatou, Basu and Lindsay (1997) for other similar examples.

Example 7.1. In the first example of this section, the true density f_m is chosen to be the mixture $(1 - \epsilon)f_2 + \epsilon f_{20}$.

In Figure 7.2 we have plotted the weighted likelihood score function $\sum_x w(\delta(x))u_\theta(x)f_m(x)$ against θ, where $\delta(x) = f_m(x)/f_\theta(x) - 1$ and $u_\theta(x)$ is the Poisson (θ) score function. The crossings of the θ-axis (X-axis) by the weighted likelihood score function represents the obtained roots. In this figure there is only one root, close to 2, for $\epsilon < 0.2$, since the function stays below the X-axis after the first crossing. For $\epsilon = 0.2$ or higher, the function is more active and additional roots appear. One of these represents the Poisson (20) component; since two downward crossings necessitate an intermediate upward crossing, there is also a third, "MLE-like" root for such values of ϵ around the center of the range. ‖

Example 7.2. In the second example we again consider a Poisson mixture represented by the density $f_m = 0.5f_6 + 0.5f_{12}$. This time the mixing proportion is kept fixed, but the weights are now raised to the power c as in Equation (7.5), and several plots of the weighted likelihood score function over θ are provided for different values of c in Figure 7.3. For small values of c, the method detects only one, "MLE-like" root. The smoothly descending

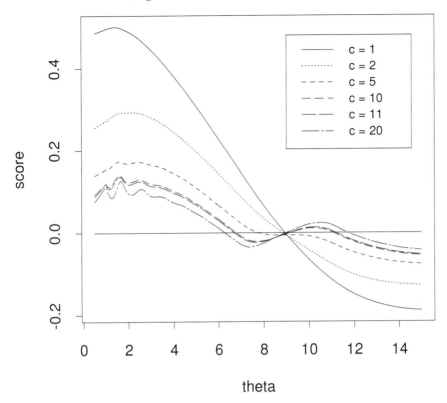

FIGURE 7.3
The weighted likelihood score function and the attained roots for the $0.5f_6 + 0.5f_{12}$ mixture, and several values of the power c.

curve gives way to a reascending redescending shape as c increases, but it requires a power larger than $c = 10$ for three properly identified roots to be clearly visible. ‖

Example 7.3. The third example is similar to the second, except that the mixture is $f_m = 0.5f_4 + 0.5f_{12}$. In this case the components are farther apart, and smaller values of the power c are successful in producing the three roots. Even at $c = 3$, the roots are well separated and clearly identified. At $c = 5$, the robust roots are very close to the means of the components. The plots are presented in Figure 7.4. ‖

If we quickly recap our findings in the above three figures, our observations may be summarized with the following views. We see that (a) if the components are sufficiently separated, there is always a robust root near the larger component; (b) the number of roots is a function of the amount of separation

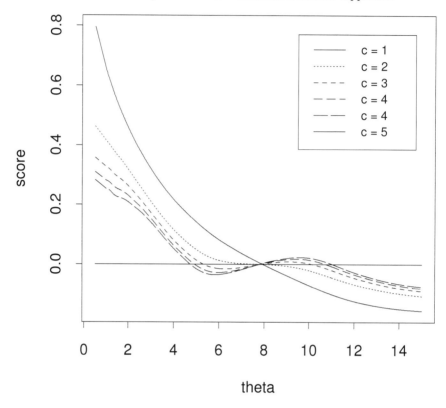

FIGURE 7.4

The weighted likelihood score function and the attained roots for the $0.5f_4 + 0.5f_{12}$ mixture, and several values of the power c.

and the degree of contamination, and (c) a higher power of the weights may, eventually, detect the (additional) robust root(s).

Example 7.4. We end this section with a real application on the Drosophila data, second experimental run, presented in Table 2.3. To determine the weighted likelihood estimator of the parameter θ under the Poisson model, we employ disparity weights based on the robustified likelihood disparity introduced in Section 2.3.6. Since the weights are identically equal to 1 on the inlier side, the only relevant parameter of the distance in the construction of the weights is α^*. The motivation of the construction of this distance was to create a weighted likelihood technique which will automatically generate the outlier deleted maximum likelihood estimator under extreme outliers. This is, in fact, what is observed with the Drosophila data, as the weighted likelihood estimator in this case produces an estimate identical to the outlier deleted maximum likelihood estimate ($\theta = 0.393$) for any α^* between 2 and 50. The final fitted weights are equal to 1 for each observation, except for the large outlying value (91) which gets a weight practically equal to zero. ‖

7.2 The Continuous Case

Our discussion in this section will primarily follow the development presented in Markatou, Basu and Lindsay (1998). In Section 7.1 we have discussed the construction of the weighted likelihood estimating equation in the case of a discrete model. To extend this approach to the continuous case, consider the kernel density estimate g_n^* and the smoothed model density f_θ^* defined in Equations (3.31) and (3.32) respectively. Let the Pearson residual $\delta(x)$ be constructed as $\delta(x) = g_n^*(x)/f_\theta^*(x) - 1$. Let $K(x, y, h)$ be the corresponding kernel function. Consider the weighted likelihood estimating equation

$$\frac{1}{n}\sum_{i=1}^n w(X_i)u_\theta(X_i) = \int w(\delta(x))u_\theta(x)dG_n(x) = 0, \qquad (7.12)$$

where $w(x) = w(\delta(x))$ is as in Equation (7.4). Then the above is a weighted likelihood estimating equation in the spirit of Equation (7.1). Such a method of selection of weights can be motivated by the minimum disparity estimation method presented by Basu and Lindsay (1994) in continuous models. The solution of Equation (7.12) will represent the weighted likelihood estimator (WLE) in the continuous case. The weight functions in (7.12) have the same interpretation as in the discrete case. Unlike the Basu–Lindsay approach to the minimum disparity estimation problem discussed in Section 3.5, the smoothing component now shows up only in the weight part, and no longer in the score part; this may be observed by contrasting Equations (3.34) and (7.12). Thus, it is intuitively clear that the asymptotic efficiency of the weighted likelihood estimators described here no longer depends on the choice of a transparent kernel.

Starting from Equation (3.34), if one uses the same manipulations used in Equations (7.2) and (7.3), one ends up with an estimating equation

$$\int w(\delta(x))\tilde{u}_\theta(x)dG_n^*(x) = 0, \qquad (7.13)$$

where $w(x) = w(\delta(x)) = (A(\delta(x))+1)/(\delta(x)+1)$ and $\tilde{u}_\theta(x) = \nabla f_\theta^*(x)/f_\theta^*(x)$. The above equation can also be solved by the iteratively reweighted estimating equation algorithm (see Section 4.10) as in the discrete case. However, G_n^* is a smoothed version of the empirical cumulative distribution; hence (7.13) is an integral over the entire support of the data and does not reduce to a sum over the observed data points. Depending on the size of the parameter, solving this estimating equation requires the numerical evaluation of several integrals at every stage of the iterative process (multiple integrals if the data are multivariate). However, an estimating equation of the form (7.12) can be obtained from (7.13) by replacing the smoothed empirical G_n^* with the ordinary (unsmoothed) empirical G_n, and $\tilde{u}_\theta$ by the ordinary score function u_θ, while keeping the weight part of the equation intact.

The following differences between the weighted likelihood methodology and the minimum disparity technique need to be highlighted in this context. The iterative solution of the minimum disparity estimating equation requires repeated implementation of numerical integrations at each step; on the other hand, the weighted likelihood technique simply requires repeated applications of the maximum likelihood algorithm. The final weights generated by the weighted likelihood method serve as natural diagnostics for the individual observations, but no such automatic diagnostic is available in the minimum disparity case. The weighted likelihood technique described here automatically leads to estimators with full asymptotic efficiency, where the minimum disparity method described by Basu and Lindsay (1994) requires a transparent kernel for full model efficiency. Even when there are multiple roots, the collection of the different roots are useful in identifying important structures in the data; it can indicate, for example, whether different parts of the data are generated by different processes.

However, one fallout of replacing the smoothed empirical with the ordinary empirical is that, unlike the discrete case, the weighted likelihood estimating equation in (7.12) no longer corresponds to the minimization of a distance. This does present a few difficulties. Firstly, the selection of roots becomes critical when the estimating equation generates multiple roots. Secondly, establishing breakdown results becomes more difficult in the absence of an objective function, as the estimators are now defined solely through the solutions of appropriate equations. And thirdly, in the absence of an objective function, defining a constrained estimator, as may be required when performing tests of composite null hypothesis, is not possible. For these reasons it is necessary to introduce concepts of parallel disparities and parallel weighted likelihoods for the satisfactory resolution of these problems. An appropriate parallel disparity in this connection could be the disparity $\rho_C(g_n^*, f_\theta^*)$, where ρ_C generates the disparity weights used in the construction of the weighted likelihood score function.

In the above, we have discussed the weighted likelihood estimation method in the spirit of the Basu–Lindsay approach, and used smoothed model densities. One could, of course, have used unsmoothed model densities as in the Beran approach to construct disparity weights. However, we believe that the fixed bandwidth consistency and the removal of the transparent kernel restriction makes the Basu–Lindsay approach a more natural one for the construction of weights.

7.2.1 Influence Function and Standard Error: Continuous Case

Unless otherwise mentioned, here we adopt the notation of Section 7.1.2. Let $G_\epsilon(x) = (1 - \epsilon)G(x) + \epsilon \wedge_y (x)$, and consider the estimating equation $\int w(\delta_\epsilon(x))u_\theta(x)dG_\epsilon(x) = 0$, where $\delta_\epsilon(x) = g_\epsilon^*(x)/f_\theta^*(x) - 1$. This equation is solved by $T(G_\epsilon) = \theta_\epsilon$, the weighted likelihood functional at G_ϵ; we can find the influence function of the weighted likelihood functional by taking derivatives of both sides of the equation

$$\int w(\delta_\epsilon(x))u_{\theta_\epsilon}(x)dG_\epsilon(x) = 0, \tag{7.14}$$

where $\delta_\epsilon(x) = g_\epsilon^*(x)/f_{\theta_\epsilon}^*(x) - 1$. The resulting equation is then evaluated at $\epsilon = 0$, and then solved for $T'(y)$. This leads to the following result.

Theorem 7.5. *The influence function of the weighted likelihood estimator for the continuous case is given by*

$$T'(y) = \frac{\partial}{\partial\epsilon}\theta_\epsilon|_{\epsilon=0} = D^{-1}N,$$

$$D = \int w'(\delta(x))(\delta(x) + 1)\tilde{u}_{\theta^g}(x)u_{\theta^g}^T(x)dG(x) + \int w(\delta(x))(-\nabla u_{\theta^g}(x))dG(x),$$

$$\begin{aligned} N &= w(\delta(y))u_{\theta^g}(y) + \int w'(\delta(x))\frac{K(x,y,h)}{f_{\theta^g}^*(x)}u_{\theta^g}(x)dG(x) \\ &- \int w'(\delta(x))(\delta(x) + 1)u_{\theta^g}(x)dG(x), \end{aligned}$$

where $\theta^g = T(G)$, $\tilde{u}_\theta(x) = \nabla \log f_\theta^*(x)$, and $\delta(x) = g(x)/f_{\theta^g}(x) - 1$. When the true distribution $G = F_\theta$ belongs to the model, we have $w(\delta(x)) = 1$ and $w'(\delta(x)) = 0$, so that the expression for the influence function reduces to $T'(y) = I^{-1}(\theta)u_\theta(y)$.

As in the discrete case, an estimate of the asymptotic variance of the estimator can thus be estimated in the "sandwich" fashion as

$$D^{-1}(G_n)\left[\frac{1}{n-1}\sum_{i=1}^n N(X_i, G_n)N(X_i, G_n)^T\right]D^{-1}(G_n).$$

A second-order analysis based on the influence function approach leads to the following analog of Theorem 4.2.

Theorem 7.6. *Assume that the model parameter is scalar, the true distribution belongs to the model, and θ is the true parameter. Let $I(\theta)$ be the corresponding Fisher information which is assumed to be finite. For an estimator defined by an estimating equation of the form (7.12), we get*

$$T''(y) = T'(y)I^{-1}(\theta)[m_1(y) + A_2m_2(y))],$$

where

$$m_1(y) = 2\nabla u_\theta(y) - 2E_\theta[\nabla u_\theta(X)] + T'(y)E_\theta[\nabla_2 u_\theta(X)]$$

and

$$m_2(y) = \frac{I(\theta)}{u_\theta(y)} \int K^2(x, y, h) u_\theta(x) \frac{f_\theta(x)}{f_\theta^{*2}(x)} dx + 2 \int u_\theta(x)\tilde{u}_\theta(x) f_\theta(x) dx$$
$$- 2\int K(x, y, h) u_\theta(x)\tilde{u}_\theta(x) \frac{f_\theta(x)}{f_\theta^*(x)} dx + T'(y)\int \tilde{u}_\theta^2(x) u_\theta(x) f_\theta(x) dx$$
$$- 2\frac{I(\theta)}{u_\theta(y)} \int \frac{K(x, y, h) u_\theta(x)}{f_\theta^*(x)} f_\theta(x) dx.$$

When the model is a one parameter exponential family model with θ being the mean value parameter, we have $m_1(y) = 0$ from Corollary 4.3. The values of $m_2(y)$ have a similar trend as those of the corresponding quantities in Theorem 4.4, and the interpretation of this result is similar to that of Corollary 4.3.

7.2.2 The Mean Downweighting Parameter

As in the discrete case, we consider the mean downweighting that occurs when the model is correct. Let $\tilde{w}$ be as defined in Equation (7.11), which is the average of the final fitted weights; notice, however, that there is a kernel involved in the construction of the residuals. For a unimodal weight function and under suitable regularity conditions, one gets

$$n(1 - \tilde{w}) \approx -\frac{A_2}{2}\sum \delta^2(X_i)$$
$$\approx -n\frac{A_2}{2}\int \left[\frac{\int K(x, y, h)dG_n(y)}{f_\theta^*(x)} - 1\right]^2 dF_\theta(x),$$

where $\delta(x) = g_n^*(x)/f_\theta^*(x) - 1$, and $A_2 = W_2 = w''(0)$. A simple calculation shows that the asymptotic mean of $n(1 - \tilde{w})$ equals

$$\Lambda = -\frac{A_2}{2}\left[\int \frac{\int K^2(x, y, h)dF_\theta(y)}{f_\theta^{*2}(x)}dF_\theta(x) - 1\right]. \tag{7.15}$$

Once again the weight function w appears in the mean downweighting formula (7.15) only through the curvature parameter A_2.

The formula above represents a complicated combination of the model and kernel, and as in all cases with continuous models, the value of the bandwidth can have a substantial effect on mean downweighting. We can get some idea of the behavior of this term for specific models and given values of the smoothing parameter. Suppose the model is normal, and the kernel is the normal density with variance h^2. It may be easily checked that the bracketed quantity in Equation (7.15) equals

$$\frac{(\sigma^2 + h^2)^{3/2}}{(3\sigma^2 + h^2)^{1/2}h^2} - 1. \tag{7.16}$$

From the above equation we see, once again, that increasing h drives the mean of $(1 - \tilde{w})$ toward zero, and hence the weights to 1, taking the equation closer to the likelihood score equation (and making it correspondingly weaker in terms of robustness). Thus, the choice of h must be moderated by the specific robustness and efficiency needs. Notice that mean downweighting also depends on the model parameter σ^2; therefore the robustness properties will vary with the true value of σ^2 even when h^2 is held fixed.

Markatou, Basu and Lindsay (1998) recommended the selection of the bandwidth in relation to the model parameter σ; in particular they suggested $h^2 = \kappa\sigma^2$ for some positive constant κ (σ will be replaced by a robust estimate $\hat{\sigma}$ of scale when it is unknown). This choice of the bandwidth reduces expression (7.16) to the constant $[(1 + \kappa)^{(3/2)}]/[\kappa(3 + \kappa)^{1/2}] - 1$, and one can then choose κ to determine the degree of downweighting. In the normal case, this also makes the weighted likelihood estimators location- and scale-equivariant. For $\kappa = 1$, the above downweighting factor equals $\sqrt{2} - 1$, which when multiplied by 0.25 (the value of the factor $-A_2/2$ for the HD or the SCS) produces a mean downweighting of only about 0.1 observations. But for κ near 0, the mean downweighting for weights generated by the HD or SCS approximately equals $0.25[1/(\kappa\sqrt{3}) - 1]$, a substantially larger number.

7.2.3 A Bootstrap Root Search

The proposed estimating equations usually have unique solutions when the data generally conform to the model or show only mild deviations. However, uniqueness of the solution is not a guaranteed issue, and the multiple root problem is very much a possibility in a given situation. Our strategy here is to search the parameter space in such a manner that all reasonable solutions are found with high "probability." For our purpose, a reasonable parameter set consists of all such values which could plausibly have been used to generate some subset of the data.

The bootstrap root search proposed by Markatou, Basu and Lindsay (1998) proceeds as follows. Let m be the minimum number of observations necessary for the maximum likelihood estimator to exist. Choose B bootstrap samples of m distinct elements of the data set. For each bootstrap sample $b = 1, \ldots, B$, let θ_b be the corresponding maximum likelihood estimator (which is thus based on m distinct elements of the original set). Use each θ_b as the starting value of the iterative reweighting algorithm on the full data, and keep track of all the roots obtained. In moderately extensive simulations, Markatou, Basu and Lindsay (1998) reported the presence of at least one, and often multiple, nondegenerate solutions where the sum of the weights exceeded half the total sample size. In a small fraction of cases, the obtained roots were degenerate, where the final sum of fitted weights were approximately m, roughly corresponding to giving weight one to each of the elements of the bootstrap sample and zero to the rest of the data.

This approach of determining multiple roots is motivated by the work of

Finch, Mendell and Thode (1989) in which a "prior" was put on the parameter space to generate the starting values. It was argued that as a byproduct one can then construct estimates of the probability that a root not yet found will be found with further searching. The particular extension provided by Markatou, Basu and Lindsay (1998) is the use of data driven starting values, designed to construct reasonable search regions automatically.

The output of the bootstrap root search procedure is a set of nondegenarate roots. If there is just one such root, as seems to be the case in most good data sets, there is little further exploration to be done. If there are multiple solutions, one has to compare the values of the parallel disparity function for each of the roots. Formally, this serves the purpose of choosing a root. Informally, it indicates the overall quality of fit or various roots, and whether there are multiple competing hypotheses which explain different subsets of the data.

7.2.4   Asymptotic Results

Following Markatou, Basu and Lindsay (1998), we present the regularity conditions needed for the existence and the asymptotic normality of the weighted likelihood estimators. While the results are true for the general multiparameter case, the conditions are presented here in case of a scalar parameter θ.

(W1) The weight function $w(\delta)$ is a nonnegative bounded and differentiable function with respect to δ.

(W2) The weight function $w(\delta)$ is regular in the sense that $w'(\delta)(\delta + 1)$ is bounded, with $w'(\delta)$ being the derivative of $w(\delta)$ with respect to δ.

(W3) For every $\theta_0 \in \Theta$, there is a neighborhood $N(\theta_0)$ such that for $\theta \in N(\theta_0)$, the quantities $|\tilde{u}_\theta(x)\nabla u_\theta(x)|$, $|\tilde{u}_\theta^2(x)u_\theta(x)|$, $|\nabla \tilde{u}_\theta(x)u_\theta(x)|$ and $|\nabla_2 u_\theta(x)|$ are bounded by $M_1(x)$, $M_2(x)$, $M_3(x)$ and $M_4(x)$, where $E_{\theta_0}[M_i(X)] < \infty$, $i = 1, 2, 3, 4$.

(W4) $E_{\theta_0}[\tilde{u}_\theta^2(X)u_\theta^2(X)] < \infty$.

(W5) The Fisher information $I(\theta) = E_\theta[u_\theta^2(X)]$ is finite.

(W6) 1. $\int |\nabla f_\theta(x)/f_\theta^*(x)|dx = \int |u_\theta(x)f_\theta(x)/f_\theta^*(x)|dx < \infty$.
      2. $\int |\tilde{u}_\theta(x)u_\theta(x)|[f_\theta(x)/f_\theta^*(x)]dx < \infty$.
      3. $\int |\nabla \tilde{u}_\theta(x)|[f_\theta(x)/f_\theta^*(x)]dx < \infty$.

(W7) The kernel $K(x, y, h)$ is bounded by a finite constant $M(h)$, which may depend on h, but not on x or y.

We refer the reader to Markatou, Basu and Lindsay (1996, 1998) for more details on the asymptotic results in this connection. Here we present the main results without proof. The actual proofs are technical, but, in essence, are modifications of the techniques of Sections 2.5 and 3.5.3.

Theorem 7.7. *Let the true distribution belong to the model, θ_0 be the true parameter, and let $\hat{\theta}_w$ be the weighted likelihood estimator. Under the above assumptions, the following results hold.*

1. *The convergence*

$$n^{1/2}|A_n - \frac{1}{n}\sum_{i=1}^{n} u_{\theta_0}(X_i)| \to 0$$

   holds in probability, where $A_n = \frac{1}{n}\sum_{i=1}^{n} w_{\theta_0}(\delta(X_i))u_{\theta_0}(X_i)$, *and* $w_{\theta_0}(\delta(X_i))$ *are weights based on* $\delta(x) = g_n^*(x)/f_{\theta_0}^*(x) - 1$.

2. *The convergence*

$$|B_n - \frac{1}{n}\sum_{i=1}^{n} \nabla u_{\theta_0}(X_i)| \to 0$$

   holds in probability where $B_n = \frac{1}{n}\sum_{i=1}^{n} \nabla(w_\theta(\delta(X_i))u_\theta(X_i))|_{\theta=\theta_0}$.

3. $C_n = O_p(1)$, *where* $C_n = \frac{1}{n}\sum_{i=1}^{n} \nabla_2(w_\theta(\delta(X_i))u_\theta(X_i))|_{\theta=\theta'}$ *where* θ' *is between* θ_0 *and* $\hat{\theta}_w$.

Using the above theorem, the consistency of the weighted likelihood estimator $\hat{\theta}_w$ follows from Serfling (1980, pp. 146–148). A straightforward Taylor series expansion of the weighted likelihood estimating equation

$$\frac{1}{n}\sum_{i=1}^{n} w_{\hat{\theta}_w}(\delta(X_i))u_{\hat{\theta}_w}(X_i) = 0$$

around $\hat{\theta}_w = \theta_0$ leads to the relation

$$n^{1/2}(\hat{\theta}_w - \theta_0) = -\frac{n^{1/2}A_n}{B_n + \frac{(\hat{\theta}_w - \theta_0)}{2}C_n},$$

which, together with the above theorem, immediately yields

$$n^{1/2}(\hat{\theta}_w - \theta_0) \xrightarrow{\mathcal{D}} Z^* \sim N(0, I^{-1}(\theta_0)).$$

7.2.5 Robustness of Estimating Equations

Many of the results about the robustness of the estimators and estimating equations of Section 4.8 carry over to the case of weighted likelihood estimators. The arguments are essentially the similar to that of Section 4.8, with the additional introduction of the parallel disparity. In particular, one can derive analogs of the convergence result in Equation (4.47), and of Theorems 4.14 and 4.16. The reader is referred to Markatou, Basu and Lindsay (1998) for a more expanded discussion of this subject.

TABLE 7.1

The estimated parameters of the different cases

Case I	Case II	Case III	Case IV	Case V	Case VI
146.19	138.27	142.14	146.32	139.32	142.22
14.10	10.09	12.05	13.97	10.30	11.67
30.15	16.38	38.77	22.86	4.79	21.97
0.75	0.90	4.84	0.83	0.94	3.59
−0.92	−0.48	7.23	−0.27	0.25	7.01

7.3 Examples

Example 7.5. Lubischew (1962) presented a data set on the characteristics of several varieties of beetles. A part of these data was previously examined by Markatou, Basu and Lindsay (1998) to illustrate the effects of our weighted likelihood technique. Here we revisit these data and present the findings in greater detail. The part of the data being considered involves two varieties of beetles, *Chaetocnema concinna* and *Chaetocnema heptapotamica*. The data considered here are bivariate, and consist of two measurements per beetle – the maximal width (measured in microns) and front angle (one unit = 7.5°) of the aedeagus (male copulative organ). It seems that the two species are not easy to distinguish visually. However, a careful look at the plotted data shows that there is a fairly clear separation. The data under study include 21 measurements from *C. concinna* (we will refer to this as Group 1) and 22 measurements form *C. heptapotamica* (Group 2 in our notation).

Fitting a bivariate normal model with $(\mu_1, \mu_2, \sigma_1^2, \sigma_2^2, \sigma_{12})$ as the vector of parameters, six sets of parameter estimates are presented in Table 7.1 which represent the following cases. (I) The maximum likelihood estimator for the 21 beetles in the first group; (II) the maximum likelihood estimates for the 22 beetles in the second group; (III) the maximum likelihood estimates for the 43 beetles in the combined group; (IV), (V) and (VI) represent the three distinct roots identified with the weighted likelihood technique when applied to the full data. In the latter three cases we used disparity weights based on the SCS measure with a bivariate normal kernel having covariance matrix $h^2 I$, where $h = 0.5$ is the smoothing parameter, and I is the two-dimensional identity matrix.

Notice that the three roots of the weighted likelihood estimating equation presented in columns (IV)–(VI) are very close to the three sets of maximum likelihood estimates presented in (I)–(III) respectively. However, unlike in Cases (I) and (II), all the roots of the weighted likelihood estimating equation are obtained for the full data. In Case (IV), the weighted likelihood clearly identifies Group 2 observations as being outliers with respect to the model and picks up a root close to that of Case (I). The same relation holds

TABLE 7.2
Fitted weights for Cases (IV)–(VI) solutions for Group 1 observations.

Group	Observation	Case IV	Case V	Case VI
1	(150, 15)	0.8685	0.0000	0.5147
1	(147, 13)	0.9396	0.0022	0.8645
1	(144, 14)	0.9972	0.0056	0.7075
1	(144, 16)	0.7061	0.0000	**0.0816**
1	(153, 13)	0.9231	0.0000	**0.1312**
1	(140, 15)	0.9351	0.0016	**0.0429**
1	(151, 14)	1.0000	0.0000	0.6923
1	(143, 14)	1.0000	0.0184	0.7788
1	(144, 14)	0.9972	0.0056	0.7075
1	(142, 15)	0.9975	0.0012	**0.1831**
1	(141, 13)	0.9715	**0.5203**	0.9689
1	(150, 15)	0.8685	0.0000	0.5147
1	(148, 13)	0.9937	0.0007	0.8895
1	(154, 15)	0.7542	0.0000	**0.2599**
1	(147, 14)	1.0000	0.0003	0.9408
1	(137, 14)	0.7447	0.0214	**0.0298**
1	(134, 15)	**0.2012**	0.0000	**0.0000**
1	(157, 14)	0.5192	0.0000	**0.0098**
1	(149, 13)	1.0000	0.0002	0.8561
1	(147, 13)	0.9396	0.0022	0.8644
1	(148, 14)	1.0000	0.0000	0.9320

between Cases (V) and (II). Case (VI) shows that the algorithm also has an "MLE-like" root, which tries to cover all the observations.

In Tables 7.2 and 7.3 we have presented the final fitted weights of the Group 1 and Group 2 observations, respectively, for each of the three roots of the weighted likelihood equation. Notice that for Case (IV), except for a few exceptions (presented in bold fonts), the root puts a reasonable weight on the Group 1 observations (Table 7.2) and weights close to zero on the Group 2 observations (Table 7.3) as it should do. The reverse is observed for Case (V). The exceptions – all marked in bold – are also easily explained through a further scrutiny of the data (plotted in Figure 7.5). For example, the three exceptions in the column corresponding to Case (V) in Table 7.3, all having width values around 130, are clearly observations which are separated from the rest of the data when considering the Group 2 observations. The 50% contour concentration ellipsoids corresponding to the three roots of the weighted likelihood equations are presented in Figure 7.5, which clearly illustrates how the different roots may be attributed to different competing hypotheses and how the set of roots helps to uncover the real structure of the data, which – at least in this example – is clearly more important than finding a single root

TABLE 7.3
Fitted weights for Cases (IV)–(VI) solutions for Group 2 observations.

Group	Observation	Case IV	Case V	Case VI
2	(145, 8)	0.0000	**0.0420**	**0.0106**
2	(140, 11)	0.0178	1.0000	0.9347
2	(140, 11)	0.0178	1.0000	0.9347
2	(131, 10)	0.0000	**0.0179**	**0.1321**
2	(139, 11)	0.0233	1.0000	0.9983
2	(139, 10)	0.0007	1.0000	0.9727
2	(136, 120)	0.1046	0.7492	0.4462
2	(129, 11)	0.0001	**0.0002**	**0.0032**
2	(140, 10)	0.0005	1.0000	0.7662
2	(137, 9)	0.0000	1.0000	0.8848
2	(141, 11)	0.0431	1.0000	1.0000
2	(138, 9)	0.0000	1.0000	0.9047
2	(143, 9)	0.0000	0.8534	**0.4016**
2	(142, 11)	0.0696	1.0000	1.0000
2	(144, 10)	0.0044	0.8463	0.7735
2	(138, 10)	0.0006	1.0000	0.9863
2	(140, 10)	0.0005	1.0000	0.7662
2	(130, 9)	0.0000	0.0023	**0.1469**
2	(137, 11)	0.0135	1.0000	0.9435
2	(137, 10)	0.0004	1.0000	0.9611
2	(136, 9)	0.0000	0.9783	0.9186
2	(140,10)	0.0005	1.0000	0.7662

which fits 50% of the data. In Figure 7.5 Group 1 observations are coded with a dot, while Group 2 observations are indicated with a star.

The individual roots corresponding to Cases (IV) and (V), although close to maximum likelihood estimates for the two groups, have their own differences with the latter as well. For example, the sign of the covariance terms are not consistent in Cases (II) and (V); as a result the concentration ellipsoid for Group 2 is very different from the one that would be obtained with the maximum likelihood estimator, and has an entirely different orientation.   ∥

Example 7.6. In this example, we fit a normal model to Newcomb data, considered earlier in Example 3.3 (Chapter 3), using weighted likelihood estimation based on disparity weights obtained from the robustified likelihood disparity. We use a normal kernel with bandwidth $h = 0.5$; and a tuning parameter of $\alpha^* = 2$, and the weighted likelihood estimates of the location and scale parameters are obtained as 27.750 and 5.044. Thus, the parameter estimates are exactly same as the outlier deleted maximum likelihood estimates, as they are expected to be in such a situation. The final fitted weights are equal to 1 for each observation other than the two large outliers, which

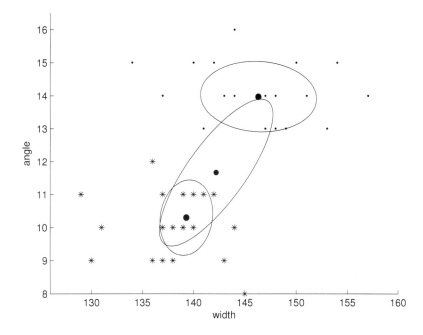

FIGURE 7.5
Group 1 and Group 2 observations, the obtained roots of the weighted likelihood equation, and the 50% contour concentration ellipsoids.

get weights practically equal to zero. The histogram of the Newcomb data with the maximum likelihood fit and the weighted likelihood fit (in effect the likelihood fits for the full data and the outlier deleted data) are presented in Figure 7.6; clearly the weighted likelihood estimate performs just as well as our robust minimum distance estimators in this case. For the sake of comparison we note that the minimum distance estimate based on the robustified likelihood disparity $\mathrm{RLD}_{-1,2}(g_n^*, f_\theta^*)$, the parallel disparity for this problem, for the same kernel and the same bandwidth turn out to be $\hat{\mu} = 27.869$ and $\hat{\sigma} = 4.995$ under the Basu–Lindsay approach. ‖

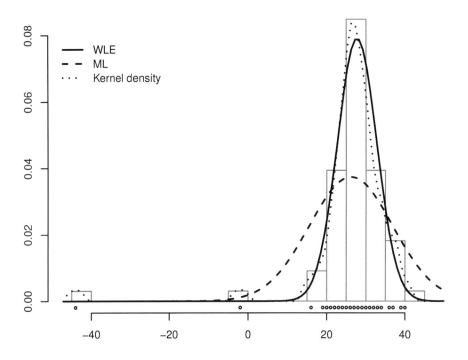

FIGURE 7.6
The weighted likelihood normal fit and the maximum likelihood normal fit
superimposed on the histogram of Newcomb data.

7.4 Hypothesis Testing

We present the extension of the weighted likelihood methodology to the case of testing of parametric hypothesis. The aim here is to construct test procedures which are asymptotically equivalent to the corresponding tests based on the maximum likelihood theory, but are more robust than the latter in preserving the level and power of the tests when actual observations are generated from a contaminated version of the true density. To fix ideas we consider a parametric model $\mathcal{F}$ as defined in Section 1.1, and a simple null hypothesis $H_0 : \theta = \theta_0$ versus the alternative $H_1 : \theta \neq \theta_0$. We present here the analogs of the Wald type, score type and likelihood ratio type tests based on the weighted likelihood methodology discussed in the previous sections of this chapter. Our discussion here follows the approach of Agostinelli and Markatou (2001).

1. Consider the version of the usual Wald type test statistic for the above null hypothesis which has the form

$$W = (\hat{\theta} - \theta_0)^T [nI(\hat{\theta})](\hat{\theta} - \theta_0),$$

where $\hat{\theta}$ is the maximum likelihood estimator of $\theta \in \Theta$. The corresponding Wald statistic based on the weighted likelihood methodology is defined as

$$W_w = (\hat{\theta}_w - \theta_0)^T \left[\sum_{i=1}^{n} w(\delta_{\hat{\theta}_w}(x_i)) I(\hat{\theta}_w) \right] (\hat{\theta}_w - \theta_0),$$

where $\delta_{\hat{\theta}_w}(x_i) = g_n^*(x_i)/f_{\hat{\theta}_w}^*(x_i) - 1$, and $\hat{\theta}_w$ is the weighted likelihood estimator of θ. In this context we will refer to it as the *unrestricted* weighted likelihood estimator, since later we have to define the corresponding restricted estimator to handle tests of hypotheses related to composite hypotheses.

A comparison of the two Wald type statistics show that in the weighted likelihood version the total sample size n is replaced by the sum of the weights at the unrestricted weighted likelihood estimator.

2. The usual score (Rao) type test function has the form

$$S = \varrho^T(\theta_0)[nI(\theta_0)]\varrho(\theta_0)$$

where $\varrho(\theta) = (\partial/\partial\theta) \log \prod_{i=1}^{n} f_\theta(x_i) = \sum_{i=1}^{n} u_\theta(x_i)$. In comparison, the score type test statistic based on the weighted likelihood methodology has the form

$$S_w = \varrho_w^T(\theta_0) \left[\sum_{i=1}^{n} w(\delta_{\hat{\theta}_w}(x_i)) I(\theta_0) \right] \varrho_w(\theta_0),$$

where $\varrho_w(\theta) = \sum_{i=1}^{n} w(\delta_{\hat{\theta}_w}(x_i))u_\theta(x_i)$, and $\hat{\theta}_w$ is the unrestricted weighted likelihood estimator of θ. Once again, the weighted likelihood version replaces the total sample size n by the sum of the weights at the unrestricted weighted likelihood estimator.

3. The usual likelihood ratio test statistic for the above hypothesis is given by

$$\Lambda = -2\left[\sum_{i=1}^{n} \log f_{\theta_0}(x_i) - \sum_{i=1}^{n} \log f_{\hat{\theta}}(x_i)\right]$$

$$= -2\sum_{i=1}^{n} \left[\log f_{\theta_0}(x_i) - \log f_{\hat{\theta}}(x_i)\right]$$

and the corresponding version based on weighted likelihood is given by

$$\Lambda_w = -2\sum_{i=1}^{n} w(\delta_{\hat{\theta}_w}(x_i))\left[\log f_{\theta_0}(x_i) - \log f_{\hat{\theta}_w}(x_i)\right].$$

Defining the tests under a composite hypothesis presents a problem since the idea of a constrained estimator is not directly applicable to the weighted likelihood situation. Just as one can use the parallel disparity in case of root selection and breakdown, here we define the parallel weighted likelihood. This may be defined as

$$\text{WL}(\theta) = \prod_{i=1}^{n} [f_\theta(x_i)]^{w(\delta_{\hat{\theta}_w}(x_i))}.$$

Here the weights are the final fitted weights for the unrestricted weighted likelihood estimator. A constrained weighted likelihood estimator under the null hypothesis can then be constructed by minimizing the above weighted likelihood over the null space, treating the weights to be fixed constants, i.e., the constrained estimator $\hat{\theta}_0$ equals

$$\hat{\theta}_0 = \arg\max_{\theta \in \Theta_0} \text{WL}(\theta) = \prod_{i=1}^{n} [f_\theta(x_i)]^{w(\delta_{\hat{\theta}_w}(x_i))}.$$

Building on the above background, Agostinelli and Markatou (2001) have shown that the Wald type, score type and likelihood ratio type tests based on the weighted likelihood methodology have the same asymptotic distribution under a correctly specified model and null hypothesis, as the corresponding tests based on ordinary likelihood theory. In this connection we present the following theorem from Agostinelli and Markatou.

Theorem 7.8. [Agostinelli and Markatou (2001, Theorem 4.1)]. *Under conditions (W1)–(W7) of Section 7.2.4 the Wald test, score test and the disparity difference test based on the weighted likelihood methodology are asymptotically equivalent to the corresponding likelihood based tests under the null and under contiguous alternatives.*

See Agostinelli and Markatou (2001) for a discussion on other aspects of hypothesis testing based on the weighted likelihood methodology.

7.5 Further Reading

In this chapter we have attempted to provide a basic flavor of the weighted likelihood methodology, and restricted ourselves to a general discussion of the technique. The technique has been adapted in many ways by several authors to deal with specialized problems. Here we give very brief descriptions of some of them to demonstrate how the method is branching out and adapting itself to such problems. In particular, the methodology has been integrated as part of the R software package (by Claudio Agostinelli) making it computationally accessible for most practitioners.

One area where the weighted likelihood methods have been explored in detail is mixture models (Markatou, 2000, 2001). In the above, problems associated with the analysis of data from a mixture of distributions in presence of data contamination and model misspecification have been considered. Numerical and algorithmic issues related to the evaluation of the weighted likelihood estimators have been studied in detail, and an algorithm, competitive with the EM algorithm is proposed. The use of the technique in the case of model selection is also studied.

See Agostinelli and Markatou (1998), and Agostinelli (2002b) for applications of the method in case of linear regression. The method has also been applied in the context of robust model selection (Agostinelli, 2002a) and robust time series analysis (Agostinelli, 2003). Also see Sarkar, Kim and Basu (1999) for a description of the test for robust homogeneity of variances based on the weighted likelihood methodology.

8

Multinomial Goodness-of-Fit Testing

8.1 Introduction

For the greater part of this book, we have attempted to give a comprehensive description of density-based minimum distance estimation using disparity measures. A second utility of these measures is that they are extremely useful in testing goodness-of-fit for discrete multivariate models. The latter is a major research area, and represents a technique which has huge applications in applied statistics. Such goodness-of-fit statistics are also routinely used by scientists in all other disciplines.

Testing goodness-of-fit with discrete multivariate data is a very old and traditional technique in statistical inference, although there have been many new developments in the field during the last few decades. There are several classic books on discrete multivariate models which represent very popular resources for this field; these include, among others, the books by Bishop, Fienberg and Holland (1975), Fienberg (1980), Agresti (1984) and Freeman (1987). Of special relevance to us is the book by Read and Cressie (1988), which is entirely on the theme of testing goodness-of-fit for discrete multivariate data. It provides an excellent and comprehensive treatment of multivariate goodness-of-fit testing using the power divergence statistics as its medium of illustration. A fair bit of work has been done in this area by Pardo and his associates as well, some of which is reported in Pardo (2006). In this chapter it will be our aim to provide a brief and useful summary of the goodness-of-fit tests in general, including some of the major contents of the above resources, and present certain angles not covered in the above references. As a special illustration we will also emphasize the applications of this method in connection with testing goodness-of-fit for the kappa statistic, a measure extensively used to quantify the amount of agreement between several raters, following the approach of Donner and Eliasziw (1992).

We emphasize again that unlike the theme of robust minimum distance estimation based on disparities, which has developed much more recently, the goodness-of-fit testing aspect of the class of disparities has been studied in more detail in the literature, and much of it has been discussed at length in several popular discrete data books. We purposely avoid the derivation of well known existing results and highlight only some new recent developments. In particular, we begin by assuming the asymptotic distributional results of the Pearson's chi-square, rather than by proving them.

8.1.1   Chi-Square Goodness-of-Fit Tests

Suppose that a random variable can take values in one of k mutually exclusive and exhaustive categories. Let the cell probabilities for these k categories be π_i, $i = 1, \ldots, k$, $\pi_i > 0$, $\sum_i \pi_i = 1$. Consider n realizations of this variable which generates the multinomially distributed random vector $X = (X_1, \ldots, X_k)$, $\sum_i X_i = n$, where X_i represents the observed frequency of the i-th category. The probability of a particular realization $(x_1, \ldots, x_k)$ can be expressed as

$$P(X_1 = x_1, \ldots, X_k = x_k) = \frac{n!}{x_1! \ldots x_k!} \prod_{i=1}^{k} \pi_i^{x_i}, \quad 0 \le x_i \le n, \ i = 1, \ldots, k.$$

Among the set of parameters $(n, \pi_1, \ldots, \pi_k)$ of this multinomial variable, the cell probabilities $\pi = (\pi_1, \ldots, \pi_k)$ are usually unknown, and performing tests of hypothesis about them is a fundamental problem of statistics.

The simple null hypothesis which will be of interest to us will be of the form

$$H_0 : \pi = \pi_0, \tag{8.1}$$

where $\pi_0 = (\pi_{01}, \ldots, \pi_{0k})$ is the hypothesized null probability vector. In this case the entire probability vector is fully specified – the null hypothesis spells out all the $k - 1$ independent components. For the more general composite hypothesis, the null is specified in terms of p unknown parameters which need to be estimated from the data.

Throughout this chapter we will use the multinomial model (or the multinomial sampling distribution) to describe random data grouped into k cells. In practice, the product multinomial and Poisson distributions are also used to describe such layouts. However, as observed by Read and Cressie (and as discussed in most elementary "Discrete Data" texts), "most of the results on modeling under these three distributions turn out to be the same," and the analysis based solely upon the multinomial model will usually be adequate to maintain the focus of this discussion.

Consider the simple null hypothesis in Equation (8.1). Given a realization $(x_1, \ldots, x_k)$ of the multinomial random variable, the most popular statistic for testing the above hypothesis is the Pearson's chi-square statistic given by

$$2n\text{PCS} = n \sum_{i=1}^{k} \frac{(p_i - \pi_{0i})^2}{\pi_{0i}} = \sum_{i=1}^{k} \frac{(x_i - n\pi_{0i})^2}{n\pi_{0i}}, \tag{8.2}$$

where $p_i = x_i/n$ represents the observed proportion or relative frequency for the i-th class and

$$\text{PCS} = \frac{1}{2} \sum_{i=1}^{k} \frac{(p_i - \pi_{0i})^2}{\pi_{0i}}. \tag{8.3}$$

The other commonly used statistic is the log-likelihood ratio chi-square statistic given by

$$2n\text{LD} = 2n \sum_i p_i \log(p_i/\pi_{0i}) = 2 \sum_i x_i \log(x_i/n\pi_{0i}). \qquad (8.4)$$

Notice that the PCS and LD measures are exactly the same as the indicated distances discussed in Chapter 2 when the vectors d_n and f_θ are replaced by $p = (p_1, \ldots, p_k)$ and $\pi_0 = (\pi_{01}, \ldots, \pi_{0k})$ respectively. We make a distinction between the chi-squares and the corresponding test statistics. Thus, the measure in (8.3) is the Pearson's chi-square, while the statistic in (8.2) is the Pearson's chi-square statistic.

Notice that the test statistics in (8.2) and (8.4) are nonnegative, and equal zero if and only if the vectors p and π_0 are identically equal. Under the null hypothesis, one would expect the value of the test statistics above to be "small," and rejection would require that the statistics exceed an appropriate threshold. The asymptotic distribution of the Pearson's chi-square statistic (Pearson, 1900) given in (8.2), one of the oldest milestones in statistical literature, is a chi-square with $k - 1$ degrees of freedom. Although the distances are, in general, different, the asymptotic null distribution of the log-likelihood ratio chi-square statistic is also the same as that of the Pearson's chi-square statistic (Wilks, 1938).

Apart from the resources mentioned in the introduction of this chapter, several other authors have compared the performance of the Pearson's chi-square statistic, the log likelihood ratio chi-square statistic as well as some of the lesser known goodness-of-fit statistics in the literature. See, e.g., Cochran (1952), Hoeffding (1965), West and Kempthorne (1972), Moore and Spruill (1975), Chapman (1976), Larntz (1978), Koehler and Larntz (1980), and Cressie and Read (1984).

8.2 Asymptotic Distribution of the Goodness-of-Fit Statistics

Cressie and Read (1984) established the asymptotic null distribution of the goodness-of-fit statistics within the power divergence family. Here we consider the more general class of test statistics for multinomial goodness-of-fit testing based on disparities; the power divergence class forms a subset of this collection. These "disparity statistics" are constructed using the general family of disparities; see Lindsay (1994), and Chapter 2 of this book. We derive the asymptotic chi-square distribution of the disparity statistics under both simple and composite null hypotheses following Basu and Sarkar (1994b); also see Zografos et al. (1990). We also investigate specific disparities within the power divergence and some other families of tests with the hope of identifying test statistics that are optimal in some sense within those families.

8.2.1   The Disparity Statistics

Consider a multinomial sample of n observations on k cells with probability vector $\boldsymbol{\pi} = (\pi_1, \ldots, \pi_k)$, $\sum_{i=1}^{k} \pi_i = 1$. Let $(x_1, \ldots, x_k)$ denote the vector of observed frequencies for the k categories, $\sum_{i=1}^{k} x_i = n$. Let C be a strictly convex, thrice differentiable function on $[-1, \infty)$ with $C(0) = 0$. We will assume that $C'''(\cdot)$, the third derivative of C, is finite and continuous at zero. Let

$$\boldsymbol{p} = (p_1, \ldots, p_k) = (x_1/n, \ldots, x_k/n), \tag{8.5}$$

and let $\boldsymbol{\pi}_0 = (\pi_{01}, \ldots, \pi_{0k})$ be a prespecified probability vector with $\pi_i > 0$ for each $1 \leq i \leq n$ and $\sum_{i=1}^{k} \pi_{0i} = 1$. Then the disparity test based on the function C for testing the null hypothesis $H_0 : \boldsymbol{\pi} = \boldsymbol{\pi}_0$ is given by the disparity statistic $D_{\rho C} = 2n\rho_C(\boldsymbol{p}, \boldsymbol{\pi}_0)$, where

$$\rho_C(\boldsymbol{p}, \boldsymbol{\pi}_0) = \sum_{i=1}^{k} C\left(\frac{p_i}{\pi_{0i}} - 1\right) \pi_{0i} \tag{8.6}$$

is the disparity, as defined in Chapter 2, between the probability vectors $\boldsymbol{p}$ and $\boldsymbol{\pi}_0$. Letting $\delta_i = (\pi_{0i}^{-1} p_i - 1)$, we see that the Pearson's chi-square, denoted by $\text{PCS}(\boldsymbol{p}, \boldsymbol{\pi}_0)$, is generated by $C(\delta) = \delta^2/2$ in (8.6). The class of disparity measures have been discussed extensively in Chapter 2 and the C functions for all our standard disparities are presented in Table 2.1. Once again we make a distinction between the disparity ρ_C, and the disparity statistic $D_{\rho C} = 2n\rho_C$.

It is useful to standardize the disparity ρ_C without changing its value so that the leading term of the disparity statistic $2n\rho_C(\boldsymbol{p}, \boldsymbol{\pi}_0) = 2n\sum C(\delta_i)\pi_{0i}$, when expanded in a Taylor series in δ_i around zero (or equivalently in p_i around π_{0i}), equals $n\sum_{i=1}^{k}\delta_i^2 \pi_{0i}$, the Pearson's chi-square statistic. In this case, establishing the asymptotic chi-square limit of the disparity statistic reduces to proving that the remainder term in the above Taylor series expansion is an $o_p(1)$ term. We will, for the rest of the chapter, assume that the C functions of the disparities considered herein satisfy conditions (2.16) and (2.27) presented in Chapter 2, which, apart from guaranteeing the above, also ensure that the summand of the disparity is itself nonnegative. Also, unless otherwise mentioned, our discussion below will refer to the case where the number of cells k is a fixed constant.

8.2.2   The Simple Null Hypothesis

Assume the setup of the previous section. Given a disparity ρ_C, we are interested in the properties of the disparity statistic $D_{\rho C} = 2n\rho_C(\boldsymbol{p}, \boldsymbol{\pi}_0)$ as a test statistic for testing the simple null hypothesis

$$H_0 : \boldsymbol{\pi} = \boldsymbol{\pi}_0, \tag{8.7}$$

where $\pi_{0i} > 0$ for all i.

Theorem 8.1. [Basu and Sarkar (1994b, Theorem 3.1)]. *The disparity statistic D_{ρ_C} has an asymptotic χ^2_{k-1} distribution under the null hypothesis (8.7) as $n \to \infty$.*

Proof. By a first-order Taylor series expansion of the disparity statistic (as a function of p_i around π_{0i}) we get

$$
\begin{aligned}
2n\rho_C(\boldsymbol{p}, \boldsymbol{\pi}_0) &= 2n \sum_{i=1}^{k} C((p_i - \pi_{0i})/\pi_{0i})\pi_{0i} \\
&= 2n \sum_{i=1}^{k} C(0)\pi_{0i} + 2n \sum_{i=1}^{k} (p_i - \pi_{0i})C'(0) \\
&\quad + n \sum_{i=1}^{k} (p_i - \pi_{0i})^2 C''(0)\pi_{0i}^{-1} \\
&\quad + \frac{n}{3} \sum_{i=1}^{k} (p_i - \pi_{0i})^3 C'''(\pi_{0i}^{-1}\xi_i - 1)\pi_{0i}^{-2} \\
&= S_1 + S_2 + S_3 + S_4
\end{aligned}
$$

where ξ_i lies between p_i and π_{0i} and C', C'' and C''' represent the derivatives of C for the indicated order. The term S_1 vanishes since $C(0) = 0$ from the disparity conditions. Since p_i and π_{0i} both sum to 1, the term S_2 also vanishes whenever $C'(0)$ is finite (here $C'(0) = 0$). As $C''(0) = 1$, the term S_3 equals the Pearson's chi-square statistic

$$n \sum_{i=1}^{k} \frac{(p_i - \pi_{0i})^2}{\pi_{0i}}.$$

Thus, to complete the proof all one needs to show is that $S_4 \to 0$ in probability as $n \to \infty$. Note that

$$
\begin{aligned}
3S_4 &= n \sum_{i=1}^{k} (p_i - \pi_{0i})^3 C''' \left(\frac{\xi_i}{\pi_{0i}} - 1 \right) \pi_{0i}^{-2} \\
&\leq \left(n \sum_{i=1}^{k} (p_i - \pi_{0i})^2 \right) \times \left(\sup_i |p_i - \pi_{0i}| \right) \\
&\quad \times \left(\sup_i \pi_{0i}^{-2} \right) \times \left(\sup_i C''' \left(\frac{\xi_i}{\pi_{0i}} - 1 \right) \right).
\end{aligned}
$$

But $\sup_i [\pi_{0i}]^{-2}$ is bounded from the given conditions and $\sup_i |p_i - \pi_{0i}| =$

$o_p(1)$; also $\sum_{i=1}^{k} n(p_i - \pi_{0i})^2 = O_p(1)$. Again since $(\xi_i - \pi_{0i}) = o_p(1)$ for every i, $C'''(\pi_{0i}^{-1}\xi_i - 1) = O_p(1)$ by the continuity and finiteness assumption on $C'''(0)$. Therefore, $3S_4 = o_p(1)$, and the general disparity statistic $2n\rho_C(p, \pi_0)$ differs from the Pearson's chi-square statistic only by an $o_p(1)$ order term. Thus, the disparity statistic has the same asymptotic distribution as the Pearson's chi-square statistic, whose asymptotic χ_{k-1}^2 distribution under the null hypothesis is well known. □

8.2.3   The Composite Null Hypothesis

Now we consider the general composite null hypothesis following Bishop, Fienberg and Holland (1975, Chapter 14) and Read and Cressie (1988, Appendix A5). Define a parameter vector $\boldsymbol{\theta} = (\theta_1, \ldots, \theta_p) \in \mathbb{R}^p$, $p < k - 1$ and the mapping $f : \mathbb{R}^p \rightarrow \Delta_k$, where Δ_k is the set

$$\left\{ \boldsymbol{t} = (t_1, \ldots, t_k) : t_i \geq 0, \quad i = 1, \ldots, k, \quad \sum_{i=1}^{k} t_i = 1 \right\}$$

such that to each parameter vector $\boldsymbol{\theta}$ there corresponds a probability vector $\boldsymbol{\pi} = (\pi_1, \ldots, \pi_k)$ satisfying $f(\boldsymbol{\theta}) = \boldsymbol{\pi}$. The hypotheses

$$H_0 : \boldsymbol{\theta} \in Q_0 \text{ and } H_0 : \boldsymbol{\pi} \in M_0 \tag{8.8}$$

are then equivalent for $M_0 = f(Q_0)$.

The minimum distance estimator of $\boldsymbol{\theta} \in Q_0$ based on the disparity ρ_C is any $\hat{\boldsymbol{\theta}} \in \bar{Q}_0$ (the closure of Q_0) which satisfies

$$\rho_C(\boldsymbol{p}, f(\hat{\boldsymbol{\theta}})) = \inf_{\boldsymbol{\theta} \in Q_0} \rho_C(\boldsymbol{p}, f(\boldsymbol{\theta})).$$

Assume that the null hypothesis (8.8) is correct so that there exists a $\boldsymbol{\theta}^* \in Q_0$, with $\boldsymbol{\pi}^* = f(\boldsymbol{\theta}^*)$ where $\boldsymbol{\pi}^*$ is the true value of $\boldsymbol{\pi}$. We will assume the regularity conditions of Birch (1964) as given in Cressie and Read (1984).

1. There is an p-dimensional neighborhood of $\boldsymbol{\theta}^*$ completely contained in Q_0.

2. $f_i(\boldsymbol{\theta}^*) > 0$, $i = 1, 2, \ldots, k$.

3. f is totally differentiable at $\boldsymbol{\theta}^*$ so that the partial derivatives of f_i with respect to each θ_j exist at $\boldsymbol{\theta}^*$.

4. The Jacobian $\partial f(\boldsymbol{\theta}^*)/\partial\boldsymbol{\theta}$ is of full rank p.

5. The inverse mapping f^{-1} is continuous at $f(\boldsymbol{\theta}^*)$.

6. The mapping f is continuous at every point $\boldsymbol{\theta} \in Q_0$.

Under these assumptions nonuniqueness and unboundedness of the minimum distance estimator occurs with probability zero as n goes to infinity.

Definition 8.1. Any estimator $\hat{\boldsymbol{\theta}} \in Q_0$, satisfying the expansion

$$\hat{\boldsymbol{\theta}} = \boldsymbol{\theta}^* + (\boldsymbol{p} - \boldsymbol{\pi}^*)D_{\boldsymbol{\pi}^*}^{-1/2}A(AA)^{-1} + o_p(n^{-1/2})$$

is called a Best Asymptotically Normal (BAN) estimator of $\hat{\boldsymbol{\theta}}$ where $D_{\boldsymbol{\pi}^*}$ is a diagonal matrix with the i-th diagonal element π_i^* and the (i,j)-th element of A is given by $(\pi_i^{*-1/2}\partial f_i(\boldsymbol{\theta}^*)/\partial\theta_j)$.

The theorem below follows in a straightforward manner using the arguments of Birch (1964) and Read and Cressie (1988, Appendix A5).

Theorem 8.2. *Under the regularity conditions above any minimum distance estimator based on disparities is BAN.*

The next theorem which applies to any BAN estimator $\hat{\boldsymbol{\theta}}$ has been proved in Bishop, Fienberg and Holland (1975, pp. 517-518).

Theorem 8.3. *Let the above regularity conditions hold and suppose that $\boldsymbol{\pi}^* \in M_0$. Then, if $\hat{\boldsymbol{\theta}}$ is any BAN estimator of $\boldsymbol{\theta}$ and $\hat{\boldsymbol{\pi}} = f(\hat{\boldsymbol{\theta}})$,*

$$n^{1/2}[(\boldsymbol{p}, \hat{\boldsymbol{\pi}}) - (\boldsymbol{\pi}^*, \boldsymbol{\pi}^*)] \rightarrow N[0, \Sigma],$$

where

$$\Sigma = \left[\begin{array}{cc} D_{\boldsymbol{\pi}^*} - \boldsymbol{\pi}^{*'}\boldsymbol{\pi}^* & (D_{\boldsymbol{\pi}^*} - \boldsymbol{\pi}^{*'}\boldsymbol{\pi}^*)L \\ L'(D_{\boldsymbol{\pi}^*} - \boldsymbol{\pi}^{*'}\boldsymbol{\pi}^*) & L'(D_{\boldsymbol{\pi}^*} - \boldsymbol{\pi}^{*'}\boldsymbol{\pi}^*)L \end{array} \right]$$

and

$$L = D_{\boldsymbol{\pi}^*}^{-1/2}A(A'A)^{-1}A'D_{\boldsymbol{\pi}^*}^{1/2}.$$

Following the proof of Theorem 8.1, we have

$$2n\rho_C(\boldsymbol{p}, \hat{\boldsymbol{\pi}}) = 2n\text{PCS}(\boldsymbol{p}, \hat{\boldsymbol{\pi}}) + o_p(1).$$

Since $2n\text{PCS}(\boldsymbol{p}, \hat{\boldsymbol{\pi}})$ converges in distribution to a χ^2_{k-p-1} random variable (see, e.g., Bishop, Fienberg and Holland, 1975, Theorem 14.9-4) under (8.8), so does every other disparity statistic. Therefore, we have the following.

Theorem 8.4. [Basu and Sarkar (1994b, Theorem 4.3)]. *Suppose the regularity conditions above hold and $\hat{\boldsymbol{\theta}}$ is any BAN estimator of $\boldsymbol{\theta}$ and $\hat{\boldsymbol{\pi}} = f(\hat{\boldsymbol{\theta}})$. Then under (8.8), $2n\rho_C(\boldsymbol{p}, \hat{\boldsymbol{\pi}})$ converges in distribution to a χ^2_{k-p-1} random variable as $n \rightarrow \infty$.*

8.2.4 Minimum Distance Inference versus Multinomial Goodness-of-Fit

In this book we have described the use of the "disparity" measures for two different purposes. Much of the book is devoted to the description of minimum distance estimation based on disparities. In this chapter our focus is on the use of disparities for the purpose of multinomial goodness-of-fit. We wish to point out here that these two issues are inherently different. In minimum distance estimation our aim is to downweight observations with large, positive, Pearson residuals (outliers) to maintain the stability of the estimators; simultaneously we wish to downweight the inliers so that the procedure has good small sample efficiency. In either case, our intention is to downplay the discrepancy in the observed pattern of the data and the pattern predicted by the model. As we have discussed in Chapter 6, there is often a need to fix up robust disparities to make them less sensitive on the inlier side. In goodness-of-fit testing, however, the discrepancy between the pattern of the observed data and the pattern predicted by the model are the items of interest to us. One would expect better power for disparities for which such discrepancies are highlighted more. Thus, while minimum distance estimation aims to downplay small differences for better results, the goodness-of-fit testing procedure requires that such differences be highlighted to achieve better results.

A further qualification to the above discussion is necessary. On the issue of downweighting small differences we have already observed that this is done differently by different disparities. Robust disparities within the power divergence family of Cressie and Read downweight outliers, but blow up the effect of inliers. Disparities like the Pearson's chi-square or the likelihood disparity control inliers appropriately but fail to contain the effect of outliers. In Chapter 6 we have discussed in detail the construction of new disparities with the aim of achieving the desired insensitivity over the entire range of the residual. Such constructions are relevant in the case of goodness-of-fit testing also, since the natural distances are rarely sensitive to both kinds of deviations. Within the power divergence family, for example, disparity goodness-of-fit tests corresponding to large positive values of λ (such as the Pearson's chi-square statistic) show superior performance in detecting outlier type discrepancies (or bump alternative type discrepancies, see Section 8.3) but largely fail to detect inlier type discrepances (or dip alternative type discrepancies). Disparities corresponding to large negative values of λ exhibit the opposite kind of behavior.

We will discuss disparity goodness-of-fit testing based on inlier modified disparities in Section 8.6.

8.3   Exact Power Comparisons in Small Samples

In statistical inference it is often the case that the small samples properties of estimators and tests cannot be determined exactly except under some very special conditions. What we do in such situations is determine the asymptotic properties of the statistical procedure in question, and hope that they provide reasonable approximations for the actual properties of the procedure in finite samples. Depending on the nature of the model and the scope of the procedure, the asymptotic results do often provide reasonable reflections of the performance of the methods in small samples, but there are situations where they poorly approximate the actual small sample scenario. In the multinomial goodness-of-fit testing situation, several such asymptotic performance indicators have been discussed in Read and Cressie (1988, Sections 4.2 and 4.3). However, the authors also point out that the approach for the extension of these conclusions to small samples is "complicated," and in some cases the conditions under which the asymptotic results are appropriate are "very restrictive in small samples" (Read and Cressie, 1988, Section 5.4).

To get a better idea of the small sample behavior of the class of disparity tests and its different subfamilies, we look at the actual exact distributions of the test statisics and their exact powers in the spirit of the example in Section 6.2.5. For illustration we choose the equiprobable null model, and consider such alternatives where one of the cell probabilities has been perturbed. This model has been utilized by Read and Cressie (1988) for the illustration of several aspects of goodness-of-fit statistics throughout their book. In a k cell multinomial, the equiprobable null $H_0 : \pi_i = 1/k$, $i = 1, 2, \ldots, k$, specifies equal cell probabilities over the k cells; the alternative hypothesis that we will consider will have the form

$$H_1 : \pi_i = \begin{cases} \{1 - \gamma/(k-1)\}/k & i = 1, \ldots, k-1 \\ \dfrac{1+\gamma}{k} & i = k \end{cases}, \qquad (8.9)$$

where $-1 \leq \gamma \leq k - 1$ is a fixed constant. Notice that the last cell is primarily responsible for the violation of the equiprobable hypothesis. A positive value of γ generates a "bump" alternative and ensures that the final cell has a larger probability, whereas the first $k - 1$ cells are equiprobable among themselves but less probable than the last one. Conversely, a negative value of γ generates the "dip" alternative. The null model corresponds to $\gamma = 0$.

Choosing such values of k and n so that the enumeration of all the samples is feasible, we focus on randomized tests of exact size 0.05. Apart from the issue of getting a feeling of the actual performance of the disparity tests in small samples, it allows us to treat the disparity tests uniformly; because of the discrete nature of the tests the actual attainable levels for the nonrandomized tests vary over the disparities, and investigating their relative performance over a common level is only possible through randomizing.

TABLE 8.1
Exact powers of the disparity tests for the equiprobable null hypothesis based on the power divergence family at several alternatives for a randomized test of size 0.05. Most of the numbers presented here have been reported previously by Cressie and Read (1984, Table 2). Here $n = 20$ and $k = 4$.

λ	γ		
	1.5	0.5	−0.9
−2.0	0.6500	0.1231	0.7434
−1.5	0.7017	0.1253	0.7428
−1.0	0.7960	0.1384	0.7342
−0.8	0.7960	0.1384	0.7342
−0.5	0.8009	0.1412	0.7263
0.0	0.8640	0.1567	0.7045
0.2	0.8640	0.1567	0.7045
0.5	0.8640	0.1567	0.7045
2/3	0.8640	0.1567	0.7045
0.7	0.8640	0.1567	0.7045
1.0	0.8745	0.1629	0.5150
1.5	0.8855	0.1682	0.3844
2.0	0.8962	0.1725	0.3291

The randomized test of size a for a disparity test given by the statistic $D_C = D_{\rho C}$ can be described as follows. Let c represent the constant such that

$$P(D_C > c|H_0) = a_1 < a \quad \text{and} \quad P(D_C \geq c|H_0) = a_2 > a.$$

Then the randomized, level a critical function $\phi(D_C)$ has the form

$$\phi(D_C) = \begin{cases} 1 & \text{for } D_C > c \\ \dfrac{a - a_1}{a_2 - a_1} & \text{for } D_C = c \\ 0 & \text{for } D_C < c, \end{cases}$$

and the exact small sample power of the randomized disparity statistic D_C at the alternative H_1 will be given by

$$b_1 + \frac{a - a_1}{a_2 - a_1}(b_2 - b_1),$$

where

$$P(D_C > c|H_1) = b_1 \quad \text{and} \quad P(D_C \geq c|H_1) = b_2.$$

Here we present the exact powers of the disparity tests for the equiprobable null hypothesis for a $k = 4$ cell multinomial with $n = 20$ for several members

TABLE 8.2
Exact powers of the disparity tests for the equiprobable null hypothesis based on the blended weight Hellinger distance family at several alternatives for a randomized test of size 0.05. Some of the numbers presented here have been reported previously by Basu and Sarkar (1994b, Table 1). Here $n = 20$ and $k = 4$.

α	γ		
	1.5	.5	$-.9$
0	.8745	.1629	.5150
.05	.8647	.1577	.6363
.10	.8640	.1567	.7045
.15	.8640	.1567	.7045
.20	.8640	.1567	.7045
.25	.8640	.1567	.7045
.30	.8640	.1567	.7045
1/3	.8640	.1567	.7045
.40	.8525	.1538	.7108
.50	.8009	.1412	.7263
.60	.7960	.1384	.7342
.70	.7353	.1291	.7410
.80	.7017	.1253	.7428
.90	.6500	.1231	.7434
1.0	.6500	.1231	.7434

of the power divergence, blended weight Hellinger distance and the blended weight chi-square families. Following Cressie and Read (1984) and Read and Cressie (1988) we choose values of $\gamma = 1.5, 0.5$ and -0.9, representing two different bump alternatives and a dip alternative. The results for the three families are presented in Tables 8.1, 8.2 and 8.3 respectively. Most of the entries of Table 8.1 are as reported in Cressie and Read (1984); parts of Tables 8.2 and 8.3 are available in Basu and Sarkar (1994b).

The results of the tables are instructive and require some discussion. For the power divergence family, the power of the disparity tests is an increasing function of λ for bump alternatives (i.e., for $\gamma > 0$). On the other hand, the power of these disparity tests is a decreasing function of λ for the dip alternative (i.e., for $\gamma < 0$). See Wakimoto, Odaka and Kang (1987) for similar observations in relation to the alternatives in Equation (8.9) for some specific disparity tests within the power divergence family. This phenomenon shows that although there has been a fair amount of research on multinomial goodness-of-fit testing and specific criteria have declared certain members to be optimal in their specific senses, one can always choose an alternative to the equiprobable simple null so that the selected "optimal" test will fare poorly at that alternative in relation to many other disparity tests in small samples. We

TABLE 8.3

Exact powers of the disparity tests for the equiprobable null hypothesis based on the blended weight chi-square family at several alternatives for a randomized test of size 0.05. Some of the numbers presented here have been reported previously by Basu and Sarkar (1994b, Table 2). Here $n = 20$ and $k = 4$.

α	γ		
	1.5	.5	$-.9$
0	.8745	.1629	.5150
.05	.8651	.1584	.5767
.10	.8647	.1577	.6363
.15	.8640	.1567	.7045
.20	.8640	.1567	.7045
.25	.8640	.1567	.7045
.30	.8640	.1567	.7045
1/3	.8640	.1567	.7045
.40	.8526	.1540	.7050
.50	.8009	.1412	.7263
.60	.7973	.1392	.7324
.70	.7960	.1384	.7342
.80	.7353	.1291	.7410
.90	.7017	.1253	.7428
1.0	.6500	.1231	.7434

will use this observation as the basis of our effort in Section 8.6 where we try to develop modified tests which have better overall power than the ordinary disparity tests.

We now consider the exact power of the disparity tests based on the blended weight Hellinger distance family (Table 8.2). From the equivalence of the power divergence family and the blended weight Hellinger distance family described in Section 6.2.5 it is seen that the relationship between the tuning parameters run in reverse – i.e., large values of λ for the power divergence family correspond to small values of α for the blended weight Hellinger distance. Notice also that the second term in the Taylor series (6.24) is increasing with λ, while the second term in the Taylor series (6.25) is a decreasing function of α. This indicates that the power of the disparity tests for bump alternatives based on the blended weight Hellinger distance family in case of the simple equiprobable null will be increasing for decreasing α, and will show the opposite trend for dip alternatives. This is indeed borne out by the exact calculations of Table 8.2.

A quick comparison of the forms of the blended weight Hellinger distance family and the blended weight chi-square family reveals that the members of the two families have the same "trend" as a function of α, and the power of the disparity tests for the two different families should proceed in the same direction as functions of α. This is also suggested by the Taylor series expansion of the blended weight chi-square disparity statistics in the spirit of Equations (6.24) and (6.25). This expansion is given by

$$2n\mathrm{BWCS}_\alpha(\boldsymbol{p}, \boldsymbol{\pi}_0) = \sum_{i=1}^{k} \frac{W_i^2}{\pi_{0i}} - \frac{\alpha}{n^{1/2}} \sum_{i=1}^{k} \frac{W_i^3}{\pi_{0i}^2} + \frac{\alpha^2}{n} \sum_{i=1}^{k} \frac{W_i^4}{\pi_{0i}^3} + O_p(n^{-3/2}). \quad (8.10)$$

Note that the expansions (6.25) and (8.10) are identical up to the cubic term W_i^3.

Both the BWHD and BWCS families generate the Pearson's chi-square for $\alpha = 0$ and the Neyman's chi-square for $\alpha = 1$. But none of the two subfamilies actually contains the log likelihood ratio chi-square. However, within each of these subfamilies there is a member (which corresponds to $\alpha = 1/3$ in either case) that is very close to the log likelihood ratio chi-square. In a Taylor series expansion the $2n\mathrm{BWHD}_{1/3}(\boldsymbol{p}, \boldsymbol{\pi}_0)$ statistic matches the log likelihood ratio test statistic $2n\mathrm{LD}(\boldsymbol{p}, \boldsymbol{\pi}_0)$ up to the first five terms, while the $2n\mathrm{BWCS}_{1/3}(\boldsymbol{p}, \boldsymbol{\pi}_0)$ statistic matches the latter up to the first four terms (counting the terms that vanish).

While most of our illustrations in this chapter involve the equiprobable null hypothesis against the general unrestricted alternative, sometimes alternatives which satisfy a suitable order restriction may be of interest. Appropriate tests for such situations which have chi-bar square type null distributions have been described by, among others, Lee (1987) and Bhattacharya and Basu (2003).

8.4 Choosing a Disparity to Minimize the Correction Terms

The detailed derivation of the first three moments of the power divergence test statistic $2n\mathrm{PD}_\lambda(\boldsymbol{p}, \boldsymbol{\pi}_0)$ under the simple null in (8.7) has been presented in Read and Cressie (1988, Appendix A11, pp. 175–181) up to $O(n^{-1})$ terms for $\lambda > -1$. The moments are based on the expansion (6.24), and are derived as functions of the number of cells k and $t = \sum_{i=1}^{k} \pi_{0i}^{-1}$. The moments are useful, among other things, in assessing the speed of convergence of the distribution of the test statistic to its asymptotic chi-square limit through the magnitude of the correction terms for the moments. For each of the three moments, the first-order term is the moment of the limiting chi-square random variable with $(k-1)$ degrees of freedom. The second-order term – or the correction term – is of order $O(n^{-1})$; value(s) of λ for which the correction terms vanish may be considered preferable in terms of the accuracy of the chi-square approximation. In particular, the closeness of the exact mean and variance of the test statistic to the asymptotic mean and variance will influence the closeness of the exact

significance level to the asymptotic significance level. See the discussion by
Read and Cressie (1988, Section 5.1).

Here we present the first two moments of the power divergence statistic
$2n\text{PD}_\lambda(\boldsymbol{p}, \boldsymbol{\pi}_0)$ for $\lambda > -1$ under the simple null. These moments are

$$
\begin{aligned}
E[2n\text{PD}_\lambda] &= (k-1) + n^{-1}[(\lambda-1)(2-3k+t)/3 \\
&\quad + (\lambda-1)(\lambda-2)(1-2k+t)/4] + o(n^{-1}) \\
&= (k-1) + n^{-1}a_\lambda + o(n^{-1})
\end{aligned}
$$

and

$$
\begin{aligned}
\text{Var}[2n\text{PD}_\lambda] &= (2k-2) + n^{-1}[(2-2k-k^2+t) \\
&\quad +(\lambda-1)(8-12k-2k^2+6t) \\
&\quad +(\lambda-1)^2(4-6k-3k^2+5t)/3 \\
&\quad +(\lambda-1)(\lambda-2)(2-4k+2t)] + o(n^{-1}) \\
&= (2k-2) + n^{-1}b_\lambda + o(n^{-1}),
\end{aligned}
$$

where

$$
a_\lambda = (\lambda-1)(2-3k+t)/3 + (\lambda-1)(\lambda-2)(1-2k+t)/4, \tag{8.11}
$$
$$
\begin{aligned}
b_\lambda &= (2-2k-k^2+t) + (\lambda-1)(8-12k-2k^2+6t) \\
&\quad + (\lambda-1)^2(4-6k-3k^2+5t)/3 + (\lambda-1)(\lambda-2)(2-4k+2t) \tag{8.12}
\end{aligned}
$$

and $t = \sum_{i=1}^{k} \pi_{0i}^{-1}$. The above moment calculations follow easily using relation (6.24) and the expressions for the quantities $E[W_i^2]$, $E[W_i^3]$ and $E[W_i^4]$ which have been derived by Read and Cressie (1988, Appendix A11).

Here our aim is to choose the tuning parameter λ so that the terms a_λ and b_λ are as close to zero as possible. Consider the special case of the equiprobable null hypothesis; in this case one gets $t = \sum_{i=1}^{k} \pi_{0i}^{-1} = k^2$. For $k \geq 2$, the quadratic equation $a_\lambda = 0$ has one solution at $\lambda = 1$, while the second solution converges to $\lambda = 2/3$ as $k \to \infty$.

The value k^2 is the minimum possible value of t under the constraint $\pi_{0i} > 0$ for each i and $\sum_{i=1}^{k} \pi_{0i} = 1$. We get $t \geq k^2$ for a completely specified arbitrary hypothesis and it may be of interest to investigate how large values of t affect the solutions of the equations $a_\lambda = 0$ and $b_\lambda = 0$. Suppose that the values of t are of order larger than k^2. The resulting equations obtained by treating the quadratics in k to be negligible compared to t yield solutions

$$
\lambda = 1 \quad \text{and} \quad \lambda = 2/3
$$

for the equation $a_\lambda = 0$, and, approximately,

$$
\lambda = 0.3 \quad \text{and} \quad \lambda = 0.61
$$

for the equation $b_\lambda = 0$. On the basis of their overall analysis, Cressie and Read

(1984) and Read and Cressie (1988) propose, together with the traditional statistic at $\lambda = 1$, the statistic at $\lambda = 2/3$ as one of the outstanding candidates for testing goodness-of-fit within their family of power divergence statistics.

We now turn our attention to the blended weight Hellinger distance family over the range $0 \leq \alpha < 1$. We have already pointed out the equivalence of this family with the power divergence family through Equations (6.24) and (6.25), under the null, up to $O_p(n^{-1})$ terms. Thus, for the BWHD family the preferred statistics can be obtained by simply using the equivalence rule $\alpha = (1 - \lambda)/3$ between the two families. The relevant moment calculations now give

$$E[2n\mathrm{BWHD}_\alpha] = (k - 1) + n^{-1}a_\alpha + o(n^{-1})$$

and

$$\mathrm{Var}[2n\mathrm{BWHD}_\alpha] = (2k - 2) + n^{-1}b_\alpha + o(n^{-1}),$$

where

$$a_\alpha = -\alpha(2 - 3k + t) + (3/4)(3\alpha^2 + \alpha)(1 - 2k + t), \qquad (8.13)$$

and

$$b_\alpha = (2 - 2k - k^2 + t) + \alpha^2(30 - 54k - 9k^2 + 33t)$$
$$+ \alpha(-18 + 24k + 6k^2 - 12t). \qquad (8.14)$$

Under large values of t the mean equation $a_\alpha = 0$ leads to two solutions $\alpha = 0$ and $\alpha = 1/9$, while the variance equation $b_\alpha = 0$ produces, approximately, the solutions $\alpha = 7/30$ and $\alpha = 0.13$. On the whole, therefore, the statistic corresponding to $\alpha = 1/9$ also emerges as an outstanding candidate for performing tests of goodness-of-fit, together with the traditional Pearson's chi-square statistic ($\alpha = 0$).

For the BWCS family, similar calculations under the null yield, for $0 \leq \alpha < 1$,

$$E[2n\mathrm{BWCS}_\alpha] = (k - 1) + n^{-1}a_\alpha + o(n^{-1})$$

and

$$\mathrm{Var}[2n\mathrm{BWCS}_\alpha] = (2k - 2) + n^{-1}b_\alpha + o(n^{-1}),$$

where

$$a_\alpha = -\alpha(2 - 3k + t) + \alpha^2(3 - 6k + 3t) \qquad (8.15)$$

and

$$b_\alpha = (2 - 2k - k^2 + t) + \alpha^2(36 - 66k - 9k^2 + 39t)$$
$$+ \alpha(-24 + 36k + 6k^2 - 18t). \qquad (8.16)$$

For large values of t the mean equation $a_\alpha = 0$ yields $\alpha = 0$ and $\alpha = 1/3$ as solutions. On the other hand, for the variance equation $b_\alpha = 0$, the solutions are, approximately, $\alpha = 0.06$ and $\alpha = 0.4$. On the whole it appears that one can recommend the BWCS statistic with $\alpha = 1/3$ as another major competitor for the list of desirable chi-square goodness-of-fit tests, together with the Pearson's chi-square test ($\alpha = 0$).

8.5   Small Sample Comparisons of the Test Statistics

All our test statistics have appropriate chi-square null distributions which are valid for large sample sizes. However, depending on the particular test being used, the chi-square approximation may be a poor one in small samples. Different authors have suggested modifications to these techniques for enhancing the performance of the methods in small samples.

We will begin with the description of the modifications suggested by Read (1984b) for the power divergence family. We will follow this up with extensions to some other families of divergences.

Let $F_E(y)$ be the exact distribution function of a disparity statistic $2n\mathrm{PD}_\lambda(p, \pi_0)$ within the power divergence family (we keep the dependence on λ implicit). It follows from the results of Cressie and Read (1984) as well as our general results that

$$F_E(y) = F_{\chi^2_{k-1}}(y) + o(1) \tag{8.17}$$

as $n \to \infty$, where $F_{\chi^2_\nu}(y)$ is the distribution function of the chi-square distribution with ν degrees of freedom evaluated at y. While $F_{\chi^2_{k-1}}(y)$ is the default approximation used for the distribution of the disparity statistic for most practitioners, here we present some other methods for the power divergence, the BWHD and the BWCS families which can provide better approximations to the exact small sample distributions and critical values of the test statistics.

8.5.1   The Power Divergence Family

Here we consider the approximations of the power divergence statistic proposed by Read (1984b). The first approximation $F_C(\cdot)$ provides a correction so that the mean and the variance of the statistic agrees with the asymptotic mean and variances up to the second order (rather than the first order). For the power divergence statistic this approximation to $F_E(y)$ is given by

$$F_C(y) = F_{\chi^2_{k-1}}(d_\lambda^{-1/2}(y - c_\lambda)),$$

where $c_\lambda = (k-1)(1 - d_\lambda^{1/2}) + n^{-1}a_\lambda$ and $d_\lambda = 1 + [n(2k-2)]^{-1}b_\lambda$. Essentially this approximation considers a moment corrected statistic

$$d_\lambda^{-1/2}(2n\mathrm{PD}_\lambda(p, \pi_0) - c_\lambda),$$

where the moment corrected statistic has mean and variance (for $\lambda > -1$) matching the chi-squared mean $(k-1)$ and variance $2(k-1)$ respectively up to $O(n^{-1})$ terms. Here a_λ and b_λ are as in Equations (8.11) and (8.12). Note that the corrected statistics are well defined even when the moments of the ordinary statistics do not exist.

A second approximation $F_S(y)$ is derived by extracting the λ-dependent second-order component from the small order term on the right-hand side of Equation (8.17). The method draws from the techniques of Yarnold (1972). The approximation may be presented as

$$
\begin{aligned}
F_S(y) = & F_{\chi^2_{k-1}}(y) + (24n)^{-1}\Big[2(1-t)F_{\chi^2_{k-1}}(y) + \{3(3t - k^2 - 2k) \\
& + (\lambda - 1)6(t - k^2) + (\lambda - 1)^2(5t - 3k^2 - 6k + 4) \\
& - (\lambda - 1)(\lambda - 2)3(t - 2k + 1)\}F_{\chi^2_{k+1}}(y) \\
& + \{-6(2t - k^2 - 2k + 1) - (\lambda - 1)4(4t - 3k^2 - 3k + 2) \\
& - (\lambda - 1)^2 2(5t - 3k^2 - 6k + 4) + (\lambda - 1)(\lambda - 2)3(t - 2k + 1)\}F_{\chi^2_{k+3}}(y) \\
& + \lambda^2(5t - 3k^2 - 6k + 4)F_{\chi^2_{k+5}}(y)\Big] \\
& + (N_\lambda(y) - n^{(k-1)/2}V_\lambda(y))e^{-y/2}/[(2\pi n)^{k-1}Q]^{1/2}.
\end{aligned}
$$

Here $N_\lambda(y)$ is the number of multinomial vectors $(x_1, \ldots, x_k)$ such that $2n\mathrm{PD}_\lambda(\boldsymbol{p}, \boldsymbol{\pi}_0) < y$, where $\boldsymbol{p} = (x_1/n, \ldots, x_k/n)$, $Q = \prod_{i=1}^k \pi_{0i}$, and

$$
\begin{aligned}
V_\lambda(y) = & \frac{(\pi y)^{(k-1)/2}}{\Gamma((k+1)/2)}Q^{1/2}\Big[1 + \frac{y}{24(k+1)n}\{(\lambda - 1)^2(5t - 3k^2 - 6k + 4) \\
& - 3(\lambda - 1)(\lambda - 2)(t - 2k + 1)\}\Big].
\end{aligned}
$$

This approximation has been discussed in Read (1984a).

The final approximation of the exact null distribution of the disparity statistic corresponds to the situation where the number of cells k increases with the sample size n. We assume that $n/k \to a$ for a fixed a $(0 < a < \infty)$. In this case the asymptotic result (8.17) no longer holds. In the case of Pearson's chi-square and the log likelihood ratio statistics Koehler and Larntz (1980) examined the applicability of the normal approximation for moderate sample sizes with moderately many cells. Using the results of Holst (1972) with

$$
f_i(x) = \begin{cases} \dfrac{2}{\lambda(\lambda + 1)}\dfrac{n}{k}\left[\left(\dfrac{kx}{n}\right)^{\lambda+1} - 1\right] & \lambda > -1, \neq 0 \\[3mm] 2x\log\left(\dfrac{kx}{n}\right) & \lambda = 0 \end{cases}
$$

and Theorem 2.4 of Cressie and Read (1984) it can be shown, under the equiprobable null hypothesis,

$$
F_E(y) = F_N(y) + o(1), \tag{8.18}
$$

where

$$F_N(y) = P\{N(0,1) < \sigma_n^{-1}(y - \mu_n)\}, \tag{8.19}$$

$N(0,1)$ represents a standard normal random variable, and

$$\mu_n = \begin{cases} 2n(\lambda(\lambda+1))^{-1}E\{(Y/a)^{\lambda+1} - 1\} & \lambda > -1, \neq 0 \\ 2nE\{(Y/a)\log(Y/a)\} & \lambda = 0, \end{cases}$$

$$\sigma_n^2 = \begin{cases} \{2a/(\lambda(\lambda+1))\}^2 k[\text{Var}\{(Y/a)^{\lambda+1}\} \\ \quad -a\text{Cov}^2\{Y/a, (Y/a)^{\lambda+1}\}] & \lambda > -1, \neq 0 \\ (2a)^2 k[\text{Var}\{(Y/a)\log(Y/a)\} \\ \quad -a\text{Cov}^2\{Y/a, (Y/a)\log(Y/a)\}] & \lambda = 0, \end{cases}$$

and Y is a Poisson (a) random variable.

The performance of the four approximations $F_{\chi_{k-1}^2}$ with respect to the power divergence statistics have been discussed extensively by Read (1984b). Similar approximations for the BWHD and BWCS families have been carried out by Shin, Basu and Sarkar (1995, 1996), which we present here.

8.5.2   The BWHD Family

For the BWHD family, the F_C and F_S corrections relate to the corresponding quantities for the power divergence family under the equivalence $\alpha = (1-\lambda)/3$. In this case we have

$$F_C(y) = F_{\chi_{k-1}^2}(d_\alpha^{-1/2}[y - c_\alpha]),$$

where $c_\alpha = (k-1)(1 - d_\alpha^{1/2}) + n^{-1}a_\alpha$ and $d_\alpha = 1 + [n(2k-2)]^{-1}b_\alpha$, where a_α and b_α are as in Equations (8.13) and (8.14) respectively.

A general derivation of the second approximation $F_S(\cdot)$ is presented later in Section 8.5.4 in terms of the derivatives of the disparity generating function C. The power divergence statistics results can also be obtained as a special case of this. Here we present the corresponding result for the BWHD family.

$$\begin{aligned}
F_S(y) = {} & F_{\chi_{k-1}^2}(y) + (24n)^{-1}\Big[2(1-t)F_{\chi_{k-1}^2}(y) + \{3(3t - k^2 - 2k) \\
& + \alpha(-27t + 18k^2 + 18k - 9) + \alpha^2(18t - 27k^2 + 9)\}F_{\chi_{k+1}^2}(y) \\
& + \{-6(2t - k^2 - 2k + 1) + \alpha(57t - 36k^2 - 54k + 33) \\
& + \alpha^2(-63t + 54k^2 + 54k - 45)\}F_{\chi_{k+3}^2}(y) \\
& + (-6\alpha + 9\alpha^2 + 1)(5t - 3k^2 - 6k + 4)F_{\chi_{k+5}^2}(y)\Big] \\
& + (N_\alpha(y) - n^{(k-1)/2}V_\alpha(y))e^{-y/2}/[(2\pi n)^{k-1}Q]^{1/2}.
\end{aligned}$$

Here $N_\alpha(y)$ is the number of multinomial vectors $(x_1, \ldots, x_k)$ such that

$2n\mathrm{BWHD}_\alpha(\boldsymbol{p}, \boldsymbol{\pi}_0) < y$, where $\boldsymbol{p} = (x_1/n, \ldots, x_k/n)$, $Q = \prod_{i=1}^k \pi_{0i}$, and

$$V_\alpha(y) = \frac{(\pi y)^{(k-1)/2}}{\Gamma\{(k+1)/2\}} Q^{1/2} \left\{1 + \frac{y[\alpha(-9t + 18k - 9) + \alpha^2(18t - 27k^2 + 9)]}{24n(k+1)}\right\}.$$

The above forms can be derived directly from the general expressions (8.22), (8.23) and (8.24) in Section 8.5.4.

The approximation $F_N(\cdot)$ works under the same setup as that of the power divergence family as in Equations (8.18) and (8.19). In this case the relevant quantities are,

$$f_i(x) = \left\{\frac{[(kx/n) - 1]}{\alpha(kx/n)^{1/2} + (1 - \alpha)}\right\}^2,$$

$$\mu_n = nE\left\{\frac{[(Y/a) - 1]}{\alpha(Y/a)^{1/2} + (1 - \alpha)}\right\}^2$$

and

$$\sigma_n^2 = a^2 k \left\{\mathrm{Var}\left(\left[\frac{(Y/a) - 1}{\alpha(Y/a)^{1/2} + (1 - \alpha)}\right]^2\right)\right.$$
$$\left. - a\mathrm{Cov}^2\left(Y/a, \left[\frac{(Y/a) - 1}{\alpha(Y/a)^{1/2} + (1 - \alpha)}\right]^2\right)\right\}.$$

8.5.3 The BWCS Family

For the BWCS family, the F_C correction is given by

$$F_C(y) = F_{\chi_{k-1}^2}(d_\alpha^{-1/2}[y - c_\alpha])$$

where $c_\alpha = (k - 1)(1 - d_\alpha^{1/2}) + n^{-1}a_\alpha$ and $d_\alpha = 1 + [n(2k - 2)]^{-1}b_\alpha$, and a_α and b_α are as in Equations (8.15) and (8.16) respectively.

The second approximation $F_S(\cdot)$ can be easily derived from the general form in Section 8.5.4. For the BWCS family this is given by

$$F_S(y) = F_{\chi_{k-1}^2}(y) + (24n)^{-1}\left[2(1 - t)F_{\chi_{k-1}^2}(y) + \{3(3t - k^2 - 2k)\right.$$
$$- 18\alpha(t - k^2) + 9\alpha^2(t - 3k^2 + 2k)\}F_{\chi_{k+1}^2}(y)$$
$$+ \{-6(2t - k^2 - 2k + 1) + 12\alpha(4t - 3k^2 - 3k + 2)$$
$$+ 18\alpha^2(-3t + 3k^2 + 2k - 2)\}F_{\chi_{k+3}^2}(y)$$
$$\left. + (9\alpha^2 - 6\alpha + 1)(5t - 3k^2 - 6k + 4)F_{\chi_{k+5}^2}(y)\right]$$
$$+ (N_\alpha(y) - n^{(k-1)/2}V_\alpha(y))e^{-y/2}/[(2\pi n)^{k-1}Q]^{1/2}.$$

Here $N_\alpha(y)$ is the number of multinomial vectors $(x_1, \ldots, x_k)$ such that

$2n\text{BWCS}_\alpha(\boldsymbol{p}, \boldsymbol{\pi}_0) < y$, where $\boldsymbol{p} = (x_1/n, \dots, x_k/n)$, $Q = \prod_{i=1}^{k} \pi_{0i}$, and

$$V_\alpha(y) = \frac{(\pi y)^{(k-1)/2}}{\Gamma\{(k+1)/2\}} Q^{1/2} \left\{ 1 + \frac{y[9\alpha^2(t - 3k^2 + 2k)]}{24n(k+1)} \right\}.$$

The approximation $F_N(\cdot)$ works under the same setup as that of the power divergence family as in Equations (8.18) and (8.19). In this case the relevant quantities are,

$$f_i(x) = \frac{[(kx/n) - 1]^2}{\alpha(kx/n) + (1 - \alpha)},$$

$$\mu_n = nE\left\{ \frac{[(Y/a) - 1]^2}{\alpha(Y/a) + (1 - \alpha)} \right\}$$

and

$$\sigma_n^2 = a^2 k \left\{ \text{Var}\left(\frac{[(Y/a) - 1]^2}{\alpha(Y/a) + (1 - \alpha)} \right) - a\text{Cov}^2\left(Y/a, \frac{[(Y/a) - 1]^2}{\alpha(Y/a) + (1 - \alpha)} \right) \right\}.$$

8.5.4   Derivation of $F_S(y)$ for a General Disparity Statistic

In this section we provide a derivation of the approximation $F_S(y)$ for a general disparity statistic based only the disparity generating function C. This description follows Shin, Basu and Sarkar (1995).

In order to expand $2n\rho_C(\boldsymbol{p}, \boldsymbol{\pi}_0)$ in a fourth order Taylor series, we assume that the fourth derivative of the disparity generating function C exists. Let $W_j = n^{1/2}(p_j - \pi_{0j})$ for $j = 1, 2, \dots, k$, and let $r = k - 1$. Consider the lattice

$$L = \{\boldsymbol{w} = (w_1, \dots, w_r) : \boldsymbol{w} = n^{1/2}(n^{-1}\boldsymbol{m} - \tilde{\boldsymbol{\pi}}_0) \text{ and } \boldsymbol{m} \in M\},$$

where $\tilde{\boldsymbol{\pi}}_0 = (\pi_{01}, \dots \pi_{0r})$ and $M = \{\boldsymbol{m} = (m_1, \dots, m_r) : m_j, j = 1, \dots, r$ are nonnegative integers satisfying $\sum_{j=1}^{r} m_j \leq n\}$. The normalized vector $\boldsymbol{W} = (W_1, \dots, W_r)$ takes values in the above lattice. Using a general asymptotic probability result for lattice random variables of Yarnold (1972), Read (1984a) derived the asymptotic expansion of the limiting distribution of the $2n\text{PD}_\lambda$ statistic under the null hypothesis (8.7). Read's result for the PD statistics contains that of Yarnold (1972) for the Pearson's chi-square and that of Siotani and Fujikoshi (1980) for the log likelihood ratio chi-square as special cases. We generalize Read's result to the entire class of disparity tests and exploit the technique of Yarnold (1972, Theorem 2), which gives a useful expression for the probability of lattice random variables belonging to an extended convex set B. See Read (1984a, Definition 2.1) for the definition of an extended convex set. Let

$$B_C(y) = \{\boldsymbol{w} = (w_1, \dots, w_r) : 2n\rho_C(n^{-1}(\boldsymbol{m}, m_k), \boldsymbol{\pi}_0) < y\}, \tag{8.20}$$

where $w_k = -\sum_{j=1}^{r} w_j$, $\boldsymbol{m} = n^{1/2}\boldsymbol{w} + n\tilde{\boldsymbol{\pi}}_0$, and $m_k = n^{1/2}w_k + n\pi_{0k}$.

Expanding $2n\rho_C(\boldsymbol{p}, \boldsymbol{\pi}_0)$ (as a function of p_i around π_{0i}) in a fourth order Taylor series, the asymptotic expansion for the distribution function $F_E(y)$ of the disparity test $2n\rho_C(\boldsymbol{p}, \boldsymbol{\pi}_0)$ is given by

$$F_E(y) = J_1^C(y) + J_2^C(y) + J_3^C(y) + O(n^{-3/2}), \tag{8.21}$$

where J_1^C, J_2^C and J_3^C are as defined by J_1, J_2 and J_3 respectively in Theorem 2.1 of Read (1984a) and $B = B_C(y)$ is as given in (8.20). The term J_1^C has the form

$$
\begin{aligned}
J_1^C(y) &= F_{\chi_{k-1}^2}(y) + \frac{1}{24n}\Big\{2(1-t)F_{\chi_{k-1}^2}(y)\\
&\quad + [3g_1 + 6g_2 + 6d_3g_1 + d_3^2g_4 - 3d_4g_3]F_{\chi_{k+1}^2}(y)\\
&\quad + [-6g_1 - 6g_3 - 6d_3g_1 - 2d_3g_4 - 2d_3^2g_4 + 3d_4g_3]F_{\chi_{k+3}^2}(y)\\
&\quad + [(2d_3 + d_3^2 + 1)g_4]F_{\chi_{k+5}^2}(y)\Big\}.
\end{aligned}
\tag{8.22}
$$

Here $d_3 = C^{(3)}(0)$ and $d_4 = C^{(4)}(0)$ are the third and fourth derivatives of the disparity generating function $C(\cdot)$ evaluated at $\delta = 0$. Also $g_1 = (t - k^2)$, $g_2 = (t - k)$, $g_3 = (t - 2k + 1)$, $g_4 = (5t - 3k^2 - 6k + 4)$. An approximation to J_2^C to the first order is given by

$$\hat{J}_2^C(y) = \{N_C(y) - n^{(k-1)/2}V_C(y)\}\{e^{-y/2}(2\pi n)^{-(k-1)/2}Q^{-1/2}\}. \tag{8.23}$$

Here $N_C(y)$ is the number of multinomial vectors $(x_1, \ldots, x_k)$ such that $2n\rho_C(\boldsymbol{p}, \boldsymbol{\pi}_0) < y$, where $\boldsymbol{p} = (x_1/n, \ldots, x_k/n)$, $Q = \prod_{i=1}^k \pi_{0i}$, and

$$V_C(y) = \left(\frac{(\pi y)^{(k-1)/2}}{\Gamma\{(k+1)/2\}}\right)Q^{1/2}\left\{1 + \frac{y(d_3^2g_4 - 3d_4g_3)}{24n(k+1)}\right\}. \tag{8.24}$$

Note that if the distribution of $2n\rho_C(\boldsymbol{p}, \boldsymbol{\pi}_0)$ was continuous, the term J_1^C could have been obtained by the multivariate Edgeworth approximation. The discontinuous nature of $2n\rho_C(\boldsymbol{p}, \boldsymbol{\pi}_0)$ statistic is accounted for by the term J_2^C. The term J_3^C is very complex, but, as pointed out by Read (1984a), any term of J_3^C that is dependent on the disparity generating function C must be an $O(n^{-3/2})$ term, and hence from the point of view of the representation (8.21), J_3^C is independent of C. As in Read (1984a) an approximation of $F_E(y)$ up to order $O(n^{-1})$ may therefore be obtained as

$$F_S(y) = J_1^C + \hat{J}_2^C.$$

From the above expressions one can directly obtain the F_N approximations for the PD, BWHD and BWCS families using the definitions of g_1, g_2, g_3, g_4 and d_1, d_2.

Also see Pardo (1998, 1999) and Pardo and Pardo (1999) for examples of similar manipulations in some other families including the Rukhin's distance and the Burbea–Rao divergence.

8.6   Inlier Modified Statistics

In Section 8.3 we have considered the equiprobable null hypothesis against the bump and the dip alternatives. We have also noticed that the hierarchy in the power of the goodness-of-fit tests within the power divergence family runs in reverse when one switches to dip alternatives from bump alternatives. Thus, tests which are more powerful against the bump alternative would be the weaker tests against the dip alternative. This phenomenon is also observed in case of the blended weight Hellinger distance and the blended weight chi-square tests.

Our concern here is that if one is not aware of the alternative against which to test the equiprobable null hypothesis, one may get stuck with a test which poorly discriminates between the null and the underlying truth. Our motivation here is to construct modified tests which perform reasonably at all alternatives although they may not be the best at any (somewhat in the spirit of uniformly most powerful unbiased tests for two sided alternatives in elementary text book examples).

For a better understanding of the way in which a disparity test functions, we further explore the structure of the test statistic

$$2n\rho_C(\boldsymbol{p}, \boldsymbol{\pi}_0) = 2n \sum_{i=1}^{k} C\left(\frac{p_i - \pi_{0i}}{\pi_{0i}}\right) \pi_{0i}.$$

The nature of the function C will determine whether the statistic will highlight outlying cells (cells for which $p_i > \pi_{0i}$) or inlying cells (corresponding to $p_i < \pi_{0i}$). In Figures 6.3 and 6.4 we have observed that within the power divergence and blended weight Hellinger distance families, the particular distances which do well in magnifying the effect of outliers show a dampened response with respect to inliers and vice versa. This appears to be the general trend in most other subfamilies of disparities. The natural distances are not simultaneously sensitive to both kinds of deviations. We have discussed this issue in detail in Chapter 6, where we have made extensive modifications to the disparities for developing distances which are insensitive to both outliers and inliers.

In this chapter we will consider similar modifications to our disparities, but now our purpose is to make the disparities sensitive to small deviations of both types. Notice that in the power divergence family a large positive value of λ leads to disparity generating functions which rise sharply on the outlier side; it is therefore intuitive that such a distance will do well in detecting bump alternatives. At the same time the slow incline of the function on the inlier side makes such a distance insensitive to dip alternatives. As one might expect, the choice of a large negative value of λ reverses the situation. Intuitively, therefore, the observations in Table 8.1 (and those in Tables 8.2 and 8.3) are as one would expect them to be.

In principle we can apply each of the five types of modifications considered in Chapter 6 to make our disparity tests more sensitive to small deviations of each type. In the first two cases (involving penalized distances and combined distance) it is necessary to provide fresh proofs for the asymptotic null distributions of the corresponding test statistics. The distribution of the test statistics in the other cases – coupled, ϵ-combined, and inlier shrunk – follow from our results in Section 8.2.

8.6.1 The Penalized Disparity Statistics

Given the null hypothesis in (8.8), consider the penalized disparity statistic

$$2n\rho_{C_h}(p, \hat{\pi}), \tag{8.25}$$

where $\hat{\pi} = f(\hat{\theta})$, f is as defined in Section 8.2.3, $\hat{\theta}$ is the minimum penalized disparity estimator of θ under the null, and C_h is the penalized disparity generating function defined in Chapter 6. The following theorem (Mandal and Basu, 2010b) will show that the penalized disparity statistic in (8.25) has the same asymptotic limit as the ordinary disparity statistic in Theorem 8.4 under the null.

Theorem 8.5. *Suppose that the disparity generating function C satisfies conditions (2.16) and (2.27) and that the third derivative C''' is finite and continuous at zero. Also suppose that the regularity conditions of Section 8.2.3 hold. Then the corresponding penalized disparity statistic given in (8.25) generated by the function C_h has an asymptotic χ^2_{k-p-1} distribution under the null hypothesis (8.8) for any fixed, positive h.*

Proof. By Theorem 8.4 the ordinary disparity statistic has an asymptotic χ^2_{k-p-1} distribution under the null hypothesis. As discussed in Section 8.2.3, the asymptotic distribution of the disparity statistic does not depend on the estimate of π, i.e., $\hat{\pi}$, as long as it is a best asymptotically normal (BAN) estimator. For the desired result it is sufficient to show that under the null

$$R_n(\theta) = 2n\rho_C(p, \hat{\pi}) - 2n\rho_{C_h}(p, \hat{\pi}) = o_p(1), \tag{8.26}$$

where $\hat{\pi}$ is the minimum penalized disparity estimator. Now

$$
\begin{aligned}
R_n(\theta) &= 2n(C(-1) - h) \sum_{i:p_i=0} \hat{\pi}_i \\
&= 2n(C(-1) - h) \sum_{i=1}^{k} \hat{\pi}_i \varsigma(p_i),
\end{aligned}
\tag{8.27}
$$

where $\varsigma(y) = 1$ if $y = 0$ and 0 otherwise. So

$$
\begin{aligned}
E\left[|R_n(\boldsymbol{\theta})|\right] &= 2n|C(-1) - h| \; E\left[\sum_{i=1}^{k} \hat{\pi}_i \varsigma(p_i)\right] \\
&\leq 2n|C(-1) - h| \sum_{i=1}^{k} E\left[\varsigma(p_i)\right] \\
&\leq 2n|C(-1) - h| \sum_{i=1}^{k} (1 - \pi_i)^n, \qquad (8.28)
\end{aligned}
$$

where π_i is the true cell probability for the i-th cell. Suppose $t_n(x) = n(1-x)^n$, where $0 < x < 1$; notice that as $t_n(x) \to 0$ for all $0 < x < 1$ as $n \to \infty$. As the number of cells k is a finite constant, we get from (8.28)

$$
E\left[|R_n(\boldsymbol{\theta})|\right] \to 0 \qquad \text{as } n \to \infty. \qquad (8.29)
$$

Hence, by Markov's inequality, Equation (8.26) holds and the desired result is established. $\qquad \square$

The asymptotic distribution of the penalized disparity statistics was derived in Park et al. (2001) for the special case of the blended weight Hellinger distance.

8.6.2   The Combined Disparity Statistics

Given the null hypothesis in (8.8), consider the combined disparity statistic

$$
2n\rho_{C_m}(\boldsymbol{p}, \hat{\boldsymbol{\pi}}), \qquad (8.30)
$$

where $\hat{\boldsymbol{\pi}} = f(\hat{\theta})$, f is as defined in Section 8.2.3, $\hat{\theta}$ is the minimum combined disparity estimator of θ under the null, and $\rho_{C_m}(\cdot, \cdot)$ is the combined disparity defined in Chapter 6 generated by the disparity generating function C_1 on the inlier side and the function C_2 on the outlier side. The following theorem will show that the above combined disparity statistic has the same limiting distribution as the ordinary disparity statistic under the null (Mandal and Basu, 2010b); see also Basu et al. (2002). From our results of Chapter 6, the minimum combined disparity estimator $\hat{\boldsymbol{\pi}}$ of $\boldsymbol{\pi}$ is a BAN estimator.

Theorem 8.6. *Suppose that the disparity generating functions C_1 and C_2 both satisfy conditions (2.16) and (2.27), and that the third derivative C_i''' is finite and continuous at zero for $i = 1, 2$. Also suppose that the regularity conditions of Section 8.2.3 hold. Then the combined disparity statistic given in (8.30) has an asymptotic χ^2_{k-p-1} distribution under the null hypothesis (8.8).*

Proof. The combined disparity statistic can be expressed as

$$2n \sum_{i=1}^{k} C_m \left(\frac{p_i - \hat{\pi}_i}{\hat{\pi}_i} \right) \hat{\pi}_i$$

$$= 2n \sum_{p_i \le \hat{\pi}_i} C_1 \left(\frac{p_i - \hat{\pi}_i}{\hat{\pi}_i} \right) \hat{\pi}_i + 2n \sum_{p_i > \hat{\pi}_i} C_2 \left(\frac{p_i - \hat{\pi}_i}{\hat{\pi}_i} \right) \hat{\pi}_i. \quad (8.31)$$

When $p_i \le \hat{\pi}_i$, a Taylor series expansion about $p_i = \hat{\pi}_i$ gives

$$2nC_m \left(\frac{p_i - \hat{\pi}_i}{\hat{\pi}_i} \right) \hat{\pi}_i$$

$$= 2nC_1(0)\hat{\pi}_i + 2n(p_i - \hat{\pi}_i)C_1'(0) + \frac{n}{\hat{\pi}_i}(p_i - \hat{\pi}_i)^2 C_1''(0)$$

$$+ \frac{n}{3\hat{\pi}_i^2}(p_i - \hat{\pi}_i)^3 C_1''' \left(\frac{\xi_{1i} - \hat{\pi}_i}{\hat{\pi}_i} \right), \quad (8.32)$$

where $p_i \le \xi_{1i} \le \hat{\pi}_i$. Since $p_i - \hat{\pi}_i$ and $\xi_{1i} - \hat{\pi}_i$ lie on the same side of zero, the expansion in (8.32) is valid. Now $C_1(0) = C_1'(0) = 0$ and $C_1''(0) = 1$. Hence, when $p_i \le \hat{\pi}_i$, Equation (8.32) reduces to

$$2nC_m \left(\frac{p_i - \hat{\pi}_i}{\hat{\pi}_i} \right) \hat{\pi}_i = \frac{n}{\hat{\pi}_i}(p_i - \hat{\pi}_i)^2 + \frac{n}{3\hat{\pi}_i^2}(p_i - \hat{\pi}_i)^3 C_1''' \left(\frac{\xi_{1i} - \hat{\pi}_i}{\hat{\pi}_i} \right). \quad (8.33)$$

A similar argument works for the case where $p_i > \hat{\pi}_i$, and we get

$$2nC_m \left(\frac{p_i - \hat{\pi}_i}{\hat{\pi}_i} \right) \hat{\pi}_i = \frac{n}{\hat{\pi}_i}(p_i - \hat{\pi}_i)^2 + \frac{n}{3\hat{\pi}_i^2}(p_i - \hat{\pi}_i)^3 C_2''' \left(\frac{\xi_{2i} - \hat{\pi}_i}{\hat{\pi}_i} \right), \quad (8.34)$$

However, from the given conditions, C_j''' is bounded in the neighborhood of zero for $j = 1, 2$; so we get, combining (8.33) and (8.34),

$$2nC_m \left(\frac{p_i - \hat{\pi}_i}{\hat{\pi}_i} \right) \hat{\pi}_i = \frac{n}{\hat{\pi}_i}(p_i - \hat{\pi}_i)^2 + o_p(1), \quad (8.35)$$

and, finally,

$$2n \sum_i C_m \left(\frac{p_i - \hat{\pi}_i}{\hat{\pi}_i} \right) \hat{\pi}_i = 2n \sum_i \frac{(p_i - \hat{\pi}_i)^2}{2\hat{\pi}_i} + o_p(1).$$

Since

$$2n \sum_i \frac{(p_i - \hat{\pi}_i)^2}{2\hat{\pi}_i}$$

has the same limiting distribution as the Pearson's chi-square statistic $2n$PCS for any BAN estimator $\hat{\pi}$ of π, the desired result holds. $\qquad\square$

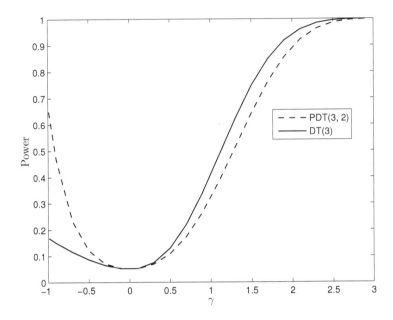

FIGURE 8.1
The exact powers of the ordinary and the penalized statistics for several values
of γ.

8.6.3   Numerical Studies

Here we present some numerical examples to illustrate the performance of our
inlier modified disparity tests in terms of their attained exact power and their
discriminatory capability over specified nulls.

We first consider a multinomial model with $n = 20$ and $k = 5$ cells. Under
this setup we test the equiprobable null hypothesis against the alternative
hypothesis in (8.9) over different values of γ in its allowable range and study
the exact power of the tests under consideration. We hope to show that the
power of the inlier modified tests give a steadier overall protection compared
to the extreme behavior of the ordinary tests. We choose the power divergence
family for illustration; the same point can also be made with other families of
disparities such as the blended weight Hellinger distance family.

The exact power at each alternative is evaluated in terms of the critical
value determined through the randomized test of exact size α. For a moment
we go back to the main philosophy of the inlier modification methods considered in Chapter 6. In that case our basic approach was to choose a disparity
with a dampened response on the outlier side, and reduce the sensitivity of
its inlier component. Our strategy for the goodness-of-fit testing problem will

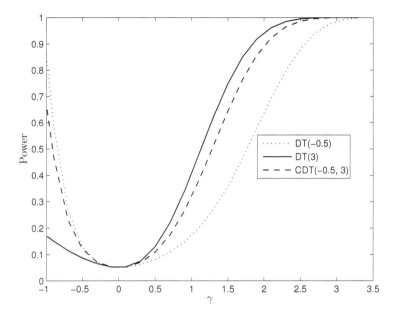

FIGURE 8.2

The exact powers of the ordinary and the combined statistics for several values of γ.

be exactly the opposite. Now we will begin with an outlier sensitive disparity; since the corresponding disparity generating function is likely to exhibit a very slow growth on the inlier side, we now apply a modification to the inlier component to increase its sensitivity.

For a specific illustration, we choose the disparity test within the power divergence family with $\lambda = 3$; in Figures 8.1–8.3 we refer to this test as DT(3). To make its inlier component more sensitive, we may apply any of our five inlier modification techniques, but in a reverse direction than the one considered in Chapter 6. For example, one such method could be to replace the weight of the empty cells by an appropriate constant h; in this case h should be a large positive value, necessarily larger than $C(-1)$. We will use $h = 2$, and in Figure 8.1 this test will be denoted by PDT(3, 2).

Alternatively one could use the combined disparity $PD_{(-0.5,3)}$, as defined in Section 6.3, to generate the desired effect. In Figure 8.2 we will denote this test by CDT(−0.5, 3). Yet another possible modification would be to consider the coupled disparity $CpPD_{(k_0,3)}$, as defined in Section 6.5. In this presentation we will use $k_0 = -2$, and denote the corresponding test in Figure 8.3 by CpDT(−2, 3). Other inlier modified tests may be similarly defined.

The exact powers of these inlier modified tests are presented in Figures 8.1–

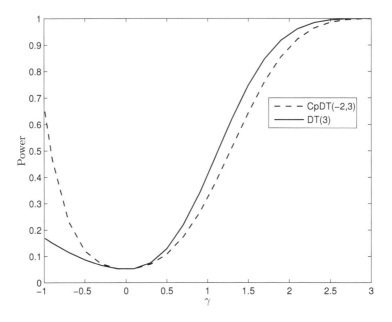

FIGURE 8.3
The exact powers of the ordinary and the coupled statistics for several values of γ.

8.3. It is immediately observed that the tests are successful in achieving our main purpose. The powers of the modified tests are significantly higher than those of the ordinary tests for dip alternatives, although they are slightly worse than the latter for bump alternatives. Clearly, the gains are far more substantial than the losses. It is also easily observed that the different inlier modified tests considered here perform very similarly.

We follow up the exact calculations above with two real data examples.

Example 8.1. This data set is taken from Agresti (1990, p. 72, Table 3.10); it has been analyzed previously by Basu et al. (2002) and in more detail by Mandal and Basu (2010b). A total of 182 psychiatric patients on drugs were classified according to their diagnosis. The breakdown of the frequencies of the different classes is as shown in Table 8.4. The sample size is fairly large, and we expect that the chi-square approximation will be reasonable for all our ordinary and modified statistics.

Consider testing a simple null hypothesis where the probability vector is given by $\pi_0 = (0.56, 0.06, 0.09, 0.25, 0.04)$. The deviations of the observed relative frequencies from the null probabilities appear to be small; the research question is whether these small deviations are enough to indicate a significant difference from the null given the large sample size. The χ^2 critical value at

TABLE 8.4
The null probabilities and the observed relative frequencies for 182 psychiatric patients.

Diagnosis	Frequency	Relative Frequency	Null Probability
Schizophrenia	105	0.577	0.56
Affective disorder	12	0.066	0.06
Neurosis	18	0.099	0.09
Personality disorder	47	0.258	0.25
Special symptoms	0	0.000	0.04

4 degrees of freedom and 5% level of significance is given by 9.488. The ordinary power divergence statistics for $\lambda = 2, 1$ and $2/3$ are, respectively, 5.273, 7.690 and 9.142, and these tests fail to reject the null hypothesis. However, for the corresponding penalized tests (note that the data set contains one empty cell) corresponding to $\lambda = 2, 1$ and $2/3$ with penalty weights $h = 2$ the test statistics are 29.540, 29.530, and 29.526, so that these statistics reject the null hypothesis comfortably.

The combined statistics for $\mathrm{PD}_{(-0.5,2)}$, $\mathrm{PD}_{(-0.5,1)}$ and $\mathrm{PD}_{(-0.5,2/3)}$, and the coupled statistics for $\mathrm{CpPD}_{(-2,2)}$, $\mathrm{CpPD}_{(-2,1)}$ and $\mathrm{CpPD}_{(-2,2/3)}$, produce statistics of similar orders and again comfortably reject the null. In this example, the outlier modified statistics clearly perceive the null to be an unlikely representation of the truth, unlike the ordinary statistics. ‖

Example 8.2. Next we consider the time passage example data (Read and Cressie, 1988, pp. 12–16) which studied the relationships between stresses in life and illnesses in Oakland, California. These data have also been analyzed by Basu et al. (2002), and in more detail by Mandal and Basu (2010b). The data are in the form of an 18-cell multinomial, with the frequencies representing the total number of respondents for each month who indicated one stressful event between 1 and 18 months before the interview. The null hypothesis $H_0 : \pi_{0i} = 1/18$, $i = 1, ..., 18$, of equiprobability is clearly an untenable one and is soundly rejected by practically all the ordinary disparity tests. However, the model fit appears to improve if we consider a log-linear time trend model $H_0 : \log(\pi_{0i}) = \vartheta + \beta i$, $i = 1, ..., 18$. Expected frequencies on the basis of estimates of ϑ and β obtained by using maximum likelihood are given in Read and Cressie (1988, Table 2.2). The test statistics are now compared with the critical value of a χ^2-statistic with 16 degrees of freedom (rather than 17), and the critical value at 5% level of significance is 26.296. All the PD_λ tests with $\lambda \in [0, 3]$ fail to reject the time trend null hypothesis; for example the PD_2 test generates a statistic equal to 23.076. However, the combined statistics corresponding to $\mathrm{PD}_{(-1,2)}$ and $\mathrm{PD}_{(-2,1)}$ are equal to 35.271 and 44.840 respectively, and in either case the null is soundly rejected. Thus, unlike the ordinary disparity tests, the time trend model fails to pass the goodness-of-fit standards of these inlier modified disparity tests. $\qquad \qquad \|$

8.7   An Application: Kappa Statistics

Cohen's kappa (Cohen, 1960) measures the agreement between two raters, each of whom classify n subjects into a number of mutually exclusive categories. This measure is given by

$$\kappa = \frac{\Pi_a - \Pi_e}{1 - \Pi_e},$$

where Π_a is the observed probability of agreement among raters, and Π_e is the hypothetical probability of chance agreement under independence. If the raters are in complete agreement then $\kappa = 1$. If there is no agreement among the raters (other than what would be expected under chance alone), one gets $\kappa = 0$. The kappa statistic is heavily used in measuring agreement between two raters, and tests of hypotheses about the kappa statistic have great practical value.

Donner and Eliasziw (1992) suggested a goodness-of-fit approach for performing tests of significance about the kappa statistic when the number of categories is 2. The approach may be described as follows. Consider two raters who independently rate n subjects on a binary scale, the ratings being either success or failure. Let η denote the probability that the rating of the jth subject, $j = 1, 2, \ldots, n$, by the kth rater, $k = 1, 2$, is a success. The probabilities for the joint responses can be expressed as a function of η and κ as

$$\text{Pr(both ratings are successes)} \quad = \quad \eta^2 + \eta(1 - \eta)\kappa \quad = EP_1,$$

$$\text{Pr}\begin{pmatrix} \text{one rating is a success} \\ \text{and one failure} \end{pmatrix} \quad = \quad 2\eta(1 - \eta)(1 - \kappa) \quad = EP_2,$$

and

$$\text{Pr(both ratings are failures)} \quad = (1 - \eta)^2 + \eta(1 - \eta)\kappa \quad = EP_3.$$

This is the common correlation model for dichotomous data. Let n_1, n_2 and n_3 represent, respectively, the number of subjects rated as successes by both raters, the number of subjects rated as successes by exactly one rater, and the number of subjects rated as failures by both of the raters, where $n = n_1 + n_2 + n_3$. The vector (n_1, n_2, n_3) is the response vector. Under the common correlation model the maximum likelihood estimator of κ is

$$\hat{\kappa} = 1 - \frac{n_2}{2n\hat{\eta}(1 - \hat{\eta})}, \qquad (8.36)$$

where

$$\hat{\eta} = \frac{2n_1 + n_2}{2n} \tag{8.37}$$

is the maximum likelihood estimator for η. Now given a null hypothesis

$$H_0 : \kappa = \kappa_0,$$

Donner and Eliasziw tested it by testing the goodness-of-fit of the data (n_1, n_2, n_3) scaled by n against the estimates of the expected probabilities EP_i, $i = 1, 2, 3$. Thus, they used the statistic

$$n \sum_{i=1}^{3} \frac{(d_i - \widehat{EP}_i)^2}{\widehat{EP}_i},$$

where $d_i = n_i/n$, and $\widehat{EP}_i$ are the quantities defined above with η replaced by $\hat{\eta}$ and κ replaced by κ_0. This statistic has an asymptotic chi-square distribution with one degree of freedom under the null hypothesis.

In general, within the power divergence family, the goodness-of-fit statistic may be expressed as

$$\frac{2n}{\lambda(\lambda + 1)} \sum_{i=1}^{3} \left[d_i \left\{ \left(\frac{d_i}{\widehat{EP}_i} \right)^{\lambda} - 1 \right\} \right],$$

or, equivalently as

$$2n \sum_{i=1}^{3} \left[\frac{d_i}{\lambda(\lambda + 1)} \left\{ \left(\frac{d_i}{\widehat{EP}_i} \right)^{\lambda} - 1 \right\} + \frac{\widehat{EP}_i - d_i}{\lambda + 1} \right],$$

as the function of the tuning parameter $\lambda \in (-\infty, \infty)$. Depending on the null and alternative hypotheses, the behavior of the test statistic and its observed power may be studied over the whole range of values for the tuning parameter λ. See Donner and Eliasziw (1992) and Basu and Basu (1995) for more details and further extensions of this approach. Also see Altaye, Donner and Klar (2001) for an extension of this method to the case of multiple raters.

9

The Density Power Divergence

Throughout this book we have emphasized on the benefits of minimum distance estimation based on disparities. At the same time, we have pointed out that in case of continuous models one is forced to use some form of nonparametric smoothing such as kernel density estimation to produce a continuous estimate of the true density. As a result, the minimum distance estimation method based on disparities inherits all the associated complications such as those related to bandwidth selection in continuous models. Basu and Lindsay (1994) considered another modification of this approach, discussed in Chapter 3, where the model is smoothed with the same kernel as the data to reduce the dependence of the procedure on the smoothing method. Although the Basu and Lindsay procedure largely reduces the effect of the bandwidth, the process still involves a nonparametric smoothing component.

In this chapter we discuss yet another family of density-based distances called the "density power divergences." This family is not directly related to the "power-divergence" family of Cressie and Read (1984), although some indirect connections may be drawn. The density power divergence family is indexed by a single nonnegative tuning parameter α which controls the trade-off between robustness and asymptotic efficiency of the parameter estimates which are the minimizers of this family of divergences. When $\alpha = 0$, the density power divergence reduces to the Kullback–Leibler divergence (Kullback and Leibler, 1951) and the resulting estimator is the maximum likelihood estimator. On the other hand, when $\alpha = 1$, the divergence is simply the integrated squared error, and the corresponding parameter estimate is highly robust (although somewhat inefficient). The degree of robustness increases with increasing α; however, larger values of α are also associated with more inefficient estimators. On the whole, several of the estimators within this class have strong robustness properties with little loss in asymptotic efficiency relative to maximum likelihood under the model. A big advantage of the method lies in the fact that the estimation method does not require any nonparametric smoothing for any value of α under continuous models.

Our description of the density power divergence family and related inference will follow the developments of Basu et al. (1998). In short, we will often refer to these divergences as the BHHJ divergences, and also refer to the corresponding estimators as the BHHJ estimators – after the names of the authors Basu, Harris, Hjort and Jones (1998). In effect we use the terms "density power divergence" and "BHHJ divergence" interchangeably. These divergences are also referred to as "type 1 divergences," in connection with a superfamily of divergences which we will introduce later in this chapter.

9.1   The Minimum L_2 Distance Estimator

To motivate the development of the density power divergence family and the corresponding estimator, we begin with one of its special cases. Consider a parametric family of distributions $\mathcal{F}$ as defined in Section 1.1; let $\{f_\theta\}$ be the associated class of model densities with respect to an appropriate dominating measure, and let $\mathcal{G}$ be the class of all distributions having densities with respect to the given measure. The L_2 distance between the densities g and f_θ is given by $[\int \{g(x) - f_\theta(x)\}^2 dx]^{1/2}$. The minimum L_2 distance functional $T_1(G)$ is defined, for every G in $\mathcal{G}$, by the relation,

$$\int \{g(x) - f_{T_1(G)}(x)\}^2 dx = \min_{\theta \in \Theta} \int \{g(x) - f_\theta(x)\}^2 dx, \qquad (9.1)$$

where g is the density of G. We will assume that the above minimum distance functional exists and is unique.

Suppose that the parametric family is identifiable in the sense of Definition 2.2. Under this assumption, the Fisher consistency of the minimum L_2 distance functional is immediate from Equation (9.1), and $T_1(F_\theta) = \theta$, uniquely. Notice, however, that the squared L_2 distance is not a member of the class of disparities, as the squared deviations are not weighted by the inverse of either the data density or the model density (or any appropriate combination of both).

Also note that given a distribution $G \in \mathcal{G}$, the squared L_2 distance between the corresponding density g and f_θ can be represented as

$$\int f_\theta^2(x) dx - 2 \int f_\theta(x) dG(x) + K,$$

where the quantity K is a constant independent of θ and can be removed from the objective function as it does not contribute to the minimization procedure. Notice that once the constant term is removed, the true density g appears linearly in this objective function (only through the $dG(x)$ term), unlike, say, the objective function for the Hellinger distance. Thus, given a random sample $X_1, \ldots, X_n$ from the true distribution with density g, one can actually minimize

$$\int f_\theta^2(x) dx - 2 \int f_\theta(x) dG_n(x) = \int f_\theta^2(x) dx - 2n^{-1} \sum_{i=1}^n f_\theta(X_i) \qquad (9.2)$$

with respect to θ, where G_n is the empirical distribution function, to obtain the minimum L_2 distance estimator of the parameter θ. Remarkably, one does not need a smooth nonparametric estimate of g for this inference process, in contrast to the works of Beran (1977), Cao et al. (1995) and others.

Suppose that the model satisfies appropriate differentiability and regularity conditions, so that the minimum L_2 distance estimator can be obtained by solving the estimating equation

$$n^{-1} \sum_{i=1}^{n} u_\theta(X_i) f_\theta(X_i) - \int u_\theta(x) f_\theta^2(x) dx = 0, \tag{9.3}$$

where $u_\theta(x) = \nabla \log f_\theta(x)$ is the maximum likelihood score function. When the true density g belongs to the family of model densities $\{f_\theta\}$, the above estimating equation is immediately seen to be unbiased.

Example 9.1 (Location Model). The above estimating equation in (9.3) gives nice insights into the robustness of the minimum L_2 distance functional. For further illustration, consider a location model $\{F_\theta\}$, and let θ be the location parameter. Notice that for this model, $\int f_\theta^2(x) dx$, the first term in Equation (9.2), is independent of θ and hence does not affect the minimization process. The minimum L_2 distance estimator is now simply the maximizer of $\sum_i f_\theta(X_i)$; this provides an interesting contrast with the maximum likelihood estimator, which maximizes the product of the model densities $\prod_i f_\theta(X_i)$, rather than their sum.

For comparison we present below the estimating equations of the minimum L_2 distance estimator and maximum likelihood estimator in this case, which are

$$\sum_i u_\theta(X_i) f_\theta(X_i) = 0 \quad \text{and} \quad \sum_i u_\theta(X_i) = 0, \tag{9.4}$$

respectively. Notice that the score function is weighted by the model density in the estimating equation of the minimum L_2 distance estimator, which may be viewed as another version of a weighted likelihood equation. This provides an automatic downweighting for the scores $u_\theta(\cdot)$ of observations that are unlikely under the model. Once again the weighting is probabilistic, rather than geometric, and the model density is directly utilized to provide the desired shrinkage.

For several parametric models such as the normal, $u_\theta(x) f_\theta(x)$ is a bounded function of x for fixed θ, which prevents any surprising observation from having a disproportionate impact on the estimation process. For a random variable X in the one parameter exponential family where θ is the mean value parameter, $u_\theta(x)$ equals $(x - \theta)/\sigma^2$, where σ^2 is the variance of X; thus the sample mean is the maximum likelihood estimator for the mean parameter in these families, suggesting the robustness problems of maximum likelihood. However, when the scores are weighted by the densities, the corresponding estimating equation generates a much more stable solution. ‖

The outlier resistant property of the minimum L_2 distance estimator and its variants have also been noticed elsewhere. Brown and Hwang (1993) used the minimum L_2 distance in determining the "best fitting normal density" to a given histogram. See also Hjort (1994), Jones and Hjort (1994), and Scott (2001). The small contribution of outliers to L_2 distance based on histograms or kernel density estimates makes this robustness intuitively apparent. Also see Shen and Harris (2004) and Durio and Isaia (2004) for some applications of the minimum L_2 method in case of mixture models.

In many cases, however, it turns out that the robustness of the minimum L_2 distance estimator requires a high price to be paid in terms of asymptotic efficiency. Numerical examples with respect to several common parametric models will be presented later in this section which will illustrate this point. In order to generate estimators which provide a better compromise between robustness and efficiency, we will now introduce the family of density power divergences, which will establish a smooth link between the efficient but non-robust maximum likelihood estimator, and the highly robust but relatively inefficient minimum L_2 distance estimator. As we will see, there are many members of this class of minimum distance estimators which exhibit strong robustness features with minimal loss in asymptotic efficiency.

9.2 The Minimum Density Power Divergence Estimator

A moment's reflection shows that the two estimating equations in (9.4) can be linked together by a generalized estimating equation of the form

$$\sum_{i=1}^{n} u_\theta(X_i) f_\theta^\alpha(X_i) = 0, \quad \alpha \in [0, 1]. \tag{9.5}$$

The last equation represents an interesting density power downweighting, and hence robustification of the usual likelihood score equation. The estimating equations in (9.4) can be recovered from (9.5) by the choices $\alpha = 1$ and $\alpha = 0$ respectively. Within the given range of α, $\alpha = 1$ will lead to the maximum downweighting for the score functions of the surprising observations; on the other extreme, the score functions will be subjected to no downweighting at all for $\alpha = 0$. Intermediate values of α provide a smooth bridge between these two estimating equations, and the degree of downweighting increases with increasing α.

For models beyond the location model, the estimating equation in (9.5) can be further generalized to generate the form

$$\frac{1}{n} \sum_{i=1}^{n} u_\theta(X_i) f_\theta^\alpha(X_i) - \int u_\theta(x) f_\theta^{1+\alpha}(x) dx = 0, \quad \alpha \geq 0, \tag{9.6}$$

which is an unbiased estimating equation when the true distribution G belongs to the model $\{F_\theta\}$. Notice that we have also replaced the range $\alpha \in [0, 1]$ with $\alpha \geq 0$, since technically there is no reason to exclude the distances with $\alpha > 1$; however, from efficiency considerations one would rarely use a distance with $\alpha > 1$ for estimating the unknown model parameter, even though such estimators would have solid robustness properties.

With this background, we are now ready to introduce the density power divergence family. Given two densities g and f, the density power divergence $d_\alpha(g, f)$ corresponding to index α between g and f is defined to be

$$d_\alpha(g, f) = \int \left\{ f^{1+\alpha}(x) - \left(1 + \frac{1}{\alpha}\right) g(x) f^\alpha(x) + \frac{1}{\alpha} g^{1+\alpha}(x) \right\} dx \qquad (9.7)$$

for $\alpha > 0$. The expression in (9.7) reduces to the L_2 distance when $\alpha = 1$. However, $d_\alpha(g, f)$ is not defined when we replace α by zero, so we define the divergence $d_0(g, f)$ as

$$d_0(g, f) = \lim_{\alpha \to 0} d_\alpha(g, f) = \int g(x) \log(g(x)/f(x)) dx. \qquad (9.8)$$

Notice that the above is a form of the Kullback–Leibler divergence, and given a class of model densities $\{f_\theta\}$, $d_0(g, f_\theta)$ is minimized over the parameter space by the maximum likelihood functional $T_0(G)$. It turns out that $d_0(g, f)$ is the only common member between the class of disparities and the family of density power divergences. When dealing with the estimators resulting from the family in (9.7) we will be primarily interested in smaller values of α close to zero, as the procedure typically becomes less and less efficient as α increases.

Our first task here is to show that the BHHJ family defined by (9.7) and (9.8) represents a family of genuine statistical distances.

Theorem 9.1. [Basu et al. (1998, Theorem 1)]. *Given two densities g and f, the divergence $d_\alpha(g, f)$ represents a genuine statistical distance for all $\alpha \geq 0$.*

Proof. For $\alpha = 0$ it is well known that the above Kullback–Leibler divergence is nonnegative and equals zero if and only if $g \equiv f$ (see the discussion on the nonnegativity of disparities in Section 2.3.1). So assume $\alpha > 0$. It is easily seen that the divergence in (9.7) is zero when the densities are identical, so all we have to show is that the divergence in (9.7) is always nonnegative. For this we will show that the integrand itself is nonnegative. We factor out the term $g^{1+\alpha}(x)/\alpha$, and let $y = f(x)/g(x)$. Then the factored integrand becomes

$$t_\alpha(y) = (\alpha y^{1+\alpha} + (1 + \alpha) y^\alpha + 1), \text{ for } y \geq 0.$$

Elementary calculus shows that the function $t_\alpha(y)$ is nonnegative and has a unique minimum at 1. Thus, the desired result holds. □

As an immediate consequence of the above theorem we see that the minimum density power divergence functional $T_\alpha(G)$, defined by the requirement

$$d_\alpha(g, f_{T_\alpha(G)}) = \min_{\theta \in \Theta} d_\alpha(g, f_\theta),$$

is Fisher consistent when the model is identifiable.

One can write the density power divergence $d_\alpha(g, f_\theta)$ as

$$d_\alpha(g, f_\theta) = \int f_\theta^{1+\alpha}(x)dx - \left(1 + \frac{1}{\alpha}\right) \int f_\theta^\alpha(x)dG(x) + K, \qquad (9.9)$$

where the term K is independent of the parameter θ. Let $X_1, \ldots, X_n$ be independent and identically distributed observations from the distribution G, not necessarily in the model, having density g. We ignore the constant and replace $G(x)$ with $G_n(x)$, the empirical distribution function, in (9.9) and obtain the minimum density power divergence estimator $\hat{\theta}$ by minimizing the resultant expression

$$H_n(\theta) = \int f_\theta^{1+\alpha}(x)dx - \left(1 + \frac{1}{\alpha}\right) n^{-1} \sum_{i=1}^n f_\theta^\alpha(X_i)$$

$$= n^{-1} \sum_{i=1}^n V_\theta(X_i), \qquad (9.10)$$

with respect to $\theta \in \Theta$, where

$$V_\theta(X_i) = \int f_\theta^{1+\alpha}(x)dx - (1 + 1/\alpha)f_\theta^\alpha(X_i). \qquad (9.11)$$

The above optimization can equivalently be thought of as the maximization of

$$\frac{1}{n\alpha} \sum_{i=1}^n f_\theta^\alpha(X_i) - \frac{1}{1+\alpha} \int f_\theta^{1+\alpha}(x)dx, \qquad (9.12)$$

with respect to θ. The quantity in Equation (9.12) is akin to a likelihood. It has been referred to as β-likelihood in the literature (see, e.g., Fujisawa and Eguchi, 2006) who denoted the tuning parameter as β and referred to the quantity in (9.7) as the β-divergence. We will follow the original formulation of Basu et al. (1998) and continue to use the term "density power divergence" (or the BHHJ divergence), and the tuning parameter α.

The restatement of the distance in terms of its empirical version as given in (9.10) without taking recourse to density estimation and nonparametric smoothing is a remarkable step. For most density-based minimum distance methods, the need for data smoothing for the creation of the distance is a default requirement. This is also the source of a substantial additional complication in the estimation process, which, unfortunately, is unavoidable in such situations. The density power divergence is a special divergence which allows an empirical version based only on the empirical distribution function and requires no data smoothing. Indeed, existing theory (e.g., De Angelis and Young, 1992) shows that in general there is little or no advantage in introducing smoothing for such functionals which may be empirically estimated using the empirical distribution function alone, except in very special cases.

For the same reason, we do not pursue the Cao, Cuevas and Fraiman (1995) approach.

Notice also that the BHHJ divergence presented in (9.7) is a special case of the so-called Bregman divergence, discussed in, among others, Jones and Byrne (1990) and Csiszár (1991). This divergence takes the form

$$\int \left[B(g(x)) - B(f(x)) - \{g(x) - f(x)\}B'(f(x)) \right] dx,$$

where B is a convex function. Taking $B(f) = f^{1+\alpha}$ gives α times the quantity in (9.7). No smoothing is needed to implement an estimation strategy based on any Bregman divergence. However, Jones et al. (2001) argue that if it is to behave like a weighted mean integrated squared error, then the divergence in (9.7) is the only statistically interesting case.

Under differentiability of the model, the minimum density power divergence estimator is obtained by solving the estimating equation obtained by setting the derivative of (9.10) to zero. Straightforward calculations show that this estimating equation is exactly the same as the one presented in (9.6).

9.2.1 Asymptotic Properties

The next task is to derive the asymptotic distribution of the minimum density power divergence estimator $\hat{\theta}$. We first present the setup and the necessary conditions, and follow it up with the proof. Let the parametric model $\mathcal{F}$ be as defined in Section 1.1, and let θ^g represent the best fitting parameter of θ.

The minimum density power divergence estimator $\hat{\theta}$ is obtained as a solution of Equation (9.6) via the minimization of $H_n(\theta)$ in (9.10). In analogy with Equation (9.10), we also define

$$H(t) = \int f_\theta^{1+\alpha}(x)dx - \left(1 + \frac{1}{\alpha} \right) \int f_\theta^\alpha(x)g(x)dx. \tag{9.13}$$

Define $i_\theta(x) = -\nabla[u_\theta(x)]$, the so called information function of the model. Also let $J = J(\theta^g)$ and $K = K(\theta^g)$ be given by

$$\begin{aligned} J = &\int u_{\theta^g}(x)u_{\theta^g}^T(x)f_{\theta^g}^{1+\alpha}(x)dx \\ &+ \int \{i_{\theta^g}(x) - \alpha u_{\theta^g}(x)u_{\theta^g}^T(x)\}\{g(x) - f_{\theta^g}(x)\}f_{\theta^g}^\alpha(x)dx \end{aligned} \tag{9.14}$$

and

$$K = \int u_{\theta^g}(x)u_{\theta^g}^T(x)f_{\theta^g}^{2\alpha}(x)g(x)dx - \xi\xi^T, \tag{9.15}$$

where

$$\xi = \int u_{\theta^g}(x)f_{\theta^g}^\alpha(x)g(x)dx. \tag{9.16}$$

For any given $\alpha \geq 0$, we will make the following assumptions:

(D1) The distributions F_θ of X have common support, so that the set $\mathcal{X} = \{x | f_\theta(x) > 0\}$ is independent of θ. The distribution G is also supported on $\mathcal{X}$, on which the corresponding density g is greater than zero.

(D2) There is an open subset of ω of the parameter space Θ, containing the best fitting parameter θ^g such that for almost all $x \in \mathcal{X}$, and all $\theta \in \omega$, the density $f_\theta(x)$ is three times differentiable with respect to θ and the third partial derivatives are continuous with respect to θ.

(D3) The integrals $\int f_\theta^{1+\alpha}(x) dx$ and $\int f_\theta^\alpha(x) g(x) dx$ can be differentiated three times with respect to θ, and the derivatives can be taken under the integral sign.

(D4) The $p \times p$ matrix $J(\theta)$, defined by

$$J_{kl}(\theta) = \frac{1}{1+\alpha} \left[E_g \left\{ \nabla_{kl} \left(\int f_\theta^{1+\alpha}(x) dx - \left(1 + \frac{1}{\alpha}\right) f_\theta^\alpha(X) \right) \right\} \right]$$

is positive definite where E_g represents the expectation under the density g. Notice that for $\theta = \theta^g$, $J = J(\theta^g)$ is as defined in (9.14).

(D5) There exists a function $M_{jkl}(x)$ such that

$$|\nabla_{jkl} V_\theta(x)| \le M_{jkl}(x) \text{ for all } \theta \in \omega,$$

where $E_g[M_{jkl}(X)] = m_{jkl} < \infty$ for all j, k, and l.

Theorem 9.2. [Basu et al. (1998, Theorem 2)]. *Under the given conditions, the following results hold:*

(a) *The minimum density power divergence estimating equation in (9.6) has a consistent sequence of roots $\hat\theta = \hat\theta_n$.*

(b) *$n^{1/2}(\hat\theta - \theta^g)$ has an asymptotic p-dimensional multivariate normal distribution with (vector) mean zero and covariance matrix $J^{-1} K J^{-1}$ where J and K are as defined in (9.14) and (9.15).*

Proof. To prove the existence, with probability tending to 1, of a sequence of solutions to an estimating equation which is consistent, we shall consider the behavior of the density power divergence on a sphere Q_a with the center at the best fitting parameter θ^g and radius a. We will show that for any sufficiently small a the probability tends to 1 that $H_n(\theta) > H_n(\theta^g)$ for all points θ on the surface of Q_a, and hence that $H_n(\theta)$ has a local minimum in the interior of Q_a. Since at a local minimum the density power divergence equations must be satisfied it will follow that for any $a > 0$ with probability tending to 1 as $n \to \infty$, the density power divergence estimating equation has a solution $\hat\theta(a)$ within Q_a.

To study the behavior of $H_n(\theta)$ on Q_a, we expand $H_n(\theta)$ around θ^g, and divide it by $(1 + \alpha)$. Thus,

$$\frac{[H_n(\theta^g) - H_n(\theta)]}{1 + \alpha} = -\frac{[\{H_n(\theta) - H_n(\theta^g)\}]}{1 + \alpha}$$

$$= \sum_j (-A_j)(\theta_j - \theta_j^g)$$

$$+ \frac{1}{2} \sum_j \sum_k (-B_{jk})(\theta_j - \theta_j^g)(\theta_k - \theta_k^g)$$

$$+ \frac{1}{6} \sum_j \sum_k \sum_l (\theta_j - \theta_j^g)(\theta_k - \theta_k^g)(\theta_l - \theta_l^g) \frac{1}{n} \sum_{i=1}^n \gamma_{jkl}(x_i) M_{jkl}(x_i)$$

$$= S_1 + S_2 + S_3$$

where

$$A_j = \frac{1}{1 + \alpha} \nabla_j H_n(\theta)|_{\theta = \theta^g}, B_{jk} = \frac{1}{1 + \alpha} \nabla_{jk} H_n(\theta)|_{\theta = \theta^g}$$

and $0 \le |\gamma_{jkl}(x)| \le 1$.

First note that by the law of large numbers

$$\begin{aligned} A_j &= \int u_{j\theta^g}(x) f_{\theta^g}^{1+\alpha}(x) dx - \frac{1}{n} \sum_{i=1}^n u_{j\theta^g}(x) f_{\theta^g}^\alpha(x) \\ &\to \int u_{j\theta^g}(x) f_{\theta^g}^{1+\alpha}(x) dx - \int u_{j\theta^g}(x) f_{\theta^g}^\alpha(x) g(x) dx \\ &= (1 + \alpha)^{-1} \nabla_j H(\theta)|_{\theta = \theta^g} \\ &= 0, \end{aligned}$$

with probability tending to 1. For any given a it therefore follows that $|A_j| < a^2$, and hence $|S_1| < pa^3$ with probability tending to 1.

Similarly, an inspection of the form of the matrix J in (9.14) reveals that $B_{jk} \to J_{jk}$ with probability tending to 1. Consider the representation

$$2S_2 = \sum \sum [-J_{jk}(\theta_j - \theta_j^g)(\theta_k - \theta_k^g)] + \sum \sum \{-B_{jk} + J_{jk}\}(\theta_j - \theta_j^g)(\theta_k - \theta_k^g).$$

For the second term in the above equation it follows from an argument similar to that for S_1 that its absolute value is less that $p^2 a^3$ with probability tending to 1. The first term is a negative (nonrandom) quadratic form in the variables $(\theta_j - \theta_j^g)$. By an orthogonal transformation this can be reduced to a diagonal form $\sum_i \lambda_i \xi_i^2$ with $\sum_i \xi_i^2 = a^2$. As each λ_i is negative, by ordering them as $\lambda_p \le \lambda_{p-1} \le \ldots \le \lambda_1 < 0$, one gets $\sum_i \lambda_i \xi_i^2 \le \lambda_1 a^2$. Combining the first and the second terms, there exist $c > 0$, $a_0 > 0$ such that for $a < a_0$, $S_2 < -ca^2$, with probability tending to 1.

Finally, with probability tending to 1, $|\frac{1}{n} \sum M_{jkl}(X_i)| < 2m_{jkl}$, and hence

$|S_3| < ba^3$ on Q_a where $b = \frac{1}{3}\sum\sum\sum m_{jkl}$. Combining these inequalities, we see that

$$\max(S_1 + S_2 + S_3) < -ca^2 + (b+s)a^3,$$

which is less than zero if $a < c/(b+s)$.

Thus, for sufficiently small a there exists a sequence of roots $\hat{\theta} = \hat{\theta}(a)$ such that $P(||\hat{\theta} - \theta||_2 < a) \to 1$ where $|| \cdot ||_2$ represents the L_2 norm. It remains to show that we can determine such a sequence independently of a. Let θ^* be the root closest to θ. This exists because the limit of a sequence of roots is again a root by the continuity of $H_n(\theta)$ as a function of θ. Then clearly $P(||\theta^* - \theta||_2 < a) \to 1$ for all $a > 0$. This concludes the proof of the existence of a sequence of consistent solutions to the density power divergence estimating equations with probability tending to 1.

We now present the proof of the asymptotic normality of the minimum density power divergence estimator $\hat{\theta}$. Let $H_n(\theta)$ be defined as in (9.10). Expanding $H_n^j(\theta) = \nabla_j H_n(\theta)$ about θ^g we obtain:

$$H_n^j(\theta) = H_n^j(\theta^g) + \sum_k (\theta_k - \theta_k^g)H_n^{jk}(\theta^g) + \frac{1}{2}\sum_k\sum_l(\theta_k - \theta_k^g)(\theta_l - \theta_l^g)H_n^{jkl}(\theta^*)$$

where θ^* is a point on the line segment connecting θ and θ^g, and H_n^{jk} and H_n^{jkl} denote the indicated second and third partial derivatives of H_n. Evaluating the above at $\theta = \hat{\theta}$, we get (since $H_n^j(\hat{\theta}) = 0$)

$$n^{1/2}\sum_k(\hat{\theta}_k - \theta_k^g)\left[H_n^{jk}(\theta^g) + \frac{1}{2}\sum_l(\hat{\theta}_l - \theta_l^g)H_n^{jkl}(\theta^*)\right] = -n^{1/2}H_n^j(\theta^g).$$

This has the form

$$\sum A_{jkn}Y_{kn} = T_{jn} \qquad (9.17)$$

with

$$Y_{kn} = n^{1/2}(\hat{\theta}_k - \theta_k^g),$$

$$A_{jkn} = \left[H_n^{jk}(\theta^g) + \frac{1}{2}\sum_{l=1}^p(\hat{\theta}_l - \theta_l^g)H_n^{jkl}(\theta^*)\right],$$

and

$$T_{jn} = -n^{1/2}H_n^j(\theta^g).$$

By Lemma 4.1 of Lehmann (1983, pp. 432–433), the solutions $(Y_{1n}, \ldots, Y_{pn})$ of (9.17) tend in law to the solutions of

$$\sum_{k=1}^n a_{jk}Y_k = T_j, \ (j = 1, \ldots, p),$$

where $(T_{1n}, \ldots, T_{pn})$ converges weakly to $(T_1, \ldots, T_p)$, and for each fixed j and k, A_{jkn} converges in probability to a_{jk} for which the matrix

$$A = ((a_{jk}))_{p \times p}$$

is nonsingular. The solution $(Y_1 \ldots, Y_p)$ is given by

$$Y_j = \sum_{k=1}^{p} b_{jk} T_k$$

where $B = ((b_{jk}))_{p \times p} = A^{-1}$.

From condition (D5) it follows that $H_n^{jkl}(\theta^*)$ is bounded with probability tending to 1, so that the consistency of $\hat{\theta}$ implies that the second term of A_{jkn} converges to zero in probability. Thus, a_{jk} is simply the limit of $H_n^{jk}(\theta^g)$. As in the proof of the consistency part, $H_n^{jk}(\theta^g) \to (1+\alpha)J_{jk}$, so that $A = (1+\alpha)J$.

Next we will show that $(T_{1n}, \ldots, T_{pn})$ has an asymptotic multivariate normal distribution with mean vector zero and covariance matrix $(1+\alpha)^2 K$ where K is as in Equation (9.15). Note that

$$T_{jn} = -n^{1/2} H_n^j(\theta^g) = -n^{1/2} \frac{1}{n} \sum_{i=1}^{n} V_{j\theta^g}(X_i). \qquad (9.18)$$

Now the best fitting parameter θ^g must minimize the quantity in Equation (9.13) and hence satisfy

$$(1+\alpha) \int f_{\theta_0}^{1+\alpha}(x) u_{j\theta_0}(x) dx - (1+\alpha) \int f_{\theta_0}^{\alpha}(x) u_{j\theta_0}(x) g(x) dx = 0,$$

for $j = 1, \ldots, p$. Thus, we have

$$E\left(V_{j\theta^g}(X)\right) = (1+\alpha) \left[\int f_{\theta^g}^{1+\alpha}(x) u_{j\theta^g}(x) dx - \int f_{\theta^g}^{\alpha}(x) u_{j\theta^g}(x) g(x) dx \right]$$

$$= 0, \qquad (9.19)$$

for $j = 1, \ldots, p$. It then follows from a simple application of the central limit theorem, together with Equations (9.18) and (9.19) that the limiting distribution of $(T_{1n}, \ldots, T_{pn})$ is multivariate normal with mean zero.

The covariance matrix of $(T_1, \ldots, T_p)$ is the covariance matrix of the vector formed by the elements

$$-(1+\alpha) f_\theta^\alpha(X) u_{j\theta}(X)$$

for $j = 1, \ldots, p$. and direct, straightforward calculations show that this covariance equals $(1+\alpha)^2 K$ where $K = K(\theta)$ is as in Equation (9.15).

Thus, the asymptotic distribution of $(T_1, T_2, \ldots, T_p)$ is multivariate normal with mean vector zero and covariance matrix

$$A^{-1}(1+\alpha)^2 K A^{-1} = [(1+\alpha)J]^{-1}(1+\alpha)^2 K[(1+\alpha)J]^{-1}$$

$$= J^{-1} K J^{-1}. \qquad (9.20)$$

This completes the proof. $\qquad\qquad\qquad\qquad\qquad\qquad\qquad\qquad\qquad\qquad \square$

When $g = f_{\theta_0}$ and $\alpha = 0$, one gets $J = K = I(\theta_0)$, the Fisher information matrix, so that the asymptotic covariance matrix in (9.20) is simply the inverse of the Fisher information matrix.

It is useful to note that the minimum density power divergence estimator solves an estimating equation of the form $\sum_{i=1}^{n} \psi(X_i, \theta) = 0$. A comparison with Equation (9.6) shows that the relevant function is given by

$$\psi(x, \theta) = u_\theta(x) f_\theta^\alpha(x) - \int u_\theta(x) f_\theta^{1+\alpha}(x) dx, \tag{9.21}$$

and hence the BHHJ estimator belongs to the class of M-estimators (Huber 1981; Hampel et al. 1986). As a result, many of the properties of this estimator can be directly obtained from the known properties of M-estimators. This also gives an independent verification of the asymptotic distribution of the minimum density power divergence estimator.

9.2.2   Influence Function and Standard Error

Let $T_\alpha(\cdot)$ represent the minimum density power divergence functional based on the tuning parameter α. An implicit differentiation of the estimating equation, or direct application of M-estimation theory, leads to an influence function of the form

$$T_\alpha'(y) = \mathrm{IF}(y, T_\alpha, G) = J^{-1} \left[u_\theta(y) f_\theta^\alpha(y) - \xi \right] \tag{9.22}$$

where $\theta = T_\alpha(y)$, and J and ξ are as in (9.14) and (9.16) evaluated at θ. If we assume that J and ξ are finite, this is a bounded function of y whenever $u_\theta(y) f_\theta^\alpha(y)$ is bounded. This is true for most standard parametric models, including, for example, the normal location-scale case. This is in contrast with density-based minimum distance estimation based on disparities, where the influence function, as we have seen throughout the earlier chapters, equals that of the maximum likelihood estimator at the model and hence is generally unbounded.

In Figure 9.1, we have presented the influence function for the estimates of the normal mean when $\sigma = 1$ for several different values of $\alpha \geq 0$. It is clear that $\alpha = 0$ represents the only case for which the influence function is unbounded. For all the other values of α, the redescending nature of the influence function is clearly apparent.

A consistent estimate of the asymptotic variance of $n^{1/2}$ times the estimator can be obtained in a sandwich fashion using the form of the influence function in Equation (9.22). Given an independently and identically distributed sample $X_1, \ldots, X_n$, let R_i, $i = 1, \ldots, n$, be the quantity $u_\theta(X_i) f_\theta^\alpha(X_i) - \xi$ evaluated at $\theta = \hat{\theta}_\alpha$, the minimum density power divergence estimator, and $G = G_n$. Similarly, let $\hat{J}(G_n)$ be the matrix in (9.14) evaluated at these estimates. Then

$$J^{-1}(G_n) \left[(n-1)^{-1} \sum_{i=1}^{n} R_i R_i^T \right] J^{-1}(G_n)$$

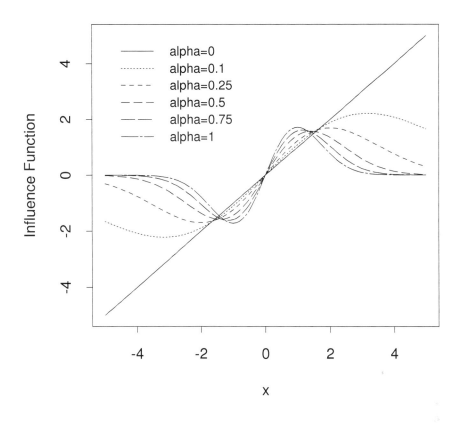

FIGURE 9.1
Influence function for the estimates of the normal mean under the $N(\mu, 1)$ model at the $N(0, 1)$ distribution.

represents a consistent estimate of the asymptotic variance of $n^{1/2}\hat{\theta}_\alpha$.

9.2.3 Special Parametric Families

Here we determine the asymptotic efficiencies of the BHHJ estimators for some common distributions. Obviously we expect better efficiencies for smaller α. Actual numerical calculations are presented later in Table 9.1 after a second family of divergences is introduced.

1. Mean of univariate normal: For a location family $\xi = 0$. Letting f_θ be

the $N(\mu, \sigma^2)$ density, σ^2 known, one gets the relevant quantities as

$$K = (2\pi)^{-\alpha}\sigma^{-(2+2\alpha)}(1+2\alpha)^{-3/2} \quad \text{and} \quad J = (2\pi)^{-\alpha/2}\sigma^{-(2+\alpha)}(1+\alpha)^{-3/2}.$$

Thus, the asymptotic variance of $n^{1/2}$ times the minimum density power divergence estimator of μ is given by

$$\left(1 + \frac{\alpha^2}{1+2\alpha}\right)^{3/2}\sigma^2 = \varsigma_\alpha\sigma^2, \tag{9.23}$$

and hence the asymptotic relative efficiency of the BHHJ estimator equals $1/\varsigma_\alpha$ in this case. For $\alpha = 0.25$, for example, this asymptotic relative efficiency equals 0.941, while the same equals 0.988 at $\alpha = 0.1$.

2. Standard deviation of univariate normal: Now consider the two parameter normal model where f_θ is the $N(\mu, \sigma^2)$ density with both parameters unknown. Direct calculations show that the 2×2 dimensional covariance matrix of the parameter estimates (location and scale) is diagonal, so that the estimates of the mean parameter and the scale parameter are asymptotically independent. The limiting distribution of the estimate of the mean parameter, therefore, remains the same as in the previous case.

For the estimation of σ, lengthy calculations show that the asymptotic variance of $n^{1/2}$ times the estimator is

$$\frac{(1+\alpha)^2}{(2+\alpha^2)^2}\left\{\frac{2Q(\alpha)}{(1+2\alpha)^{5/2}} - \alpha^2\right\}\sigma^2$$

where $Q(\alpha) = 1 + 3\alpha + 5\alpha^2 + 7\alpha^3 + 6\alpha^4 + 2\alpha^5$. This expression can be compared with that of the variance of the maximum likelihood estimator for the computation of asymptotic efficiency.

3. Poisson distribution: Actual closed form solutions are not available, but asymptotic efficiencies can be easily calculated numerically. Specific cases are presented in Table 9.1.

4. Exponential distribution: For the density $f_\theta(x) = \theta^{-1}exp(-x/\theta)$, $x > 0$, the quantities K and J in the asymptotic variance of $n^{1/2}$ times the minimum density power divergence estimator of θ turn out to be

$$K = \left\{\frac{1+4\alpha^2}{(1+2\alpha)^3} - \frac{\alpha^2}{(1+\alpha)^4}\right\}\theta^{-(2+2\alpha)} \quad \text{and} \quad J = \frac{1+\alpha^2}{(1+\alpha)^3}\theta^{-(2+\alpha)}.$$

The asymptotic variance is then given by

$$\frac{(1+\alpha)^2 P(\alpha)\theta^2}{(1+\alpha^2)^2(1+2\alpha)^3}$$

where $P(\alpha) = 1 + 4\alpha + 9\alpha^2 + 14\alpha^3 + 13\alpha^4 + 8\alpha^5 + 4\alpha^6$. Asymptotic relative efficiency is easily obtained by noting that the variance of $n^{1/2}$ times the maximum likelihood estimator is θ^2.

5. Mean of multivariate normal: For the $N_p(\mu, \Sigma)$ family, the limiting covariance matrix of $n^{1/2}$ times the minimum density power divergence estimator of μ (whether or not Σ is known) can be shown to be

$$\left(1 + \frac{\alpha^2}{1 + 2\alpha}\right)^{p/2+1} \Sigma.$$

Thus, one loses efficiency for increasing p if α is kept fixed.

9.3 A Related Divergence Measure

In this section we consider an alternative class of density-based minimum distance estimators and compare it with the BHHJ family of estimators. This development is closely related to a robust model fitting method of Windham (1995), who considered weighting the data using weights "proportional to a power of the density," and constructed weighted moment equations. In this sense Windham's approach also comes under the broad general idea of density-power downweighting. The class of estimators introduced in this section is equivalent to those of Windham when utilizing the likelihood score function in place of general moment choices. However, the key step in Windham's approach is the construction of the moment equation; his model fitting or parameter estimation is not linked to an optimization problem in this context, and Windham does not pursue a minimum distance method. In this section we will present this as a minimum distance problem. Having an associated divergence facilitates the asymptotic analysis as well as the interpretation of the method in many ways. The approach in this section follows that of Jones et al. (2001); the corresponding divergence and class of estimators will be referred to as the JHHB divergence and the JHHB estimators respectively. We will see that the BHHJ and JHHB divergences (and the corresponding estimators) are very closely related, yet in general different.

9.3.1 The JHHB Divergence

Let $\{F_\theta\}$ be the parametric family of distributions, $\theta \in \Theta$, with corresponding distributions denoted by $F_\theta(x)$. Let $\mathcal{G}$ be the class of distributions having densities with respect to the dominating measure, and let $T : \mathcal{G} \to \Theta$ be a statistical functional. For given G in $\mathcal{G}$, let $G_{\beta,t}$ be defined as:

$$dG_{\beta,t}(x) = \frac{f_t^\beta(x)}{E_G[f_t^\beta(X)]} dG(x).$$

Then, given G and the model $\{F_\theta\}$, Windham's estimator θ satisfies:

$$T\{(F_\theta)_{\beta,\theta}\} = T(G_{\beta,\theta}). \tag{9.24}$$

Thus, for the normal variance with zero mean, the motivating case considered by Windham, the scale functional T is the value of θ satisfying

$$\frac{1}{\int \phi^{\beta+1}(x;t)} \int x^2 \phi^{\beta+1}(x;t)dx = \frac{1}{\sum_i \phi^\beta(x;t)} \sum X_i^2 \phi^\beta(X_i;t),$$

where $\phi(\cdot,t)$ is the normal density with mean 0 and scale parameter t, and $X_1,\ldots,X_n$ is an independently and identically distributed sample from the true distribution. When one utilizes the likelihood score function as the choice of the moment functional, Windham's (1995) approach reduces to choosing the estimator $\hat{\theta}_\beta$ to solve the following equation in θ:

$$\frac{\sum_{i=1}^n f_\theta^\beta(X_i)u_\theta(X_i)}{\sum_{i=1}^n f_\theta^\beta(X_i)} = \frac{\int f_\theta^{1+\beta}u_\theta\,dx}{\int f_\theta^{1+\beta}\,dx}. \tag{9.25}$$

The choice of the functional may be justified by arguing that if one wishes to relate a general parameter estimation method to likelihood estimation, one should incorporate the likelihood score function in some appropriate way. Equation (9.25) may also be written as

$$\int f_\theta^{1+\beta}\,dx\, n^{-1}\sum_{i=1}^n f_\theta^\beta(X_i)u_\theta(X_i) - \int f_\theta^{1+\beta}u_\theta\,dx\, n^{-1}\sum_{i=1}^n f_\theta^\beta(X_i) = 0. \tag{9.26}$$

Equation (9.26) also represents an unbiased estimating equation when $g = f_\theta$.

When f_θ represents a location family, then Equations (9.6) and (9.26) both reduce to

$$\sum_{i=1}^n f_t^\gamma(X_i)u_t(X_i) = 0,$$

where γ is α or β, and thus for this special case both approaches are identical. In general, however, the estimating equations are distinct.

We now attempt to link the above estimating equations to appropriate divergences or objective functions. Equation (9.26) is the estimating equation resulting from the maximization of

$$\left\{n^{-1}\sum_{i=1}^n f_\theta^\beta(X_i)\right\}^{1+\beta} \Big/ \left(\int f_\theta^{1+\beta}\,dz\right)^\beta$$

with respect to θ, or the minimization of the negative of its logarithm

$$\beta\log\left(\int f_t^{1+\beta}\,dz\right) - (1+\beta)\log\left\{n^{-1}\sum_{i=1}^n f^\beta(X_i)\right\}. \tag{9.27}$$

Equation (9.27) is a direct analog of Equation (9.10), when the identity function and the log functions are interchanged. Thus, just as the BHHJ estimator $\hat{\theta}_\alpha$ minimizes

$$\alpha \int f_\theta^{1+\alpha} dx - (1+\alpha) \int f_\theta^\alpha g \, dx,$$

the JHHB estimator $\hat{\theta}_\beta$ minimizes

$$\beta \log \left(\int f_\theta^{1+\beta} dx \right) - (1+\beta) \log \left(\int f_\theta^\beta g \, dx \right).$$

It is not difficult to convert the objective function (9.27) into a genuine discrepancy. One version of this is given by

$$\frac{1}{\beta} \left\{ \int g^{1+\beta} dx - \frac{\left(\int f_t^\beta g \, dx \right)^{1+\beta}}{\left(\int f_t^{1+\beta} dx \right)^\beta} \right\}. \tag{9.28}$$

That (9.28) has the required properties of being positive for all $\beta > 0$ with equality if and only if $f_\theta = g$ follows as a special case of Hölder's inequality. Indeed, alternative versions of this discrepancy also yield the exact same $\hat{\theta}_\beta$. For instance, one might prefer

$$\frac{1}{\beta} \{1 - \rho_\beta(g, f_t)\} = \frac{1}{\beta} \left\{ 1 - \frac{\int f_t^\beta g \, dx}{\left(\int g^{1+\beta} dx \right)^{1/(1+\beta)} \left(\int f_t^{1+\beta} dx \right)^{\beta/(1+\beta)}} \right\},$$

where the ρ_β measure has a "correlation" nature, with maximal value 1 only if f_t agrees fully with g. A further form of the divergence which is similar to (9.7) is

$$\varrho_\beta(g, f_\theta) = \log \left(\int f_\theta^{1+\beta} dx \right) - \left(1 + \frac{1}{\beta}\right) \log \left(\int g f_\theta^\beta dx \right) + \frac{1}{\beta} \log \left(\int g^{1+\beta} dx \right). \tag{9.29}$$

Both (9.7) and (9.29) belong to a more general family of divergences, given by

$$\phi^{-1} \left(\int f_t^{1+\gamma} dx \right)^\phi - \frac{1+\gamma}{\gamma} \phi^{-1} \left(\int f_t^\gamma g \, dx \right)^\phi + \frac{1}{\gamma} \phi^{-1} \left(\int g^{1+\gamma} dx \right)^\phi. \tag{9.30}$$

That (9.30) is a divergence can be shown via the Hölder inequality. Selecting $\phi = 1$ immediately gives (9.7), while the limit $\phi \to 0$ recovers the distance in (9.29). Since both divergences are embedded in the same family, comparisons between (9.7) and (9.29) really represent comparisons between $\phi = 1$ and $\phi = 0$ in (9.30). One can develop an estimation method from (9.30) for any

TABLE 9.1
Asymptotic relative efficiencies of type 1 and type 0 minimum divergence estimators, given as ratios of limiting variances, in percent, for various values of $\alpha = \beta$.

Model	Estimator	$\gamma = \alpha = \beta$						
		0.00	0.02	0.05	0.10	0.25	0.50	1.00
Normal μ	both	100	99.9	99.7	98.8	94.1	83.8	65.0
Normal σ	type 1	100	99.9	99.3	97.6	88.8	73.1	54.1
	type 0	100	99.9	99.3	97.5	88.5	70.6	43.3
Exponential	type 1	100	99.8	99.1	96.8	85.8	68.4	50.9
	type 0	100	99.8	99.1	96.7	85.1	63.2	33.8
Poisson ($\theta = 3$)	type 1	100	99.9	99.7	98.8	94.4	85.0	67.9
	type 0	100	99.9	99.7	98.8	94.2	84.2	65.3
Poisson ($\theta = 10$)	type 1	100	99.9	99.7	98.8	94.1	84.0	65.6
	type 0	100	99.9	99.7	98.8	94.0	83.8	64.9
Geometric ($\theta = 0.1$)	type 1	100	99.8	99.1	96.8	85.9	68.4	51.1
	type 0	100	99.8	99.1	96.7	85.1	63.3	33.9
Geometric ($\theta = 0.9$)	type 1	100	99.9	99.4	98.0	92.0	84.1	82.2
	type 0	100	99.9	99.4	98.0	91.7	81.7	71.3

nonnegative value of ϕ. However, for values of ϕ other than 0 and 1, the corresponding estimating equations are not unbiased for finite n, but are so for the asymptotic case. We have focused on the two cases $\phi = 1$ and $\phi = 0$ only. For any other value of ϕ the method can not be represented as an M-estimation method, but consistency and asymptotic normality can be established under mild regularity conditions.

In summary, the JHHB divergence produces another smoothing free class of minimum distance procedures in the same sense as that of the BHHJ divergences. As indicated by the respective tuning parameters, the BHHJ divergence will also be called the type 1 divergence and the JHHB divergence will be called the type 0 divergence with reference to the superfamily of divergences represented by Equation (9.30).

9.3.2   Formulae for Variances

As in case of the type 1 estimator, the type 0 estimator also has an asymptotic normal distribution, which can be easily derived by appealing to M-estimation theory. Let $\hat{\theta}_\beta$ represent the type 0 estimator corresponding to tuning parameter β. Then the asymptotic variance of $n^{1/2}\hat{\theta}_\beta$ has the form $(J_\beta^{(0)})^{-1}K_\beta^{(0)}(J_\beta^{(0)})^{-1}$. To distinguish between the type 0 and type 1 estimators, we will denote the asymptotic variance of the latter by $J_\alpha^{-1}K_\alpha J_\alpha^{-1}$ as done in Jones, Hjort, Harris, and Basu (2001). We will present the formulae

TABLE 9.2
Estimated parameters for the Newcomb data under the normal model for the BHHJ and the JHHB methods.

$\gamma = \alpha = \beta$	0.00	0.02	0.05	0.10	0.25	0.50	1.00
location (type 1)	26.21	26.74	27.44	27.60	27.64	27.52	27.29
location (type 0)	26.21	26.74	27.44	27.60	27.64	27.51	27.25
scale (type 1)	10.66	8.92	5.99	5.39	5.04	4.90	4.67
scale (type 0)	10.66	8.92	5.99	5.38	5.01	4.82	4.34

TABLE 9.3
Estimated parameters for the Drosophila data under the Poisson model for the BHHJ and the JHHB methods.

$\gamma = \alpha = \beta$ Estimator	0.00	0.001	0.01	0.02	0.05	0.10	0.25	0.50	1.00
Type 1	3.059	2.506	0.447	0.394	0.393	0.392	0.386	0.375	0.365
Type 0	3.059	2.506	0.447	0.394	0.392	0.390	0.381	0.362	0.330

for both through the following unified notation. Let

$$L_{j,\gamma} = \int g^j f_\theta^\gamma dx, \quad M_{j,\gamma} = \int g^j f_\theta^\gamma u_\theta dx \quad \text{and} \quad N_{j,\gamma} = \int g^j f_\theta^\gamma u_\theta u_\theta^T dx,$$

where $j = 0, 1$ and γ is positive; the integrals are evaluated at the type 1 or type 0 estimators depending on which kind is being discussed. Under model conditions, $L_{0,1+\gamma} = L_{1,\gamma}$, and so on. For the type 1 estimator, K_α and J_α may be expressed as $K_\alpha = N_{1,2\alpha} - M_{1,\alpha} M_{1,\alpha}^T$ and

$$J_\alpha = \int f_\theta^\alpha (g - f_\theta)(i_\theta - \alpha u_\theta u_\theta^T) dx + N_{0,1+\alpha},$$

where $i_\theta(x) = -\nabla\{u_\theta(x)\}$ is the (positive definite) information function of the model. Under model conditions, these matrices may be presented in the simple form

$$J_\alpha = N_{0,1+\alpha}, \quad \text{and} \quad K_\alpha = N_{0,1+2\alpha} - M_{0,1+\alpha} M_{0,1+\alpha}^T.$$

Let $\xi_\beta = M_{1,\beta}/L_{1,\beta} = M_{0,1+\beta}/L_{0,1+\beta}$. It then turns out

$$K_\beta^{(0)} = \frac{1}{L_{1,\beta}^2} \int g f_\theta^{2\beta} (u_\theta - \xi_\beta)(u_\theta - \xi_\beta)^T dx$$

$$= \frac{1}{L_{1,\beta}^2} (N_{1,2\beta} - \xi_\beta M_{1,2\beta}^T - M_{1,2\beta} \xi_\beta^T + \xi_\beta \xi_\beta^T L_{1,2\beta}).$$

Under model conditions,

$$K_\beta^{(0)} = (1/L_{0,1+\beta}^2)(N_{0,1+2\beta} - \xi_\beta M_{0,1+2\beta}^T - M_{0,1+2\beta}\xi_\beta^T + \xi_\beta\xi_\beta^T L_{0,1+2\beta}).$$

Also,

$$
\begin{aligned}
J_\beta^{(0)} &= \frac{N_{0,1+\beta}}{L_{0,1+\beta}} - \frac{M_{0,1+\beta}}{L_{0,1+\beta}}\left(\frac{M_{0,1+\beta}}{L_{0,1+\beta}}\right)^T + \beta\left(\frac{N_{0,1+\beta}}{L_{0,1+\beta}} - \frac{N_{1,\beta}}{L_{1,\beta}}\right) \\
&+ \frac{1}{L_{1,\beta}}\int g f_\theta^\beta i_\theta dx - \frac{1}{L_{0,1+\beta}}\int f_\theta^{1+\beta} i_\theta dx,
\end{aligned}
$$

which reduces under model conditions to $J_\beta^{(0)} = N_{0,1+\beta}/L_{0,1+\beta} - \xi_\beta\xi_\beta^T$.

Let $\alpha = \beta \to 0$, and write K for K_α or $K_\beta^{(0)}$, and J for J_α or $J_\beta^{(0)}$. Then we get

$$K \to N_{1,0} = \int g u_\theta u_\theta^T dx, \quad J \to N_{0,1} + \int (g i_\theta - f_\theta i_\theta)dx = \int g i_\theta dx,$$

which is as one would expect from the limiting behavior of the maximum likelihood methods at arbitrary distributions outside the model.

9.3.3   Numerical Comparisons of the Two Methods

Here we present a brief numerical comparison of the two methods through efficiency comparisons under standard models and some real data examples.

Efficiency calculations based on the asymptotic variance formulae derived above lead to the numbers presented in Table 9.1, which gives an idea of the performance of the methods under model conditions. For values of the tuning parameter close to zero, the performances of the two methods are practically identical. For larger values of the parameter, typically, the type 1 method is more efficient; however, part of it could be due to the incorrect calibration of the methods for $\alpha = \beta$.

Tables 9.2 and 9.3 illustrate the robustness of the techniques for the Newcomb data (under the normal model) and the Drosophila data, second experimental run (under the Poisson model), also analyzed in this book through several other model fitting techniques. The examples clearly show how the tuning parameters (α or β) control the trade-off between efficiency and robustness.

9.3.4   Robustness

The primary source of the robustness of the BHHJ and JHHB estimators is the density power downweighting philosophy and the effect of such downweighting has been observed in the influence function of the estimators and real data performances. Here we mention some additional points about the robustness of the type 1 and type 0 estimators.

Both from heuristic considerations and empirical observations, the BHHJ and JHHB techniques are extremely close to each other in performance for small values of the tuning parameter ($\alpha = \beta$). As $\alpha = \beta$ increases, some differences between the estimators in terms of their influence function and other attributes become visible (e.g., Figure 1, Jones et al. 2001). The differences are, in general, small. In calculations of mean square error under contamination, the type 0 estimator tended to slightly outperform the type 1 estimator, although in some exceptional cases the type 1 estimator performed significantly better.

In breakdown calculations it has been observed that for the normal case with both parameters unknown, the breakdown point for the mean parameter is $\alpha/(1+\alpha)^{3/2}$ for the type 1 estimator. However, similar calculations show that for the type 0 estimator the corresponding bound is zero; this was indicated as a possibility by Windham (1995). This gives the type 1 error a small edge in robustness.

9.4   The Censored Survival Data Problem

In this section we will present an adaptation of the density power divergence approach (the BHHJ method) of Basu et al. (1998), as described by Basu, Basu and Jones (2006), which is useful in the context of censored survival data problems. In this case the data are not fully observed, and their analysis is complicated by various censoring mechanisms that come into play. As we have discussed before, a major advantage of the BHHJ method is that it does not require any nonparametric smoothing to produce estimates of the true underlying density function. In Section 9.2, the empirical distribution itself has been shown to be sufficient for constructing an estimate of the divergence. For the right censored situation, where the data are not fully observed, the Kaplan–Meier estimator (Kaplan and Meier, 1958) provides a consistent estimate of the distribution function; thus one can still obtain an estimate of the BHHJ divergence without introducing a nonparametric density estimate, and the minimum density power divergence estimators can then be obtained in a routine manner.

The BHHJ divergence has already been introduced in Equation (9.7), and its empirical version, presented in Equation (9.10) may be expressed, in case of independently and identically distributed data, as

$$\int f_\theta^{1+\alpha}(x)dx - \left(1 + \frac{1}{\alpha}\right) \int f_\theta^\alpha(x)dG_n(x). \qquad (9.31)$$

In the right censoring problem, each random observation X_i from the target distribution G is associated with a random C_i from the censoring distribution H (assumed to be independent of G); one does not observe X_i directly, and has information only on $Y_i = \min(X_i, C_i)$ and δ_i (the indicator of the event $X_i < C_i$). Under appropriate regularity conditions and the assumption that the target distribution G and the censoring distribution H have no common points of discontinuity, the Kaplan–Meier estimator of the survival function $\hat{S}_n(x)$ converges almost surely to the true survival function $S(x)$. Thus, $\hat{G}_n = 1 - \hat{S}_n$ is a consistent estimator of the true distribution function in this context and the empirical estimate of the density power divergence can now be constructed by replacing G_n with $\hat{G}_n$ in (9.31). The parameter estimate is the minimizer of

$$\int f_\theta^{1+\alpha}(x)dx - \left(1 + \frac{1}{\alpha}\right) \int f_\theta^\alpha(x)d\hat{G}_n(x) = \int V_\theta(x)d\hat{G}_n(x)$$

where $V_\theta(x)$ is as in Equation (9.11). The minimum density power divergence estimators of the parameter θ are obtained as the solution of the equation

$$\int \psi_\theta(x)d\hat{G}_n(x) = 0,$$

where the elements of $\psi_\theta(\cdot)$ represent the partial derivatives of $V_\theta(x)$ with respect to the components of θ.

To determine the asymptotic distributions of the BHHJ estimators in the case of censored survival data, a law of large numbers and central limit theorem type results for general functionals $\int \phi(x)d\hat{G}_n(x)$ of the Kaplan–Meier estimator are needed. This theory has been established in a series of publications by Stute and Wang (Stute and Wang, 1993; Stute, 1995; Wang, 1995, 1999). The above results show that the BHHJ estimators for the right censoring case have appropriate limiting normal distributions. The asymptotic covariance matrix depends, among other things, on the true distribution G and the censoring distribution H. See Basu, Basu and Jones (2006) for more details.

9.4.1   A Real Data Example

We consider a real data application of this technique based on an example taken from Efron (1988). The analysis presented here has been reproduced from Basu, Basu and Jones (2006). The data were generated from a study comparing radiation therapy alone (arm A) and radiation therapy and chemotherapy (arm B) for the treatment of head and neck cancer. A total of 51 patients were assigned to arm A of the study; 9 of them were lost to follow-up and, therefore, censored. On the other hand, 45 patients were assigned to arm B of the study of which 14 were lost to follow-up. The censoring levels are quite high, ranging approximately between 20% and 30%. Efron analyzed the data from various angles, and his analysis suggests that radiation and chemotherapy (as in arm B) is more effective in terms of survival time. We check the appropriateness of the BHHJ fits of the Weibull model to these data.

The maximum likelihood estimates and the BHHJ estimates of the two

TABLE 9.4

Analysis of Efron data under the Weibull model: Arm A.

	α	Scale, $\hat{a}$	Shape, $\hat{b}$
ML	0	399.24	0.91
BHHJ	0.001	418.18	0.98
	0.01	417.72	0.98
	0.1	412.72	0.99
	0.2	402.51	1.00
	0.25	395.31	1.02
	0.5	321.90	1.16
	0.75	252.85	1.44
	1.0	249.47	1.47

TABLE 9.5

Analysis of Efron data under the Weibull model: Arm B.

	α	Scale, $\hat{a}$	Shape, $\hat{b}$
ML	0	925.45	0.76
BHHJ	0.001	789.23	0.91
	0.01	790.07	0.91
	0.1	791.81	0.90
	0.2	789.26	0.90
	0.25	785.13	0.90
	0.5	726.72	0.93
	0.75	551.53	1.03
	1.0	343.07	1.31

Weibull parameters a and b are given for various values of the tuning parameter α in Tables 9.4 and 9.5. There are significant changes in the values of both parameters as the tuning parameter α moves from 0 to 1. In particular, estimate of the shape parameter b moves to the other side of 1 as α increases.

Figures 9.2 and 9.3 illustrate the results for Arms A and B, respectively. On each figure is shown:

(a) a kernel density estimate obtained by smoothing the Kaplan–Meier estimator (e.g., Wand and Jones, 1995, Section 6.2.3), where the bandwidth was subjectively chosen to avoid oversmoothing;

(b) the maximum likelihood fit to the Weibull model;

(c) the BHHJ fit to the Weibull model for $\alpha = 1$, and

(d) the maximum likelihood fit to the two component Weibull mixture model. It is clear that in each case the major part of the data are con-

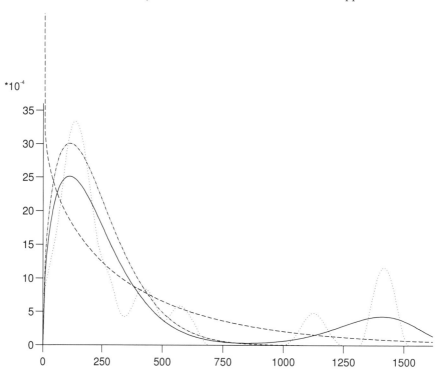

FIGURE 9.2
Kernel density estimate (dotted line), the maximum likelihood fit to the Weibull model (dashed line), the BHHJ ($\alpha = 1$) fit to the Weibull model (dot-dashed line) and the maximum likelihood fit to the two component Weibull mixture (solid line) for Arm A of the Efron (1988) data.

centrated on the left, together with a few long-lived individuals to the right.

The likelihood fits to the Weibull model are monotone decreasing since $\hat{b} < 1$. These fits try to accommodate the main body as well as the long tail, and end up failing to provide a good model for either. The robust BHHJ fit corresponding to $\alpha = 1$ appears to model the main body of the data adequately, while essentially ignoring the long tail. In this sense, the performance of the robust estimators is consistent with the robustness philosophy as appropriate for this case.

In cases where the contamination proportion is substantial, the contaminating component may also be of interest and could be modeled. The maximum likelihood fits to the two component Weibull mixtures are thus also presented in Figures 9.2 and 9.3; the associated parameter estimates, in an obvious notation, are reported in Table 9.6. In Figure 9.2, the Weibull mixture confirms the robust fit to be appropriate for the main body of data; the second

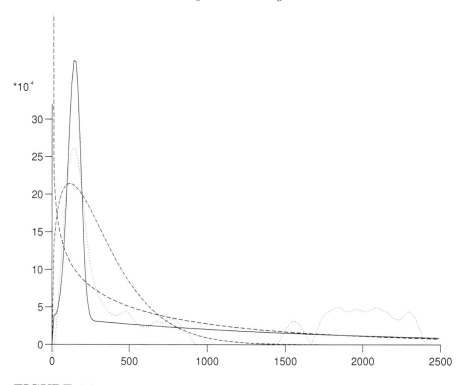

FIGURE 9.3
Kernel density estimate (dotted line), the maximum likelihood fit to the Weibull model (dashed line), the BHHJ ($\alpha = 1$) fit to the Weibull model (dot-dashed line) and the maximum likelihood fit to the two component Weibull mixture (solid line) for Arm B of the Efron (1988) data.

TABLE 9.6
Maximum likelihood parameter estimates for Efron data assuming two component Weibull mixture model.

		Arm A	Arm B
	$\hat{p}$	0.82	0.35
First Component	scale, $\hat{a}$	241.53	156.00
	shape, $\hat{b}$	1.47	4.08
Second Component	scale, $\hat{a}$	1428.11	1800.00
	shape, $\hat{b}$	9.17	0.90

component covers the smaller tail. In Figure 9.3, the mixture has a different form, showing a narrow peak to the left and a long flat tail to the right. It appears that further modeling based on models with heavier tails is necessary for the arm B data.

9.5   The Normal Mixture Model Problem

In this section we give a brief description of the application of the minimum density power divergence method in case of normal mixture models following Fujisawa and Eguchi (2006). The latter class of models represents a very flexible class that can fit a wide variety of shapes and structures. The maximum likelihood is the most common model fitting technique used in this connection; the popularity of the method is due in a large part to the availability of the EM algorithm of Dempster, Laird and Rubin (1977). However, this flexibility can also make the model behave in undesirable ways because of its lack of robustness. The likelihood can grow without bound when one of the component means coincides with an observed data point, while the corresponding variance is allowed to go to zero; the sensitivity of the maximum likelihood estimator to outlying values can generate unduly inflated variances leading to incorrect inference including incorrect estimation of mixing proportions. Even when the likelihood does not explode, the method can fit contaminating clusters as separate components. The likelihood may be made bounded by adding additional conditions such as the equality of variances, but in that case the model loses much of its flexibility. While we acknowledge all the positives of the normal mixture model fitted by the method of maximum likelihood, there is clearly a need for a more robust fitting method.

Consider the normal mixture model

$$f_\theta(x) = \sum_{i=1}^{k} \pi_i \phi(x, \mu_i, \sigma_i^2),$$

where k is the number of components, π_i, $i = 1, \ldots, k$ represent the mixing proportions, and $\phi(\cdot, \mu, \sigma^2)$ represents a normal density with mean μ and variance σ^2. Using the density power divergence, the objective function (or the "α-likelihood") to be maximized for estimation of the parameters of the mixture model is given in Equation (9.12). Fujisawa and Eguchi (2006) argue that the bound for the gross error sensitivity of the procedure – the supremum of the absolute value of the influence function – is of the order of $O(1/\alpha)$. On the other hand, $\text{Var}(n^{1/2}\hat{\theta}_\alpha) = \text{Var}(n^{1/2}\hat{\theta}) + O(\alpha^2)$ as $n \to \infty$ and $\alpha \to 0$, where $\hat{\theta}$ is the maximum likelihood estimator, so that the asymptotic relative efficiency of the estimator approaches 1 in the order of $O(\alpha^2)$. Thus, appropriate small values of α may be expected to perform well both in terms of robustness and efficiency.

Fujisawa and Eguchi (2006) show that the α-likelihood is bounded above for all $\alpha > 0$ under standard regularity conditions. In particular, the condition

$$\min_j \pi_j \geq \phi(\alpha)/n$$

is needed, where $\phi(\alpha) = (1 + \alpha)^{3/2}/\alpha$. For example, when $\alpha = 0.2$, and $n = 100$, the above condition reduces to $\min_j \pi_j \geq 0.066$.

The authors also considered the selection of the appropriate tuning parameter. This is a difficult but important problem, since the performance of the estimator can critically depend on the value of α. Heuristically they stated their preference for values of α in the neighborhood of $\alpha = 0.2$. They also suggested a data based choice by creating an objective function over α and then choosing α to minimize it. In particular, they considered the empirical Cramér von-Mises distance between the vector $\{(i - 0.5)/n\}_{i=1,\ldots,n}$ and the estimated distribution function values at the ordered $\{x_i\}$ vector, estimated at $\theta = \hat{\theta}_\alpha$, but using a "leave-one-out strategy" in the spirit of cross validation. Thus $F_{\hat{\theta}_\alpha}(x_i)$ is obtained by estimating the parameter using the reduced data leaving out x_i. To implement the "leave-one-out strategy," another approximation based on density power divergence is used, sparing the experimenter the trouble of having to go through n optimizations for each α.

Finally, to estimate the minimum density power divergence estimators of the parameters of the normal mixture model, Fujisawa and Eguchi (2006) suggest an iterative algorithm similar to the EM algorithm. The algorithm reduces to the ordinary EM algorithm for $\alpha = 0$, and, like the EM algorithm, is locally contractive in the neighborhood of an appropriate local minimum for small α. Detailed calculations are given in the Appendix of Fujisawa and Eguchi (2006).

9.6 Selection of Tuning Parameters

As is clear from the discussion of the previous sections, the choice of the tuning parameter is an important component in the performance of the minimum density power divergence estimator. The problem is a difficult one. However, there have been some general attempts to choose data driven tuning values for the implementation of the method. Hong and Kim (2001) suggested the minimization of the asymptotic variance for the choice of the tuning parameter. Warwick and Jones (2005) estimated the mean squared error of the model parameters via their asymptotic approximations and chose the tuning parameters to minimize the estimated mean squared error. The approach of Hong and Kim, in a sense, is a special case of Warwick and Jones where one ignores the bias component. Here we give a brief description of approach of Warwick and Jones, and summarize their findings.

Warwick and Jones (2005) considered the summed mean square error $E[(\hat{\theta}_\alpha - \theta^*)^T (\hat{\theta}_\alpha - \theta^*)]$ where θ^* is an appropriate target value and $\hat{\theta}_\alpha$ is the minimum density power divergence estimator. The summed mean square error is represented as the sum of the bias component and the variance component given by $(\theta_\alpha - \theta^*)^T (\theta_\alpha - \theta^*)$ and $n^{-1} tr[J_\alpha^{-1}(\theta_\alpha) K_\alpha(\theta_\alpha) J_\alpha^{-1}(\theta_\alpha)]$ respectively. Here θ_α is the asymptotic mean of $\hat{\theta}_\alpha$ so that $(\theta_\alpha - \theta^*)$ is the asymptotic bias. The matrices J_α and K_α are as in (9.14) and (9.15) with α subscripts attached (since dependence on α is the issue here). Also $tr(\cdot)$ represents the trace of the matrix.

Estimating the variance component is the easier of the two, at least in principle, if not numerically. For this one simply replaces G with the empirical G_n and θ_α by $\hat{\theta}_\alpha$. To estimate the asymptotic bias one can replace θ_α by $\hat{\theta}_\alpha$, but the issue of what value to employ for θ^* is a difficult one. Several "pilot" choices were tried out by Warwick and Jones (2005). Denoting by $\hat{\theta}(P)$ the estimator obtained with estimated α and pilot estimator P, the authors declare their overall preference for $\hat{\theta}(L_2)$, i.e., the final estimator with estimated α that uses the minimum L_2 estimator as the pilot estimator. This judgment is given on the basis of a large scale numerical study; $\hat{\theta}(L_2)$ has several competitors, but seems to be the best choice on the whole. The authors also demonstrate that the effect of the pilot estimator on the final data based choice of α can be considerable.

9.7   Other Applications of the Density Power Divergence

In this chapter we have provided a compact description of the minimum density power divergence and related methods. Although the method is not very old, it has been followed up by several authors and it has seen wide application in several other areas. Here we provide a brief description of some such applications. While this is nowhere close to an exhaustive description, at least it will give a flavor of the general usefulness of the divergence.

We have already described the application of the density power divergence method in case of Gaussian mixture models as considered by Fujisawa and Eguchi (2006). These authors and other associates of Shinto Eguchi have extensively explored the density power divergence in many settings, where they have referred to the objective function as β-divergence (or, alternatively, referred to the negative of the quantity as the β-likelihood). In particular, Fujisawa and Eguchi have considered the use of the density power divergence method against heavy contamination (Fujisawa and Eguchi, 2008). Minami and Eguchi (2002) dealt with robust blind source separation based on the density power divergence for recovering original independent signals when their linear mixtures are observed. Mollah, Minami and Eguchi (2006, 2007) have applied the density power divergence in the area of Independent Component Analysis (ICA). Also see Eguchi and Kato (2010) and Mollah et al. (2010) for other various applications of the density power divergence method.

Lee and Na (2005) have considered the problem of testing for a parameter change in terms of the cusum test constructed using the density power divergence measure, while Miyamura and Kano (2006) used the density power divergence for robust Gaussian graphical modeling.

Juárez and Schucany (2004) have considered the estimation of the shape and scale parameters of the generalized Pareto distribution based on the density power divergence; the method appears to have been integrated in standard softwares for generalized Pareto distributions. The authors have also provided an alternative proof of the asymptotic normality of the minimum density power divergence estimator under conditions that are different from those originally considered by Basu et al. (1998).

The implementation of the BHHJ measure of divergence in statistical model selection has also been considered; it has been investigated in Mattheou, Lee and Karagrigoriou (2009) and Karagrigoriou and Mattheou (2009). The authors have considered a model selection criterion called the divergence information criterion (DIC), which appears to perform well, especially in regression models. The authors have explored the performance of the method in detail, and also investigated the mean squared error of prediction. A modification of the DIC criterion was introduced in Mantalos, Mattheou and Karagrigariou (2010a) to make it appropriate for the time series model, and the authors used the modified criterion for the determination of the order of an AR process; the same criterion was used in the context of forecasting ARMA processes by Mantalos, Mattheou and Karagrigoriou (2010b) and in the context of vector AR processes by Mantalos, Mattheou and Karagrigoriou (2010c).

The application of the BHHJ divergence in goodness-of-fit testing has also been explored by several authors, and a generalized BHHJ measure has been proposed (Mattheou and Karagrigoriou, 2009). The consistency and the asymptotic normality of a properly defined estimator of the generalized BHHJ measure has been established in Karagrigoriou and Mattheou (2010). Several related aspects including the asymptotic distribution of the goodness-of-fit tests under the null hypothesis and under contiguous alternatives have been studied in Mattheou and Karagrigoriou (2009, 2010). Vonta and Karagrigoriou (2010) have combined the Csiszár family of measures and the BHHJ measure and studied generalized distances of residual lifetimes and past lifetimes for reliability and survival models, including the Cox proportional hazards models and the frailty or transformation models. Also see Harris and Basu (1997) for a generalized divergence involving disparities and density power divergences.

10

Other Applications

This book has attempted to give an extensive exposition of the use of density-based minimum distance methods in statistics and other applied sciences, with particular emphasis of the use of χ^2-type distances. In terms of the scope and utility of the methods, this represents a huge research area, and to keep a clear focus in our discussion we have primarily restricted ourselves to parametric inference problems where an independently and identically distributed sample is available from the true distribution. Within this setup, our description has been fairly comprehensive.

However, real world problems often involve models and descriptions that go beyond the one sample inference scenario for independently and identically distributed data. In the last two decades there have been several applications of density-based minimum distance methodology for solving more specialized problems that require further extension of the general techniques described in this book. In this chapter we summarize some important research works in these directions. When presented in full detail, this topic can be a whole new book by itself; our intention here is to provide a brief introduction to some of these techniques without getting into the deeper technical aspects. The discussion will be brief, and we will emphasize the results and avoid the proofs.

For specific illustrations we will focus the application of density-based minimum distance methods in problems of survival data, grouped data, mixture models and semiparametric problems. We will also indicate some of the references in allied areas involving minimum distance methods. In this chapter we will occasionally deviate from our standard notation to present the description in the notation of the author(s) who first developed it.

10.1 Censored Data

10.1.1 Minimum Hellinger Distance Estimation in the Random Censorship Model

Consider the random censorship model, where $X_1, \ldots, X_n$ are independently and identically distributed random variables with cumulative distribution function F supported on $[0, \infty)$, and $Y_1, \ldots, Y_n$ are independently and identically distributed censoring variables independent of the X_i's having distribution function G supported on $[0, \infty]$. The distribution G is allowed to have positive mass at ∞; if G is degenerate at ∞, we are able to observe the complete data $X_1, \ldots, X_n$. In the random censorship model, the pair $\{\min(X_i, Y_i), \chi(X_i \leq Y_i)\}$ is observed, where $\chi(A)$ is the indicator function for the set A. F is assumed to have density f with respect to the Lebesgue measure λ, which is modeled by the parametric family of densities $\{f_\theta : \theta \in \Theta \subset \mathbb{R}^p\}$. For a generic distribution function H, we denote $\bar{H} = 1 - H$. For any (sub-) distribution function D, let $D^{-1}(t) = \inf\{u : D(u) \geq t\}$, and $\tau_D = D^{-1}(1)$. Also let $\tau = \min(\tau_F, \tau_G)$.

Yang (1991) considered minimum Hellinger distance estimation of the parameter θ under the above model where the data are randomly censored. Direct construction of a kernel density estimate using the ordinary empirical distribution of $X_1, \ldots, X_n$ is not possible since the data are not fully observed. Yang suggested the use of its product-limit estimator (Kaplan and Meier, 1958) which we will denote by $\hat{F}_n$. The nonparametric kernel density estimate of the true density f is thus constructed as

$$f_n(x) = h_n^{-1} \int K((x - y)/h_n) d\hat{F}_n(y),$$

for some suitable kernel function K and bandwidth h_n. Under appropriate conditions the kernel density estimate f_n converges to the true density in the Hellinger metric, i.e., $\|f_n^{1/2} - f^{1/2}\|_2 \to 0$ in probability as $n \to \infty$. However, the product-limit estimator $\hat{G}_n$ estimates a function which is identical with G only on $[0, \tau]$.

Let $\tilde{X}_i = \min(X_i, Y_i)$ and $\delta_i = \chi(X_i \leq Y_i)$. Yang considers the joint density function of the observed data; $(\tilde{X}_1, \delta_1)$ has a density $\bar{F}^{1-y}(x) f^y(x)$ at (x, y) with respect to the measure μ_G on $\mathbb{R} \times \{0, 1\}$, where μ_G is defined by the relation

$$\int m \, d\mu_G = \int m(x, 0) dG(x) + \int m(x, 1) \bar{G} dx$$

for any nonnegative measurable function m on $\mathbb{R} \times \{0, 1\}$. We define, for any subdensity d on $\mathbb{R}$ with respect to λ, a subdensity $L(d)$ on $\mathbb{R} \times \{0, 1\}$ with respect to μ_G by

$$L(d)(x, y) = \bar{D}^{1-y}(x) d^y(x),$$

where D is the (sub-) distribution function of d. For a (sub-) distribution function G and a subdensity function d, the minimum Hellinger distance functional $\Psi(d, G)$ is the element of the parameter space provided one exists, that minimizes the Hellinger distance between $L(f_\theta)$ and $L(d)$, so that

$$\|[L(f_{\Psi(d,G)})]^{1/2} - [L(d)]^{1/2}\|_G = \inf_{\theta \in \Theta} \|[L(f_\theta)]^{1/2} - [L(d)]^{1/2}\|_G$$

where $\| \cdot \|_G$ denotes the L_2 norm in $L_2(\mu_G)$. The above definition needs a modification due to the fact that one can only estimate $G(t)$ for t up to τ. So, for $0 \leq \gamma \leq \infty$, define $\Psi(\cdot, G, \gamma)$ similarly where the integration is restricted to $x \in (-\infty, \gamma]$; it is easily seen that $\Psi(\cdot, \cdot, \infty) = \Psi(\cdot, \cdot)$.

Under the above setup, Yang extends the results of Beran to the case of a random censorship model. To simplify notation, we only look at the case when the parameter is one dimensional. The minimum Hellinger distance estimator of $\psi(f, G, \tau)$ is defined by $\hat{\theta}_n = \psi(f_n, \hat{G}_n, T)$, where $T = \max\{\tilde{X}_i\}$. Under additional conditions to accommodate the censoring mechanism, he shows that

$$\psi(f_n, \hat{G}_n, T) \to \psi(f, G, \tau)$$

in probability, while under $L(f)$, $n^{1/2}(\hat{\theta}_n - \Psi(f, G, \tau))$ converges weakly to a normal distribution with mean 0 and finite variance. In particular, under $L(f_\theta)$, $n^{1/2}(\hat{\theta}_n - \Psi(f, G, \tau))$ converges weakly to $N(0, 1/I)$, where

$$I = E\left[\frac{\partial}{\partial \theta} \left(\ln L(f_\theta)(\tilde{X}, \delta) \right) \right]^2,$$

is the Fisher information matrix.

10.1.2 Minimum Hellinger Distance Estimation Based on Hazard Functions

Another interesting minimum distance approach has been presented by Ying (1992), where a class of minimum distance estimators for the parameters of the distribution have been presented, where the data are subject to a possible right censorship. The primary difference of this work from that of Yang (1991) is that Ying (1992) uses a Hellinger type distance function for the hazard rate function rather than for the density functions themselves. Ying discusses several advantages of using the hazard rate function. The martingale integral approximation is a key to this approach, and Ying argues that this gives the experimenter the flexibility to trim off the "tail" data without introducing significant bias; this is an important advantage in survival analysis. The approach also allows the use of weaker conditions than those employed by Beran (1977) and Yang (1991). The approach may also be extended to handle multiplicative counting processes. The approach can handle both left truncated and right censored data.

Ying derives the consistency and asymptotic normality of a restricted minimum Hellinger type distance estimator which only uses the information pertaining to those observations falling into a fixed finite interval, and incorporates an appropriate weight function into the distance. The resulting minimum Hellinger distance estimator is consistent and asymptotically normal under regularity conditions satisfied by common parametric models, and has the same efficiency as the maximum likelihood estimator in these situations.

10.1.3 Power Divergence Statistics for Grouped Survival Data

Chen, Lai and Ying (2004) have provided an interesting adaptation of power divergence statistics for grouped survival data. The minimum distance estimators derived here under parametric assumptions are shown to be asymptotically equivalent to the maximum likelihood estimators. The measures are also used for testing goodness-of-fit.

Suppose that there are n study subjects whose failures and censoring times are denoted by X_j and Y_j, $j = 1, \ldots, n$, so that observations consist of $\tilde{X}_j = \min(X_j, Y_j)$ and $\delta_j = \chi(X_j < Y_j)$. Suppose that the followup period of the survival study is the interval between 0 and β, partitioned into k subintervals $(\beta_{i-1}, \beta_i]$, $i = 1, \ldots, k$, where $0 = \beta_0 < \beta_1 < \cdots < \beta_k = \beta$. Let n_i represent the number of subjects at risk at the beginning of the i-th interval and let d_i be the number of failures during the interval. It will be assumed that censoring occurs only at the beginning of each interval. Let X_j be independent with a common distribution function F. If censoring occurs only at β_i, $i = 1, \ldots, k$, the distribution of d_i given $\{d_1, \ldots, d_{i-1}, n_1, \ldots, n_{i-1}\}$ (more technically, given the σ-field $\mathcal{F}_i$ generated by $\{d_1, \ldots, d_{i-1}, n_1, \ldots, n_{i-1}\}$), has the binomial distribution

$$P(d_i = y | \mathcal{F}_i) = \binom{n_i}{y} h_i^y (1 - h_i)^{n_i - y}, y = 0, 1, \ldots, n_i, \tag{10.1}$$

where $h_i = (F(\beta_i) - F(\beta_{i-1}))/(1 - F(\beta_{i-1}))$. Then the power divergence statistic for grouped survival data may be presented as

$$\frac{2}{\lambda(\lambda + 1)} \sum_{i=1}^{k} \left\{ d_i \left[\left(\frac{d_i}{n_i h_i} \right)^\lambda - 1 \right] + d_i^c \left[\left(\frac{d_i^c}{n_i h_i^c} \right)^\lambda - 1 \right] \right\}.$$

The authors show that the above statistic converges to a chi-square with k degrees of freedom under (10.1) as $n_i \to \infty$ for each i. The statistic above is an interesting analog of the likelihood used to describe this scenario. Here $d_i^c = n_i - d_i$ and $h_i^c = 1 - h_i$.

Consider a parametric model which specifies $h_i = h_i(\theta)$ where h_i are twice continuously differentiable. For each λ one can define the minimum distance estimators for the parameters by minimizing the quantity

$$\frac{2}{\lambda(\lambda + 1)} \sum_{i=1}^{k} \left\{ d_i \left[\left(\frac{d_i}{n_i h_i(\theta)} \right)^\lambda - 1 \right] + d_i^c \left[\left(\frac{d_i^c}{n_i h_i^c(\theta)} \right)^\lambda - 1 \right] \right\}.$$

Denoting the minimum distance estimator by $\hat{\theta}_\lambda$, the asymptotic distribution of $n^{1/2}(\hat{\theta}_\lambda - \theta_0)$ is normal with mean zero and a variance which is independent of λ. Thus the asymptotic distribution of the minimum distance estimators are all identical.

10.2 Minimum Hellinger Distance Methods in Mixture Models

We have discussed estimation of the parameters in a Gaussian mixture model based on the density power divergence in Chapter 9. In this section we consider minimum Hellinger distance estimation under different parametric mixture model settings. In particular, we will consider, in some detail, the works of Woodward, Whitney and Eslinger (1995) and Cutler and Cordero-Braña (1996). We will also point out the application of the minimum Hellinger distance method in some specific contexts.

Woodward, Whitney and Eslinger (1995) considered the mixture of two known densities where the mixing proportion is the only unknown quantity. The authors used the Epanechnikov kernel and obtained the appropriate bandwidth based on a method of Parzen (1962). Notice that in this case the parameter space $[0, 1]$ is compact, facilitating the use of the results of Beran (1977) and Tamura and Boos (1986). Under appropriate regularity conditions, the authors derived an asymptotic model optimality result for the minimum Hellinger distance estimator in the spirit of Theorem 4.1 of Tamura and Boos (Theorem 3.4 of this book). They also discussed the practical feasibility of employing the minimum Hellinger distance estimator in this setting, and presented detailed empirical results; the theoretical and empirical results suggest that the minimum Hellinger distance estimator attains full asymptotic efficiency at the model, and is competitive with the minimum Cramér–von-Mises distance estimator (Woodward et al. 1984) under this setting. Woodward, Whitney and Eslinger also empirically investigated the performance of the minimum Hellinger distance estimator in case of normal mixtures where all the involved parameters are unknown.

Cutler and Cordero-Braña (1996) considered the problem of minimum Hellinger distance estimation over a more expanded class of models in the mixture setting. They chose the model

$$g_\theta(x) = \sum_{i=1}^{k} \pi_i f_{\phi_i}(x),$$

where ϕ_i, $i = 1, \ldots, k$ represent s dimensional parameters, and

$$\theta = (\pi_1, \ldots, \pi_{k-1}, \phi_1^T, \ldots, \phi_k^T)^T$$

is the full parameter vector of interest. Cutler and Cordero-Braña consider the continuity of the minimum Hellinger distance functional with respect to the Hausdorff metric. Under appropriate conditions on the parameter space and suitable identifiability conditions for the model, the authors show that the minimum Hellinger distance functional $T(G)$ exists, is (Hausdorff) continuous in the Hellinger topology at G, and is essentially unique under model conditions. A Hausdorff consistency condition in place of the usual consistency criterion leads to the asymptotic normality result along the lines of Tamura and Boos (1986).

Cutler and Cordero-Braña also empirically studied the α-influence function of the minimum Hellinger distance functional and the breakdown properties of the latter under normal mixtures. In either case, the minimum Hellinger distance estimator is shown to be much more resilient to data contamination than the corresponding maximum likelihood estimator.

As mentioned before, the popularity of the maximum likelihood estimators in a mixture model is in a large part due to the presence of a useful algorithm like the EM algorithm. Cutler and Cordero-Braña presented an algorithm called the HMIX which is a helpful iterative tool in determining the minimum Hellinger distance estimator in this situation. The algorithm successively maximizes over the parameter θ, and over the (normalized) weights of the k components, keeping the other variables fixed. The algorithm, like the EM algorithm, has its own convergence problems, so appropriate starting values are very important. Also, the procedure leads to an automatic adaptive density estimate.

See Karlis and Xekalaki (1998, 2001) and Lu, Hui and Lee (2003) for the application of minimum distance methods in finite mixtures of Poissons and Poisson regression models.

10.3   Minimum Distance Estimation Based on Grouped Data

In real life, grouped data is common. Let $\mathcal{F} = \{F_\theta : \theta \in \Theta \subseteq \mathbb{R}^p\}$ be the parametric model that describes the distribution of the underlying data. When grouped data are available, but the parameter is estimated without taking the grouping into consideration, the estimates of the parameter θ may be inefficient or biased. In the following, we will provide a description of robust and efficient estimation under data grouping using an approximate minimum Hellinger distance estimation following Lin and He (2006). In this approach it is assumed that the true population density f_θ is continuous.

When grouped data are available, one may consider the exact distribution $F_{n\theta}^*$ of grouped data, so that the sample may be viewed as having been drawn from $F_{n\theta}^*$. A minimum distance estimator can then be obtained by constructing a distance between the grouped data and $F_{n\theta}^*$. However, the evaluation of $F_{n\theta}^*$ involves integrating the density over the appropriate ranges, so that the computational complexity of the process becomes a matter of concern. Lin and He (2006) describe an approximation, leading to the approximate minimum Hellinger distance estimator, which avoids the use of a nonparametric kernel and, under certain conditions, is consistent and asymptotically as efficient as the maximum likelihood estimator.

We have discussed the minimum Hellinger distance estimator under several settings in this book. The distance has been defined in Equation (2.52). In discrete models the estimator can be seen to maximize

$$\phi_{n,\theta} = \sum_x d_n^{1/2}(x) f_\theta^{1/2}(x),$$

(Simpson 1987), where $d_n(x)$ is the relative frequency at the point x and $f_\theta(x)$ is the model density. The convenience of using the empirical density function $d_n(x)$ is an appealing feature of minimum distance estimation in discrete models. The approximate Hellinger distance estimator is motivated by this convenience. One could, of course, group the data even when they are actually generated from a continuous distribution. However, in that case the integrated form of the empirical density $f_{n\theta}^*$ for the discretized bins becomes a drag on the practical implementation of the method. In the derivation of the approximate minimum Hellinger distance estimator one replaces $f_{n\theta}^*$ by its first-order approximation; as a consequence the quantity can be easily computed, and the method settles back on the minimum Hellinger distance estimation process for discrete models.

Let $y_1, \ldots, y_n$ represent n independent and identically distributed realizations from the true probability density function f_θ (which is assumed to be continuous). The observed data are grouped into $k_n + 2$ classes

$$(-\infty, L_n), \ [L_n, L_n + h_n), \ [L_n + h_n, L_n + 2h_n), \ \ldots, \ [U_n - h_n, U_n), \ [U_n, \infty),$$

where the k_n central intervals have the same width $h_n = (U_n - L_n)/k_n$. As $n \to \infty$, $k_n \to 0$ at an appropriate rate. Let $x_1^*, \ldots, x_{k_n}^*$ represent the mid-points of the k_n central intervals. The grouped data can then be viewed as a random sample from a discrete distribution supported on $(L_n, x_1^*, x_2^*, \ldots, x_{k_n}^*, U_n)$. The density function for the above may be represented as

$$f_{n\theta}^*(x) = \begin{cases} F_\theta(L_n) & \text{if } x = L_n \\ F_\theta\left(x_i^* + \dfrac{h_n}{2}\right) - F_\theta\left(x_i^* - \dfrac{h_n}{2}\right) & \text{if } x = x_i^* \ (i = 1, \ldots, k_n), \\ 1 - F_\theta(U_n) & \text{if } x = U_n. \end{cases}$$

Let the empirical density $d_n(x)$ be the observed relative frequency of the of the grouped data for $x \in \{x_0^*, x_1^*, \ldots, x_{k_n}^*, x_{k_n+1}^*\}$ where $x_0^* = L_n$ and $x_{k_n+1}^* = U_n$. A minimum Hellinger distance estimator of θ can be obtained by maximizing

$$\sum_{i=0}^{k_n+1} d_n^{1/2}(x_i^*) f_{n\theta}^{*1/2}(x_i^*).$$

Given a density g, if g_n^* represents the density function of the grouped data, the minimum Hellinger distance functional at the corresponding distribution minimizes $\mathrm{HD}_n^* = 2\|g_n^{*1/2} - f_{n\theta}^*\|_2^2$, where $\|\cdot\|_2$ is the L_2 norm, provided such a minimum exists.

Under the identifiability of the family $f_{n\theta}^*$, which generally follows from the identifiability of the original family when $k_n > p$, where p is the dimension of the parameter vector, Lin and He (2006) have extended the results of Beran and proved the analog of Lemma 2.6 (Theorem 1, Beran, 1977) in the grouped data case, under similar conditions but including modifications to accommodate the grouped data part. The theorem establishes the existence of the functional, the continuity of the functional in the Hellinger metric, as well as its Fisher consistency for a sufficiently large n. The determination of the asymptotic distribution is facilitated by the result

$$-\nabla \mathrm{HD}_n^*(d_n, f_{n\theta}^*) = n^{-1}\left[2\sum_{i=1}^{n} \dot{s}_{n\theta}^*(X_i)f_{n\theta}^*(X_i)^{-1/2}\right] + o_p(n^{-1/2}),$$

which is the analog of Simpson's (1987) result for the discrete case given in Theorem 2.10. Here $\dot{s}_{n\theta}^*(x)$ represents the derivative of $s_{n\theta}^*(x) = f_{n\theta}^{*1/2}(x)$. Next, assuming that $f_\theta(x)$ is bounded in (L_n, U_n), consider $\tilde{f}_{n\theta}$, the first-order approximation of $f_{n\theta}^*$ based on the linear expansion. This approximation is given by

$$\tilde{f}_{n\theta}(x) = \begin{cases} 0 & \text{if } x = L_n \\ h_n f_\theta(x_i^*) & \text{if } x = x_i^* \ (i = 1, \ldots, k_n), \\ 0 & \text{if } x = U_n. \end{cases} \tag{10.2}$$

The approximate minimum Hellinger distance estimator, therefore, is obtained by minimizing $\mathrm{HD}(d_n, \tilde{f}_{n\theta}) = 2\|d_n^{1/2} - \tilde{f}_{n\theta}\|_2^2$, or maximizing

$$\sum_{i=0}^{k_n+1} d_n^{1/2}(x_i^*)\tilde{f}_{n\theta}^{1/2}(x_i^*).$$

Under appropriate regularity conditions described by Lin and He (2006) the approximate minimum Hellinger distance estimator has the same asymptotic distribution under the model as the maximum likelihood estimator (Lin and He, 2006, Theorem 4). Based on our knowledge of the properties of the ordinary minimum Hellinger distance estimator and its variants, we expect the approximate minimum Hellinger distance estimator to have strong robustness features. For example, in the Gaussian location case, Lin and He argue that under proper choices of L_n and U_n the finite sample breakdown point of the estimator is approximately $\frac{1}{2}$.

In another approach, Victoria-Feser and Ronchetti (1997) have considered a version of the minimum power divergence estimator for grouped data having bounded influence function, with the bounds controlling the degree of robustness. In the process they obtain a "more robust" version of the minimum Hellinger distance estimator.

Among other related work concerning the grouped data problem, see Menendez et al. (2001). Also see Bassetti, Bodini and Regazzini (2007) for some consistency results for minimum distance estimators based on grouped data.

10.4   Semiparametric Problems

With the growth of computing power over the last few decades, numerically involved statistical techniques have become more feasible. This has led to a greater interest in nonparametric and semiparametric models. In the present section we describe some applications of the minimum Hellinger distance methodology in some semiparametric problems. Our discussion in this section follows Wu and Karunamuni (2009), Karunamuni and Wu (2009), and Wu, Karunamuni and Zhang (2010). Much of the material forms the Ph.D. dissertation work of Wu (2007).

Following the above authors, we describe two specific problems. The first problem is that of statistical analysis in a two population mixture model. We have already discussed some applications of the minimum Hellinger distance methodology in the mixture model problem (Woodward, Whitney and Eslinger, 1995; Cutler and Cordero-Braña, 1996). However, in those cases the component distributions were fully parametric. The discussion here extends the previous approaches to the case where the component distributions are entirely unknown, and are treated as nuisance parameters; the main parameter of interest is the mixing proportion.

The second problem involves another two-sample semiparametric model, where the log ratio of the two underlying density functions are in the form of a regression model. This setup is closely related to the two-sample location model, and includes the two-sample location-scale model as a special case.

10.4.1   Two-Component Mixture Model

Here it is of interest to estimate the mixing proportion θ in a two-component mixture model of the form $\theta F + (1-\theta)G$ where the distributions F and G are completely unknown. The parametric family of densities is then defined as

$$h_\theta(x) = \theta f(x) + (1-\theta)g(x),$$

where f and g are the respective densities, and θ is the parameter of interest. The problem, simply stated as above, is not well defined; to make it more meaningful we assume that training samples are available from the respective densities. Thus, the data structure involves three independent and identically distributed samples: (i) $X_1, \ldots, X_{n_0}$ from F; (ii) $Y_1, \ldots, Y_{n_1}$ from G and (iii) $Z_1, \ldots, Z_{n_2}$ from $\theta F + (1-\theta)G$. For full identifiability of the problem, it is further assumed that the L_1 distance between the densities f and g is strictly positive.

Three kernel density estimates, f^*, g^* and h^* are computed from the three given samples. For any $t \in [0, 1]$ define

$$\tilde{h}_t(x) = tf^*(x) + (1-t)g^*(x).$$

Thus, $\tilde{h}_t$ is a parametric family where the parameter t is the only unknown. Then the proposed minimum Hellinger distance estimator of the mixture proportion θ is

$$\hat{T} = \arg \min_{t \in [0,1]} \|\tilde{h}_t^{1/2} - h^{*1/2}\|_2$$

where $\|\cdot\|_2$ represents the L_2 norm. Since the parameter space $[0, 1]$ is compact, this minimizer exists from Lemma 2.6.

Suppose $n_i/n \to \rho_i$, $i = 1, 2, 3$ where ρ_i are positive constants in $(0, 1)$ and $n = n_0 + n_1 + n_2$ is the total sample size. Then under appropriate regularity conditions it is shown that

$$n^{1/2}(\hat{\theta} - \theta) \to Z^* \sim N(0, V),$$

where θ is the true value of the parameter and the asymptotic variance V is given by

$$\frac{\dfrac{\theta^2}{\rho_0} \text{Var}\left[\dfrac{\partial \log h_\theta(X_1)}{\partial \theta}\right] + \dfrac{(1 - \theta)^2}{\rho_1} \text{Var}\left[\dfrac{\partial \log h_\theta(Y_1)}{\partial \theta}\right] + \dfrac{1}{\rho_2} \text{Var}\left[\dfrac{\partial \log h_\theta(Z_1)}{\partial \theta}\right]}{\left\{\text{Var}\left[\dfrac{\partial \log h_\theta(Z_1)}{\partial \theta}\right]\right\}^{-2}}.$$

See Wu and Karunamuni (2009) and Karunamuni and Wu (2009) for more details on the method.

10.4.2   Two-Sample Semiparametric Model

Suppose that $X_1, \ldots, X_{n_0}$, and $Z_1, \ldots, Z_{n_1}$ be independently and identically distributed random samples of observations from distribution having density function g and h respectively. The samples are independent of each other. The two unknown density functions are linked by an "exponential tilt" relation

$$h(x) = g(x) \exp[\alpha + r(x)\beta].$$

The set $\theta = (\alpha, \beta)$ is the parameter of interest, so that the density on the left-hand side of the above equation may be written as $h_\theta(x)$ to indicate its dependence on the parameters. For $r(x) = x$, the model covers many standard distributions such as two exponential distributions with different means, and two normal distributions with common variance but different means. The model $r(x) = (x, x^2)$ has extensive applications in logistic discrimination and logistic regression analysis, and in case control studies.

Consider kernel density estimates g^* and h^* based on $X_1, \ldots, X_n$ and $Z_1, \ldots, Z_n$ respectively. Let $\mathcal{G}$ be class of all distributions having densities with respect to the Lebesgue measure. Then, given a distribution $F \in \mathcal{G}$, the minimum Hellinger distance functional $T(F)$ may be defined as

$$T(F) = \arg \min_{\theta \in \Theta} \|h_\theta^{1/2} - f^{1/2}\|_2,$$

where f is the density of F, and $\|\cdot\|_2$ represents the L_2 norm. The functional $T(F)$ is Fisher consistent if the family $\{h_\theta\}$ is identifiable. However, the functional cannot be directly computed since g is unknown. The approach then is to replace g by its estimate g^* so that

$$\hat{h}_\theta(x) = \exp[(1, r(x))\theta]g^*(x).$$

Then the minimum Hellinger distance estimator of θ is obtained as

$$\hat{\theta} = \arg\min_{\theta \in \Theta} = \|\hat{h}_\theta^{1/2} - h^{*1/2}\|_2.$$

Wu (2007) has established the asymptotic normality of this estimator, and studied its robustness properties.

10.5   Other Miscellaneous Topics

Subsequent to the development of the theory by Beran (1977) and others, several authors have numerically investigated the performance of the minimum distance estimators, which has further illustrated the robustness properties of the latter and demonstrated the scope of their applicability. Eslinger and Woodward (1991) investigated the use of the minimum Hellinger distance estimator for the normal model with unknown location and scale, and demonstrated the robustness of the minimum Hellinger distance estimator beyond the limits that are theoretically predicted. Basu and Sarkar (1994c) demonstrated similar results for the Basu–Lindsay approach, and further illustrated the advantages of model smoothing.

See Cressie and Pardo (2000, 2002) and Cressie, Pardo and Pardo (2003) for some interesting applications of the minimum ϕ-divergence method to the case of loglinear models. Much of the work done by Leandro Pardo and his associates, including the above, has been described adequately in the book by Pardo (2006). In some of their recent work, Menendez, Pardo and Pardo (2008) have developed preliminary test estimators based on ϕ-divergence estimators in generalized linear models with binary data. Martin and Pardo (2008) have considered some new families of estimators and tests in log linear models, while Pardo and Martin (2009) considered tests of hypothesis for standardized mortality ratios based on the minimum power divergence estimators.

Several authors have worked on the dual form of the distance in order to obtain robust estimators which have a high degree of efficiency. See Broniatowski (2003), Liese and Vajda (2006), Broniatowski and Keziou (2006, 2009) and Toma and Broniatowski (2010). Between them, these papers provide a fairly comprehensive description of the duality technique. The duality form of the Kullback Leibler divergence has been considered by Broniatowski (2003) on sets of measures satisfying moment conditions. Broniatowski and Keziou

(2006) have provided the structure of the dual representation of divergences between a probability distribution and a set of signed finite measures. Broniatowski and Keziou (2009) have presented the asymptotic results for parametric inference through duality and divergences. Toma and Broniatowski (2010) have explored the dual ϕ-divergence estimators with respect to their robustness through the influence function approach. Their paper also studies the balance between the robustness and asymptotic relative efficiency for location-scale models, and also proposes and studies some tests of hypotheses based on the dual divergence criterion. Toma and Leoni-Aubin (2010) have also studied robust tests based on dual divergence estimators and saddlepoint approximations.

Toma (2008) considered minimum Hellinger distance estimators for multivariate distributions from the Johnson system. These estimators are obtained by applying a transform to the data, computing the minimum Hellinger distance estimator for the obtained Gaussian data and transforming those estimators back using the relationship between the parameters of the two distributions. Toma (2009) has also developed new robustness measures and developed some optimal M-estimators based on the power divergence as well as the density power divergence families.

Among other interesting applications of minimum distance methods, Chen and Kalbfleisch (1994) considered the use of a Hellinger distance in inverse problems in fractal construction; the method does not require full knowledge of the image to be approximated and can eliminate unwanted locally optimal approximations.

Pak (1997) and Pak and Basu (1998) have considered the extension of minimum disparity estimation to the case of linear regression models. The Ph.D. thesis by Alexandridis (2005) represents an application of minimum distance method based on disparities in ranked set sampling.

See Dey and Birmiwal (1994) and Peng and Dey (1995) for some applications of distance based techniques in the Bayesian context.

11

Distance Measures in Information and Engineering

In this section and the next we somewhat detach ourselves from the discussion of the previous chapters and consider the use of the distance measures described in this book in the context of information science and engineering. For this purpose we use the established notation in this field, even if that may be at variance with the notation of the previous ten chapters. In this chapter we will refer to a probability density function defined on an appropriate set X as a "probability function."

11.1   Introduction

A distance is a measure of discrepancy between two elements of a given space, and it is important to use the topological definition of the given mathematical space. The term "information divergence" is used in information theory and statistics, and it denotes the discrimination between two probability distributions. Some important statistical uses of such divergence measures have been discussed in the earlier chapters of this book. Here we start with the origin of information measures and retrace a part of their history.

The "Shannon Entropy," an entropy based on the logarithmic function, is one of the most important information measures for a single probability distribution presenting the information complexity (Shannon, 1948). The "Kullback–Leibler" divergence (Kullback and Leibler, 1951) is a typical information divergence related to the Shannon entropy. A class of order-α entropies was proposed by Shützenberger (1954). Rényi (1960) further elaborated on order-α entropies including the Shannon entropy, and pointed out the importance of this type of measure in information theory. The divergence measures based on order-α entropies are called order-α divergences. The fundamental idea of such entropies and the associated divergences is derived from the old measures presented by Bhattacharyya (1943) and Jeffreys (1948). Tsallis (1988) introduced a new entropy in the field of statistical physics which is often called the Tsallis entropy.

Csiszár (1963) introduced the generalized class of "f-divergences" that includes the Kullback–Leibler divergence, the Hellinger discrimination (Hellinger, 1909), and the variation distance among others. Later it was independently introduced by Ali and Silvey (1966). These generalized divergences are often referred to as ϕ-divergences in the field of statistics and other disciplines. As a result, both names are widely used. Some important theoretical work on f-divergences has been reported in Csiszár (1967b), Vajda (1972), Csiszár (1972), Kafka et al. (1991), and Leise and Vajda (2006) among others. Another type of information divergence using convex functions was introduced by Bregman (1967) in terms of the discrimination between two points in a convex space, and it is consequently called the Bregmen divergence.

An interesting case of a modified information divergence based on f-divergences is one in which a function f in the definition of the f-divergence is not convex, that is, one in which the f-divergence can be defined using a non-convex function f. One such case was presented by Shioya and Da-Te (1995). Their proposed divergence measure is based on Hermite-Hadamard's inequality; see, e.g., Pečarić (1992) and Pečarić et al. (1992). This non-convex f-divergence measure was renamed the integral mean divergence. Some mathematical properties and inequalities in this connection have been presented in Barnett and Dragomir (2002), Barnett et al. (2002), and Cerone and Dragomir (2005).

For another generalization of information divergences we will consider the class of information divergence measures between two nonnegative functions with finite volumes. Csiszár was the first to introduce such a measure; it is called the I-divergence and is used in Chapter 12 of this book. We will end this chapter with a study of some mathematical properties of information divergences.

11.2   Entropies and Divergences

Shannon (1948) has considered the entropy

$$H(X) = \sum_{x \in \mathcal{X}} -p(x) \log p(x),$$

in connection with the random variable X with probability function $p(x)$ on the finite set $\mathcal{X}$. The optimal code length of the elementary prefix coding using an independently and identically distributed sequence from the corresponding probability distribution has also been discussed. Two encoders P and Q are considered with probability functions $p(x)$ and $q(x)$ respectively. Then the following inequality provides the optimality of the average code length of P:

$$\sum_{x \in \mathcal{X}} -p(x) \log p(x) \leq \sum_{x \in \mathcal{X}} -p(x) \log q(x),$$

This establishes the nonnegativity of $\sum_{x\in\mathcal{X}} p(x)\log\{p(x)/q(x)\}$. If we consider another random variable Y on the finite set $\mathcal{Y}$, then we obtain the mutual information for the channel coding theorem for information transmission (Shannon, 1948) by the Kullback–Leibler divergence between the joint probability function $p(x,y)$ of X and Y, and the product of the individual probability functions.

When we extend the Shannon entropy, we get the order-α entropy (often called the Rényi entropy) given by

$$H_\alpha(X) = \frac{1}{1-\alpha} \log \sum_{x\in\mathcal{X}} p^\alpha(x),$$

where $0 < \alpha < 1$ or $\alpha > 1$. This was first introduced by Shützenberger (1954). In the limiting case we get $\lim_{\alpha\to 1} H_\alpha(X) = H(X)$. Some relationships between order-α entropies and the source coding theorem were presented by Campbell (1965).

Based on the order-α entropy, the gain in information of order α when the probability function q is replaced by p is given by

$$R_\alpha(p\|q) = \frac{1}{\alpha-1} \log \sum_{x\in\mathcal{X}} p^\alpha(x)q^{1-\alpha}(x).$$

By applying Jensen's inequality, the relation $R_\alpha(p\|q) \geq 0$ is found to be true for $\alpha > 1$ or $0 < \alpha < 1$, and its equality is established if and only if $p(x) = q(x)$ for all x. The measure $R_\alpha(p\|q)$ is called the order-α divergence. These order-α entropies and divergences were extensively studied in Rényi (1960).

To control systems with long-range interaction or long-time memory, Tsallis (1988) proposed a generalized entropy in the areas of thermodynamics and statistical physics given by

$$T_\beta(X) \stackrel{\text{def}}{=} \frac{1 - \sum_{x\in\mathcal{X}} p^\beta(x)}{\beta - 1},$$

where β is a real constant. This is called the Tsallis entropy and $\lim_{\beta\to 1} T_\beta(X) = H(X)$ holds true.

To make the notation simple and uniform, for the rest of this chapter we will let X denote the set of interest, rather than the random variable.

11.3   Csiszár's f-Divergence

11.3.1   Definition

We start with the fundamental setting for probability measures. Let $\mathcal{U}$ be a σ-finite algebra of the set X, and let $(X,\mathcal{U})$ be a measurable space. Let $\mathcal{M}_{\text{fin}}$

be the set of all finite measures on $(X, \mathcal{U})$, that is, for all $\mu \in \mathcal{M}_{\text{fin}}$ and for all $S \in \mathcal{U}$, $\mu(S) < \infty$ and $\mu(X) < \infty$ are satisfied. Let $\mathcal{M}_{\text{prob}}$ be the set of all the probability measures on $(X, \mathcal{U})$, that is, $\mathcal{M}_{\text{prob}} \subset \mathcal{M}_{\text{fin}}$, and $P(X) = 1$ is satisfied for all $P \in \mathcal{M}_{\text{prob}}$.

For all $P \in \mathcal{M}_{\text{prob}}$, a finite measure $\mu \ (\in \mathcal{M}_{\text{fin}})$ satisfying $P \prec \mu$ (μ is the dominating measure of P) exists. The probability function $p(x)$ corresponding to P is uniquely obtained from the Radon-Nikodym theorem as

$$p(x) = \frac{dP(x)}{d\mu(x)}.$$

We define the set of all the probability functions using the dominating measure μ as

$$\mathbb{P} = \left\{ p \ \middle| \ \int_X p(x) d\mu(x) = 1, \ p(x) \geq 0 \ \forall x \in X \right\}.$$

Csiszár (1963) introduced a generalized divergence measure, called f-divergence, between probability distributions P and Q as

$$D_f(P, Q) \stackrel{\text{def}}{=} \int_X q(x) f\left(\frac{p(x)}{q(x)}\right) d\mu(x), \qquad (11.1)$$

where $f(u)$ is a convex function on $(0, \infty)$, is strictly convex at $u = 1$, and satisfies $f(1) = 0$. For $p(x) = 0$ or $q(x) = 0$ in Equation (11.1), the following convention is used:

$$0 f\left(\frac{0}{0}\right) = 0, \ \ 0 f\left(\frac{a}{0}\right) = \lim_{\epsilon \to +0} \epsilon f\left(\frac{a}{\epsilon}\right) = a \lim_{u \to \infty} \frac{f(u)}{u}.$$

The function $f^*(u) \ (= u f(1/u))$ is also a convex function on $(0, \infty)$, and is strictly convex at $u = 1$. Also $f^*(1) = 1$ is satisfied. The function f^* gives the reverse f-divergence defined by $D_{f^*}(P, Q) = D_f(Q, P)$.

The integral (11.1) is an effective measure in the sense of indirect observation, that is, let $\bar{P}$ and $\bar{Q}$ be the reconstructions of P and Q, respectively, to any sub-σ-algebra $\bar{\mathcal{U}}$ of $\mathcal{U}$, we have

$$D_f(\bar{P}, \bar{Q}) \leq D_f(P, Q).$$

Some related results of indirect observations based on f-divergences are presented in detail in Csiszár (1967a). We often focus on only the probability functions p and q, and as a convention write $D_f(p\|q)$ for $D_f(P, Q)$.

Many different information divergences are obtained by formulating the f-divergence using various convex functions. Table 11.1 presents the relationships between the convex functions and well-known information divergences. The α-divergence is a parameterized divergence that includes the Hellinger discrimination ($\alpha = 0.5$), the Kullback–Leibler divergence ($\alpha = 0$), and χ^2-divergence ($\alpha = -1$). In addition, the integral of α-divergence corresponds to the Chernoff bound for the two hypothesis decision problem (Chernoff,

TABLE 11.1

Several well-known information divergences and the corresponding convex functions.

Convex function f	Information divergence
$f(u) = u \log u$	Kullback–Leibler divergence
	$D_k(p\|q) = \int_X p(x) \log \dfrac{p(x)}{q(x)} d\mu(x)$
$f(u) = -\log u$	Reverse Kullback–Leibler divergence
	$D_k(q\|p) = \int_X q(x) \log \dfrac{q(x)}{p(x)} d\mu(x)$
$f(u) = \|u - 1\|$	Variation distance
	$D_v(p\|q) = \int_X \|p(x) - q(x)\| d\mu(x)$
$f(u) = 2(1 - \sqrt{u}) + u - 1$	Hellinger discrimination
	$D_h(p\|q) = \int_X (p^{1/2}(x) - q^{1/2}(x))^2 d\mu(x)$
$f(u) = (u - 1)^2$	χ^2-divergence
	$D_{\chi^2}(p\|q) = \int_X \dfrac{(p(x) - q(x))^2}{q(x)} d\mu(x)$
$f(u) = \dfrac{1}{u} - 1$	Reverse χ^2-divergence
	$D_{\chi^2}(q\|p) = \int_X \dfrac{(p(x) - q(x))^2}{p(x)} d\mu(x)$
$f(u) = \dfrac{u - u^{1-\alpha}}{\alpha(1 - \alpha)}$	α-divergence
	$D_\alpha(p\|q) = \dfrac{1 - \int_X p^{1-\alpha}(x) q^\alpha(x) d\mu(x)}{\alpha(1 - \alpha)}$

1952). We use the name α-divergence to indicate the α power integral (Cressie and Read, 1984; Amari and Nagaoka, 2000). We distinguish between the α-divergences and the order-α divergences. Rényi's order-α divergence does not directly correspond to an f-divergence.

11.3.2 Range of the f-Divergence

Let $\mathbb{F} = \{f\}$ be the set of all convex functions on $(0, \infty)$ satisfying the required conditions for defining an f-divergence ($f(1) = 0$ is included in the conditions). For all $c \in \mathbb{R}$, the transformation from f to f_c defined by $f_c(u) \overset{\text{def}}{=} f(u) + c(u-1)$ gives the invariance property of the f-divergence, that is, the following equality holds:

$$D_{f_c}(p\|q) = \int_X q(x) f\left(\frac{p(x)}{q(x)}\right) d\mu(x) + c\left\{\int_X p(x) d\mu(x) - \int_X q(x) d\mu(x)\right\}$$
$$= D_f(p\|q).$$

Csiszár (1967a) presented some inequalities between the f-divergence and the variation distance. As a consequence it has been pointed out that the upper bound of the f-divergence is determined by $|f(0)|$ and $|f^*(0)|$. Later Vajda (1972) presented a concrete upper bound for the f-divergence given by

$$D_f(p\|q) \leq f(0) + f^*(0). \qquad (11.2)$$

Under the transformation from f to f_c, $f(0) + f^*(0)$ is invariant. When $f(0) + f^*(0) < \infty$, the equality in Equation (11.2) is established if and only if $p \perp q$, i.e., p and q are singular. In determining an upper bound for the f-divergence the following relation with the variation distance is useful.

Lemma 11.1. *The f-divergence and the variation distance satisfy the relation*

$$D_f(p\|q) \leq \frac{f(0) + f^*(0)}{2} D_v(p\|q). \qquad (11.3)$$

Proof. If $f(0) + f^*(0) < \infty$, we use the convex function

$$g(u) = f(u) + \frac{f(0) - f^*(0)}{2}(u - 1).$$

Note that $g \in \mathbb{F}$. For $0 < u \leq 1$, the following is observed:

$$g(1 \times u + 0 \times (1 - u)) \leq u\, g(1) + (1 - u)\, g(0) = (1 - u)\frac{f(0) + f^*(0)}{2}. \qquad (11.4)$$

Using the transformation from $g(t)$ to $g^*(t)$, we have

$$g^*(t) = f^*(t) + \frac{f(0) - f^*(0)}{2}\left(\frac{1}{t} - 1\right)t.$$

In the same way as in Equation (11.4), we have , for $0 < t \leq 1$

$$g^*(1 \times t + 0 \times (1 - t)) \leq t\, g^*(1) + (1 - t)\, g^*(0) = (1 - t)\frac{f(0) + f^*(0)}{2}.$$

Using $g^*(t) = tg(1/t)$ we get, for $0 < t \leq 1$.

$$g\left(\frac{1}{t}\right) \leq \left(\frac{1}{t} - 1\right)\frac{f(0) + f^*(0)}{2}. \qquad (11.5)$$

By using Equations (11.4) and (11.5), we have

$$g(u) \leq \frac{f(0) + f^*(0)}{2}|u - 1|, \text{ for } u \geq 0.$$

By substituting $p(x)/q(x)$ for u, we have

$$q(x)g\left(\frac{p(x)}{q(x)}\right) \leq \frac{f(0) + f^*(0)}{2}|p(x) - q(x)|. \qquad (11.6)$$

Taking the integral with respect to x on X, we have

$$D_g(p\|q) \leq \frac{f(0) + f^*(0)}{2} D_v(p\|q).$$

By the invariance property of f-divergence ($D_g = D_f$), we have Equation (11.3). $\qquad\square$

Vajda (1972) presented the same inequality between the f-divergence and the variation distance as in Equation (11.2); however, the description of that proof is not as simple as the one given above. The inequality Equation (11.2) represents the condition for the finiteness of the upper bound of the f-divergence. For example, the variation distance and Hellinger discrimination are finite for any two probability distributions, and their maxima are easily obtained. However, the Kullback–Leibler divergence and χ^2-divergence are not bounded.

11.3.3 Inequalities Involving f-Divergences

Inequalities of information divergences are useful for investigating a mathematical relationship of them. The following are some classical inequalities involving D_k, D_h, and D_v.

$$D_h(p\|q) \leq D_v(p\|q), \tag{11.7}$$

$$\frac{D_v(p\|q)^2}{2} \leq D_k(p\|q), \tag{11.8}$$

$$\frac{D_v(p\|q)^2}{4} \leq D_h(p\|q) \leq D_k(p\|q). \tag{11.9}$$

See Kraft (1955) for Equation (11.7), Kullback (1967, 1970) for Equation (11.8), and Pitman (1979) and Barron and Cover (1991) for the first and second parts of Equation (11.9), respectively.

The Hellinger discrimination is a member of the class of α-divergences. A generalized version of the second part in Equation (11.9) using α-divergences is given by

$$D_\alpha(p\|q) \leq \frac{1}{1-\alpha} D_k(p\|q), \quad \alpha \neq 1,$$

with equality if and only if $\alpha \to 0$ or $p = q$. This is easily obtained by noting $u - u^{1-\alpha} \leq \alpha u \log u$, for $u > 0$ and $\alpha \in [0, 1]$.

Csiszár (1972) also suggested the following inequality. A function $\Psi(x)$ ($x \in \mathbb{R}$) exists depending on f with $\lim_{x \to 0} \Psi(x) = 0$, such that

$$D_v(p\|q) \leq \Psi(D_f(p\|q)).$$

This is regarded as a generalized version of Equation (11.7).

11.3.4 Other Related Results

Kafka et al. (1991) considered powers of f-divergence that define a distance. For example, the square root of the Hellinger discrimination defines a distance. Csiszár (1967a) analyzed the topological properties of f-divergences which revealed the relationship between finite f-divergences and the topological structure of the set of all probability distributions.

For other fundamental properties, a class of symmetric f-divergences was investigated by Kumar and Chhina (2005). For other cases of Csiszár type divergence using non-convex functions, Shioya and Da-Te (1995) presented the divergence

$$D_{HH}^f(p\|q) \overset{\text{def}}{=} \int_X q(x) \left[\frac{\int_1^{\frac{p(x)}{q(x)}} f(t)dt}{\frac{p(x)}{q(x)} - 1} \right] d\mu(x),$$

where $f(u) \in \mathbb{F}$. The divergence measure proposed by them is based on Hermite-Hadamard's inequality

$$f\left(\frac{a+b}{2}\right) \leq \frac{\int_a^b f(t)dt}{b-a} \leq \frac{f(a) + f(b)}{2},$$

where f is a convex function on an appropriate domain and $a < b$. Some mathematical results in this connection have been presented in Barnett and Dragomir (2002), Barnett et al. (2002), and Cerone and Dragomir (2005).

11.4 The Bregman Divergence

Bregman (1967) introduced the generalized discriminant measure

$$B_f(p\|q) = \int_X \{f(p(x)) - f(q(x)) - f'(q(x))(p(x) - q(x))\} d\mu(x), \quad (11.10)$$

where p and q are two probability functions on X, f is a differentiable and a strictly convex function on $(0, \infty)$, and f' is the derivative function of f. The concept of this divergence is based on the following inequality presenting the relationship between the convex function f and its support function,

$$f(a) \geq f'(b)(a - b) + f(b), \quad \forall a, b \in (0, \infty),$$

where there is equality if and only if $a = b$ (because f is strictly convex). By substituting a and b with $p(x)$ and $q(x)$, respectively, and by integrating over X, Equation (11.10) is obtained.

The class of Bregman divergences includes various well-known information divergences generated by specific convex functions. The choice $f(u) = u \log u - u + 1$, B_f gives the Kullback–Leibler divergence, and for $f(u) = u - \log u - 1$ a reverse Kullback–Leibler divergence is obtained. The following convex function f_α is introduced in terms of the generalized projections of Csiszár (1995).

$$f_\alpha(u) = \begin{cases} u^\alpha - \alpha u + \alpha - 1 & \text{if } \alpha > 1 \text{ or } \alpha < 0 \\ -u^\alpha + \alpha u - \alpha + 1 & \text{if } 0 < \alpha < 1 \\ u \log u - u + 1 & \text{if } \alpha = 1 \\ u - \log u - 1 & \text{if } \alpha = 0. \end{cases}$$

When $\alpha = 2$, B_{f_α} gives the squared Euclidean distance.

Originally, the Bregman divergence was proposed as a measure of discrepancy between two points in $\mathbb{R}^d$, and it was used for finding the solution to problems in convex programming (Bregman, 1967).

In more recent times this divergence has been used for several other purposes; for example it has been used to introduce and analyze an algorithm for a set of weak-learning machines (Murata et al., 2004), as well as for a progressive-data-clustering method (Banerjee et al., 2005) in the field of machine learning. Thus, there are many applications using the Bregman divergence in the fields of engineering and information science. The class of Bregman divergences also includes the squared loss function criterion.

11.5 Extended f-Divergences

11.5.1 f-Divergences for Nonnegative Functions

So far we have looked at f-divergences as discriminant measures between two probability distributions. By expanding the scope to real valued nonnegative functions on the set X, we can define f-divergences for nonnegative functions with finite volumes.

Let $\mathcal{U}_X$ be a σ-algebra on set X. Let $\mathcal{M}_{\text{fin}}$ be the set of all finite measures on the measurable space $(X, \mathcal{U}_X)$. For all $F \in \mathcal{M}_{\text{fin}}$, the dominating measure μ of F satisfying $F \prec \mu$ exists, and $\rho(x) = \frac{dF(x)}{d\mu(x)}$ uniquely exists. We then define the set of all such functions as

$$\mathcal{P}_{\text{fin}} = \left\{ \rho \Big| \int_X \rho(x) d\mu(x) < \infty, \ \rho(x) \geq 0 \ \forall x \in X \right\}.$$

The following I-divergence between ρ and τ $(\in \mathcal{P}_{\text{fin}})$ was introduced by Csiszár (1991) and has been used in various applications.

$$I(\rho \| \tau) = \int_X \rho(x) \log \frac{\rho(x)}{\tau(x)} d\mu(x) + \int_X \tau(x) d\mu(x) - \int_X \rho(x) d\mu(x).$$

These applications show that the I-divergence discriminant criterion is quite effective for measuring discrimination and is competitive with or better than the usual squared error loss.

The following α-divergence has been applied in the context of information geometry (Amari and Nagaoka, 2000).

$$I_\alpha(\rho\|\tau) = \frac{1}{\alpha(1-\alpha)} \int_X \left\{(1-\alpha)\tau(x) + \alpha\rho(x) - \tau^{1-\alpha}(x)\rho^\alpha(x)\right\} d\mu(x), \ \alpha \in \mathbb{R}.$$

In order to define a discriminant measure between two real and nonnegative functions with finite volumes in the spirit of an f-divergence, the following restricted subset of $\mathbb{F}$ must be used:

$$\bar{\mathbb{F}} = \{f|f \in \mathbb{F}, \ f(u) \geq 0, \ \forall u \geq 0\}.$$

A real number m satisfying $f_m(u) = f(u) + m(u-1) \in \bar{\mathbb{F}}$ and $f_m(x) \geq 0$ for all $u \geq 0$ exists for all $f \in \mathbb{F}$.

For all $\rho, \tau \in \mathcal{P}_{\text{fin}}$, Csiszár (1995) defined the extended f-divergences with $f \in \bar{\mathbb{F}}$ as

$$D_f(\rho\|\tau) = \int_X \tau(x)f\left(\frac{\rho(x)}{\tau(x)}\right) d\mu(x).$$

It is immediately seen that $D_f(\rho\|\tau) \geq 0$ with equality if and only if $\rho = \tau$. It is also easy to show that the f-divergences (for $f \in \bar{\mathbb{F}}$) can be used as discriminants between two nonnegative real functions with finite volumes. For all $f \in \bar{\mathbb{F}}$, the relation $f^* \in \bar{\mathbb{F}}$ automatically holds, and the reverse divergence of $D_f(\rho\|\tau)$ is given by

$$D_{f^*}(\rho\|\tau) = \int_X \tau(x)f^*\left(\frac{\rho(x)}{\tau(x)}\right) d\mu(x) = \int_X \rho(x)f\left(\frac{\tau(x)}{\rho(x)}\right) d\mu(x) = D_f(\tau\|\rho).$$

The following theorem gives the bound for $D_f(\rho\|\tau)$ as a function of the volumes of ρ and τ.

Theorem 11.2. *For all $f \in \bar{\mathbb{F}}$ satisfying $f(0) + f^*(0) < \infty$, the upper bound for the f-divergence for nonnegative functions is given as*

$$D_f(\rho\|\tau) \leq f(0) \int_X \tau(x)d\mu(x) + f^*(0) \int_X \rho(x)d\mu(x), \qquad (11.11)$$

with equality if and only if $\rho \perp \tau$.

To prove the above result, we use the following inequality between the f-divergence $D_f(\rho\|\tau)$ and the variation distance $D_v(\rho\|\tau)$.

Lemma 11.3. *We have the inequality*

$$D_f(\rho\|\tau) \leq \frac{f(0) + f^*(0)}{2} D_v(\rho\|\tau)$$
$$+ \frac{f(0) - f^*(0)}{2} \left(\int_X \tau(x)d\mu(x) - \int_X \rho(x)d\mu(x)\right), \qquad (11.12)$$

with equality if and only if $f(u) = |u-1|$ or $\rho(x) = \tau(x)$ for $\forall x$.

Proof. Let Y_1 and Y_2 be the sets satisfying $X \cap Y_1 = \phi$, $X \cap Y_2 = \phi$ and $Y_1 \cap Y_2 = \phi$. Let Y and Z be $Y_1 \cup Y_2$ and $X \cup Y$, respectively. Let $\mathcal{U}_Z$ be a σ-algebra on Z satisfying $Y_1, Y_2 \in \mathcal{U}_Z$ and $s \in \mathcal{U}_Z$ for all $s \in \mathcal{U}_X$. Let $\tilde{\mathcal{M}}_{\text{fin}}$ be the set of all the finite measures of $(X, \mathcal{U}_Z)$. Then $\tilde{\mu} \in \tilde{\mathcal{M}}_{\text{fin}}$ exists satisfying $\tilde{\mu}(s) = \mu(s)$ for all $s \in \mathcal{U}_X$. We define the following set on Z with the dominating measure $\tilde{\mu}$.

$$\tilde{\mathcal{P}}_{\text{fin}} = \left\{ \omega \mid \int_Z \omega(z) d\tilde{\mu}(z) < \infty, \ \omega(z) \geq 0 \ \forall z \in Z \right\}.$$

We use the f-divergences between two elements of $\tilde{\mathcal{P}}_{\text{fin}}$ by

$$\tilde{D}_f(\nu \| \xi) = \int_Z \xi(z) f\left(\frac{\nu(z)}{\xi(z)}\right) d\tilde{\mu}(z),$$

where $\nu, \xi \in \tilde{\mathcal{P}}_{\text{fin}}$ and $f \in \bar{\mathbb{F}}$.

The following probability distributions on Z are defined in terms of $A_\tau = \int_X \tau(x) d\mu(x)$ and $A_\rho = \int_X \rho(x) d\mu(x)$:

$$\tilde{\rho}(z) = \begin{cases} \dfrac{\rho(z)}{A_\rho + A_\tau} & z \in X \\[2mm] \dfrac{\omega_\tau(z)}{A_\rho + A_\tau} & z \in Y_1 \\[2mm] 0 & z \in Y_2 \end{cases},$$

$$\tilde{\tau}(z) = \begin{cases} \dfrac{\tau(z)}{A_\rho + A_\tau} & z \in X \\[2mm] 0 & z \in Y_1 \\[2mm] \dfrac{\omega_\rho(z)}{A_\rho + A_\tau} & z \in Y_2 \end{cases},$$

where $\omega_\tau, \omega_\rho \in \tilde{\mathcal{P}}_{\text{fin}}$ satisfies $\int_{Y_1} \omega_\tau(z) d\tilde{\mu}(z) = A_\tau$ and $\int_{Y_2} \omega_\rho(z) d\tilde{\mu}(z) = A_\rho$, respectively.

Then, introducing $D_f(\tilde{\rho} \| \tilde{\tau})$ as the f-divergence between $\tilde{\rho}$ and $\tilde{\tau}$ on Z, we have

$$D_f(\rho \| \tau) = (A_\tau + A_\rho) \int_X \frac{\tau(x)}{A_\tau + A_\rho} f\left(\frac{\frac{\rho(x)}{A_\tau + A_\rho}}{\frac{\tau(x)}{A_\tau + A_\rho}}\right) d\mu(x)$$

$$= (A_\tau + A_\rho) \left\{ \tilde{D}_f(\tilde{\rho} \| \tilde{\tau}) - \int_Y \tilde{\tau}(z) f\left(\frac{\tilde{\rho}(z)}{\tilde{\tau}(z)}\right) d\tilde{\mu}(z) \right\}$$

$$= (A_\tau + A_\rho) \left\{ \tilde{D}_f(\tilde{\rho} \| \tilde{\tau}) - f^*(0) \int_{Y_1} \tilde{\rho}(z) d\tilde{\mu}(z) - f(0) \int_{Y_2} \tilde{\tau}(z) d\tilde{\mu}(z) \right\},$$

where $\tilde{D}_f(\tilde{\rho}\|\tilde{\tau})$ is the f-divergence between two probability distributions $\tilde{\rho}$ and $\tilde{\tau}$ on Z. We then have

$$\tilde{D}_f(\tilde{\rho}\|\tilde{\tau}) \leq \frac{f(0) + f^*(0)}{2}\tilde{D}_v(\tilde{\rho}\|\tilde{\tau}) \tag{11.13}$$

$$= \frac{f(0) + f^*(0)}{2(A_\tau + A_\rho)}\left\{\int_X |\rho(z) - \tau(z)|d\tilde{\mu}(z) + A_\tau + A_\rho\right\}$$

$$= \frac{f(0) + f^*(0)}{2(A_\tau + A_\rho)}\left\{D_v(\rho\|\tau) + A_\tau + A_\rho\right\},$$

where Equation (11.13) is obtained by using Equation (11.3). Then, we have

$$D_f(\rho\|\tau) \leq \left(\frac{f(0) + f^*(0)}{2}\right)D_v(\rho\|\tau) + \left(\frac{f(0) - f^*(0)}{2}\right)(A_\tau - A_\rho). \tag{11.14}$$

$\square$

Using Equation (11.12), we have

$$\left(\frac{f(0) + f^*(0)}{2}\right)D_v(\rho\|\tau) + \left(\frac{f(0) - f^*(0)}{2}\right)(A_\tau - A_\rho)$$

$$\leq f(0)\int_X \tau(x)d\mu(x) + f^*(0)\int_X \rho(x)d\mu(x), \tag{11.15}$$

where equality is attained if and only if $D_v(\rho\|\tau) = \int_X \rho(x)d\mu(x) + \int_X \tau(x)d\mu(x)$, that is, $\rho\perp\tau$. This proves Theorem 11.2.

Equations (11.14) and (11.15) provide generalizations for the upper bound of the f-divergence between two probability distributions. For $D_f = I_\alpha$, for $\alpha \in (0,1)$ we have

$$I_\alpha(\rho\|\tau) \leq \frac{1}{2\alpha(1-\alpha)}D_v(\rho\|\tau)$$

$$+ \frac{1}{2}\left(\frac{1}{\alpha} - \frac{1}{1-\alpha}\right)\left(\int_X \tau(x)d\mu(x) - \int_X \rho(x)d\mu(x)\right)$$

$$\leq \frac{1}{\alpha}\int_X \tau(x)d\mu(x) + \frac{1}{1-\alpha}\int_X \rho(x)d\mu(x).$$

For $\alpha \notin (0,1)$, I_α is not bounded for $\forall\rho, \tau \in \mathcal{P}_{\text{fin}}$.

Corollary 11.4. *For $f \in \bar{\mathbb{F}}$, assume that $f = f^*$ and $f(0) < \infty$. Then we have*

$$D_f(\rho\|\tau) \leq f(0)D_v(\rho\|\tau),$$

with equality if and only if $f(u) = |u - 1|$ or $\rho = \tau$.

As an example of the above case, for $\alpha \in (0,1)$ we have

$$\frac{1}{\alpha(1-\alpha)}\int_X \left(\tau(x) + \rho(x) - \tau^{1-\alpha}(x)\rho^\alpha(x) - \tau^\alpha(x)\rho^{1-\alpha}(x)\right)d\mu(x)$$

$$\leq \frac{1}{\alpha(1-\alpha)}D_v(\rho\|\tau).$$

11.5.2 Another Extension of the f-Divergence

The density power divergence family was introduced by Basu et al. (1998) who studied the corresponding statistical properties. However, in its usual form this divergence does not belong to the f-divergence class. Therefore, we must consider another extension for the discrimination between two probability distributions using the f-divergence between two positive finite measures $D_f(\rho\|\tau)$.

Using the modification $g(p(x))$ for the probability function $p(x)$ on X, where g is a strictly increasing function on $[0, \infty)$ satisfying $\int_X g(p(x))d\mu(x) < \infty$ for all $p \in \mathbb{P}$, the following information divergence between p and q for $f \in \bar{\mathbb{F}}$ using $g(p)$ was presented in Uchida and Shioya (2005):

$$
\begin{aligned}
D_{f,g}(p\|q) &= D_f(g(p)\|g(q)), \\
&= \int_X g(q(x))f\left(\frac{g(p)}{g(q)}\right)d\mu(x).
\end{aligned}
$$

The density power divergence is derived by using the extended f-divergence, $D_{f,g}(p\|q)$. By setting $f_\beta(u) = (1-\beta)\{(\frac{u-u^{1-\beta}}{\beta}) - (u-1)\}$ and $g_\beta(v) = v^{\frac{1}{1-\beta}}$ for $\beta \in [0,1)$, we have

$$
D_{f_\beta,g_\beta}(p\|q) = \int_X \left[\frac{1-\beta}{\beta}\left(p^{\frac{1}{1-\beta}}(x) - q^{\frac{\beta}{1-\beta}}(x)p(x)\right)\right.
$$

$$
\left. - (1-\beta)\left(p^{\frac{1}{1-\beta}}(x) - q^{\frac{1}{1-\beta}}(x)\right)\right]d\mu(x).
$$

Using $\alpha = \beta/(1-\beta)$, we have the density power divergence D_{pw}^α as

$$
D_{\text{pw}}^\alpha(p\|q) = \int_X \left\{\frac{1}{\alpha}p(x)\left(p^\alpha(x) - q^\alpha(x)\right) - \frac{1}{1+\alpha}\left(p^{1+\alpha}(x) - q^{1+\alpha}(x)\right)\right\}d\mu(x).
$$

It is easy to check that $D_{\text{pw}}^\alpha(p\|q)$ is equal to the Kullback–Leibler divergence for the case $\alpha \to 0$.

However, f_β is an element of $\bar{\mathbb{F}}$. Therefore, using Lemma 11.3, the following is true in case of the Tsallis entropy, $T_{1+\alpha}(p)$ for the probability function p:

$$
\begin{aligned}
D_{\text{pw}}^\alpha(p\|q) &\leq \frac{1}{2\alpha}D_v(p\|q) + \frac{\alpha-1}{2\alpha(1+\alpha)}\left(\int_X q^{1+\alpha}(x)d\mu(x) - \int_X p^{1+\alpha}(x)d\mu(x)\right) \\
&= \frac{1}{2\alpha}D_v(p\|q) + \frac{\alpha-1}{2(1+\alpha)}\left(T_{1+\alpha}(p) - T_{1+\alpha}(q)\right).
\end{aligned}
$$

The maximum of this information divergence is easily obtained by using Equation (11.11) as

$$
D_{\text{pw}}^\alpha(p\|q) \leq \frac{1}{\alpha(1+\alpha)}\int_X p^{1+\alpha}(x)d\mu(x) + \frac{1}{1+\alpha}\int_X q^{1+\alpha}(x)d\mu(x).
$$

11.6   Additional Remarks

There are many applications of information measures in the fields of computer science and engineering. Some purely mathematical results involving information divergences have been presented in terms of differential geometry by (Nagaoka and Amari, 1982) and Amari and Nagaoka (2000). Some new applications using the density power divergence have recently been presented in the fields of materials science (Shioya and Gohara, 2006) and network engineering (Uchida, 2007). From the statistical physics viewpoint, various computational applications in the fields of engineering and science have been provided in Tsallis (2009).

12

Applications to Other Models

In this chapter, various applications of minimum distance methods to other scientific disciplines will be provided.

12.1 Introduction

The minimum distance approach is widely used for estimating unknown target parameters from given data. Fundamentally, the axiom of a distance D on the set X consists of the following three conditions: $D(x, y) \geq 0$ (nonnegativity), $D(x, y) = D(y, x)$ (symmetry), and $D(x, y) + D(y, z) \geq D(x, z)$ (the triangle inequality) for all $x, y, z \in X$. The nonnegativity condition represents the consistency between two elements. We will consider "incomplete" distances satisfying the nonnegativity condition, where the measure is zero if and only if the arguments are identically equal. We have referred to such measures as statistical distances in previous chapters, which are widely used for estimating different target parameters.

In terms of the origin of information measures, the discrimination between two probability distributions has been expressed through "information divergences." The Kullback–Leibler divergence (Kullback and Leibler, 1951), the Hellinger discrimination (Hellinger, 1909) and the variation distance are well known information divergences. In the first ten chapters of this book we have studied pure statistical applications of such divergences. In this chapter, the focus is on information theory, engineering, and other fields.

The minimization of the Kullback–Leibler divergence has been effectively applied to various problems in information science and systems engineering. Statistically this produces the maximum likelihood estimator, the origin of which goes at least as far back as Gauss (1821) in case of independently and identically distributed data from the normal distribution. This is related to the maximum entropy method using prior information by Jaynes (1968). Beran (1977) considered minimum Hellinger distance estimation with a robustness objective. By focusing on the geometrical structure of the manifold generated by a set of probability distributions, Nagaoka and Amari (1982) introduced the concept of "information geometry." In this literature, the Kullback–Leibler divergence is treated as the distance of geodesics for the manifold of an exponential family. Various applications based on information geometry have also been investigated (Amari and Nagaoka, 2000).

In the field of machine learning, the architecture of input/output systems is heavily used. Minimization methods are effectively employed for training procedures for given data. The minimization of the Kullback–Leibler divergence based on the Gaussian model gives the least squares method. This method has been found to be effective even for optimization with many parameters in complex engineering systems. In research involving brain mechanisms and neurons, the artificial neural network architecture has been proposed as a mathematical model. An efficient training procedure for such networks has been theoretically developed (Amari, 1967) through the steepest descent method. The well known back propagation method was presented by Rumelhart et al. (1986) as a training procedure for multi-layered perceptrons. The training procedure for neural networks was originally based on the least squares method and is thus a direct application of the minimization of an information measure.

As in the non-probabilistic case, several entropy measures have been introduced in the field of fuzzy set theory. The concept of the fuzzy set was proposed by Zadeh (1965), and different applications have been considered by several authors. Based on the formula of the Shannon entropy, the complexity measure for a fuzzy set (fuzzy entropy) was introduced and discussed by De Luca and Termini (1972), Kosko (1986), and others. A divergence measure between two fuzzy sets was introduced by Bhandari and Pal (1993), and another type of divergence was introduced by Fan and Xie (1999). Such measures have been used in fuzzy-processing-based applications.

The ease of construction of such a distance between two probability distributions (or two nonnegative functions) has led to various applications of the minimum distance method in a variety of engineering and scientific fields. As a novel application which involves the minimization of a statistical distance in materials science, a phase retrieval algorithm based on information divergence measures was presented by Shioya and Gohara (2006). An iterative Fourier phase retrieval method using intensity measurements for recovering the phase has been used in a diverse range of fields, including electron microscopy, astronomy, crystallography, synchrotron X-ray, and others. In this chapter, above models using minimization of statistical distances are considered. The phase retrieval algorithm based on information divergences is presented here as a typical illustration.

12.2   Preliminaries for Other Models

A statistical distance is used as a measure of discrimination between two probability distributions. Such a measure is called an "information divergence." The fundamental framework of information divergences are presented here as a prerequisite for later applications. Let $\mathbb{P}$, the set of all probability functions on X, be given by

$$\mathbb{P} \overset{\text{def}}{=} \{p \mid \int_X p(x)dx = 1,\ p(x) \geq 0\ \forall x \in X\}.$$

If X is discrete, a sum replaces the integral. An information divergence is a discriminant measure between two probability functions $p, q \in \mathbb{P}$. As a general representation, let $D(p\|q)$ be a function satisfying

$$D(p\|q) \geq 0, \qquad \forall p, q \in \mathbb{P},$$
$$D(p\|q) = 0 \qquad \Leftrightarrow p(x) = q(x),\ \forall x \in X.$$

The following are some commonly used information divergences satisfying the above constraints.

- KULLBACK–LEIBLER DIVERGENCE (Kullback and Leibler, 1951)

$$D_k(p\|q) \overset{\text{def}}{=} \int_X p(x) \log\Big(\frac{p(x)}{q(x)}\Big)dx. \tag{12.1}$$

- VARIATIONAL DISTANCE

$$D_v(p\|q) \overset{\text{def}}{=} \int_X |p(x) - q(x)|dx.$$

- HELLINGER DISCRIMINATION (Hellinger, 1909; Beran, 1977)

$$D_H(p\|q) \overset{\text{def}}{=} \int_X \Big(p^{1/2}(x) - q^{1/2}(x)\Big)^2 dx.$$

- MATUSITA DISTANCE (Matusita, 1955, 1964)

$$D_M^\alpha(p\|q) \overset{\text{def}}{=} \int_X |p^\alpha(x) - q^\alpha(x)|^{1/\alpha}dx, \quad 0 < \alpha \leq 1. \tag{12.2}$$

The Kullback–Leibler divergence in Equation (12.1) is divided into the terms of Shannon's entropy and $\log|X|$ if $|X|$ is finite and q is a uniform distribution on X. The distance in (12.2) is called the Matusita distance. The case $\alpha = 1$ gives the variation distance and the case $\alpha = 1/2$ gives the Hellinger discrimination.

As a generalization of the Kullback–Leibler divergence for nonnegative real functions, the following measure was presented by Csiszár (1991).

$$I_k(g\|h) \overset{\text{def}}{=} \int_X g(x) \log\Big(\frac{g(x)}{h(x)}\Big)dx + \int_X h(x)dx - \int_X g(x)dx,$$

where $g(x) \geq 0$, $h(x) \geq 0$ for all $x \in X$, $\int_X g(x)dx < \infty$, and $\int_X h(x)dx < \infty$. This divergence defines the discriminant measure between two nonnegative real functions with finite volumes, i.e., $I_k(g\|h) \geq 0$ and $I_k(g\|h) = 0$ if and only if $h(x) = g(x)$, identically.

We have extensively discussed the statistical approach to minimum distance inference in the earlier chapters. The parameters of a statistical model are estimated by minimizing a suitable statistical distance between a nonparametric data density (appropriately smoothed if necessary) and the parametric model density over the parameter space. In this chapter some applications of similar techniques to other models and disciplines will be discussed.

12.3   Neural Networks

12.3.1   Models and Previous Works

Information technologies need to be developed for worldwide networks in order to extract or find interesting or important information from the huge amount of data in the world. In the field of machine learning, the focus has been on computational algorithms and related analysis.

Artificial neural networks are a kind of mathematical brain model consisting of neurons. The architecture of these networks is composed of nodes and connections; a weight parameter is assigned to each connection, and its value corresponds to the information flow between two connected nodes. A feed-forward architecture and a multilayer perceptron (MLP) model is useful in this scenario. In the training procedure for MLPs, the δ rule (a gradient descent learning rule) can be used and a linear-separation problem can be solved by using this procedure. However, such an algorithm is not effective for nonlinear separation problems. Important theoretical work on the statistical steepest descent method has been presented by Amari (1967), while the back propagation (BP) method has been discussed in Rumelhart et al. (1986) as an effective method for neural networks. Feed-forward neural networks have been applied to various problems for estimating the nonlinear relationship between inputs and outputs. More recently, an integrated predictor consisting of an ensemble of multiple trained predictors, called ensemble learning, has been used as one of the developed learning methods. The bagging method was presented by Breiman (1996), and the boosting method was discussed in Schapire (1990). To obtain each predictor for ensemble learning, a normal training BP method is usually used.

12.3.2   Feed-Forward Neural Networks

Feed-forward neural networks are one of the commonly used neural network architectures. The network has three kinds of layers: input, hidden, and output layers. Each input node in the input layer connects to all the hidden nodes, and each hidden node connects to all the output nodes. The direction of information flow is from the input layer to the hidden layer and then from the hidden layer to the output layer. Each node has a threshold value and receives the linear combination of the input elements and weight parameters. The output value of a node is determined by a transfer function, such as the identity function, the sigmoid function and the tanh function, among others. In particular, the sigmoid function and the tanh function are useful for formulating the training procedure of networks for handling data having nonlinearity.

The settings for the three-layer feed-forward neural network are described next. Let x_i be the variable of the ith input node $(i = 1, \cdots, I)$, h_j be the variable of the jth hidden node $(j = 1, \cdots, J)$, and y_k be the variable of the kth output node $(k = 1, \cdots, K)$; $I - J - K$ architecture is then defined. Let g be the transfer function for the hidden and output nodes, $v_{i,j}$ $(\in \mathbb{R})$ be the weight parameter between the ith input and jth hidden nodes, and $w_{j,k}$ $(\in \mathbb{R})$ be the weight parameter between the jth hidden and kth output nodes. The threshold values of the hidden and output nodes are $(v_{0,1}, \cdots, v_{0,J})$ and $(w_{0,1}, \cdots, w_{0,K})$, respectively. The output variable of the jth hidden node h_j is $g(v_{0,j} + \sum_{i=1}^{I} v_{i,j} x_i)$, and the output variable of the kth output node y_k is $g(w_{0,k} + \sum_{j=1}^{J} w_{j,k} h_j)$.

Figure 12.1 shows a three-layer feed-forward neural network with $I - J - K$ architecture. The output y_k depends on the parameters $\{v_{i,j}\}_{i=1,\cdots,I;j=1,\cdots,J}$, $\{w_{j,k}\}_{j=1,\cdots,J;k=1,\cdots,K}$, and the input elements of the layer $\{x_i\}_{i=1,\cdots,I}$. We suppose that the transfer function g for the input and parameter vectors is a differentiable function for the hidden and output nodes. By use of the parameters $\mathbf{x} = (x_1, \cdots, x_I)$, $\mathbf{v} = (v_{0,1}, v_{1,1}, \cdots, v_{I,J})$, and $\mathbf{w} = (w_{0,1}, w_{1,1}, \cdots, w_{J,K})$, the kth output element, $k = 1, \ldots, K$, is given as

$$f_k(\mathbf{v}, \mathbf{w}|\mathbf{x}) = g\left(w_{0,k} + \sum_{j=1}^{J} w_{j,k} g\left(v_{0,j} + \sum_{i=1}^{I} v_{i,j} x_i\right)\right)$$

and the output of the network is given as

$$\mathbf{f}(\mathbf{v}, \mathbf{w}|\mathbf{x}) = (f_1(\mathbf{v}, \mathbf{w}|\mathbf{x}), \cdots, f_K(\mathbf{v}, \mathbf{w}|\mathbf{x})).$$

12.3.3   Training Feed-Forward Neural Networks

We assume that the input $\mathbf{x}_s$ and desired output $\mathbf{y}_s$ presented by the data set $d_s = (\mathbf{x}_s, \mathbf{y}_s)$ are given, where s is the index of the data $(s = 1, \cdots, N)$. Our aim is to obtain a suitable input/output system using the feed-forward neural network satisfying the given data constraints. Each $\mathbf{y}_s$ is often called the "teacher datum" or "teacher signal" of the input $\mathbf{x}_s$. D_N is the teacher data set of $\{d_1, \cdots, d_N\}$. The problem is to estimate the parameters $\mathbf{v}$ and $\mathbf{w}$ satisfying $\mathbf{y}_s = \mathbf{f}(\mathbf{v}, \mathbf{w}|\mathbf{x}_s)$ for $s = 1, \cdots, N$.

The architecture of the feed-forward neural network is a kind of deterministic system based on the weight parameters and the settled transfer functions.

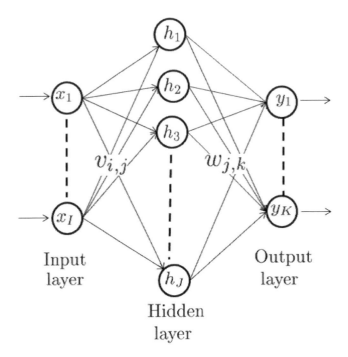

FIGURE 12.1
Three-layer feed-forward neural network with $I - J - K$ architecture. The
input and the output vectors are $(x_1, \cdots, x_I)$ and $(y_1, \cdots, y_K)$, respectively,
$v_{i,j}$ is the weight parameter between the ith input and the jth hidden nodes,
and $w_{j,k}$ is the weight parameter between the jth hidden and the kth output
nodes.

Moreover, no probabilistic mechanism is introduced in the network architecture, so the architecture does not have a statistical aspect. Thus, a direct
fitting method is needed for the training procedure of the architecture. In
contrast, we often assume that the set of teacher data includes noise factors.

We then assume that the teacher data D_N is an independently and identically distributed sample, and the stochastic relationship between the inputs
and outputs of the neural network is presented by the Gaussian model having
the conditional probability density function

$$P_{\mathrm{NN}}(\mathbf{y}_s|\mathbf{x}_s) \propto \exp\{-c\|\mathbf{y}_s - \mathbf{f}(\mathbf{v}, \mathbf{w}|\mathbf{x}_s)\|^2\},$$

where $\|\cdot\|$ is the L_2 norm and c is a positive constant. Maximum likelihood
estimation based on $\prod_{s=1}^{N} P_{\mathrm{NN}}(\mathbf{y}_s|\mathbf{x}_s)$ leads to the minimization of

$$E_{\mathbf{v},\mathbf{w}} = \sum_{s=1}^{N} \|\mathbf{y}_s - \mathbf{f}(\mathbf{v}, \mathbf{w}|\mathbf{x}_s)\|^2.$$

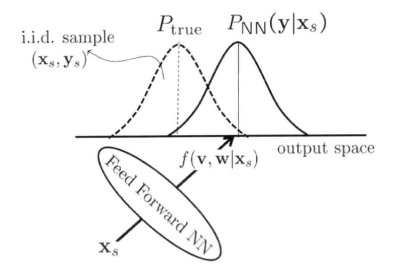

FIGURE 12.2
Relationship between a feed-forward neural network with weight parameters
$\mathbf{v}, \mathbf{w}$ and the conditional Gaussian density function $P_{\mathrm{NN}}(\mathbf{y}|\mathbf{x}_s)$.

With respect to the connected domain $\mathbf{z} = (\mathbf{x}, \mathbf{y})$, this estimation is the same
as the minimization of the Kullback–Leibler divergence between unknown true
density function P_{true} (generating the teacher data) and the Gaussian model
density function P_{NN} with the network parameters. Figure 12.2 shows the
relationship between the feed-forward neural network with weight parameters
$\mathbf{v}, \mathbf{w}$ and the conditional Gaussian density function. The estimator $(\hat{\mathbf{v}}, \hat{\mathbf{w}})$
based on the Kullback–Leibler divergence is given by

$$(\hat{\mathbf{v}}_k, \hat{\mathbf{w}}_k) = \arg\min_{\mathbf{v}, \mathbf{w}} D_k(P_{\mathrm{true}} \| P_{\mathrm{NN}}).$$

In the practical aspect of this computation, the rule for updating the
weights from $(\mathbf{v}^{\mathrm{old}}, \mathbf{w}^{\mathrm{old}})$ to $(\mathbf{v}^{\mathrm{new}}, \mathbf{w}^{\mathrm{new}})$ are presented as

$$\mathbf{v}^{\mathrm{new}} = \mathbf{v}^{\mathrm{old}} - \epsilon_1 \left.\frac{\partial E_{\mathbf{v}, \mathbf{w}}}{\partial \mathbf{v}}\right|_{\mathbf{v} = \mathbf{v}^{\mathrm{old}}}$$

and

$$\mathbf{w}^{\mathrm{new}} = \mathbf{w}^{\mathrm{old}} - \epsilon_2 \left.\frac{\partial E_{\mathbf{v}, \mathbf{w}}}{\partial \mathbf{w}}\right|_{\mathbf{w} = \mathbf{w}^{\mathrm{old}}},$$

where ϵ_1 and ϵ_2 are suitable coefficients. The process eventually provides a
suitable fit to the teacher data generated by an unknown nonlinear function.

From the viewpoint of using various information divergence measures, if we use another divergence (e.g., Hellinger discrimination, χ^2-divergence, reverse Kullback–Leibler divergence, and so on) for the minimization, a suitable update rule for the weight parameters is needed with respect to the gradient term of the information divergence between P_{true} and P_{NN}.

12.3.4 Numerical Examples

Two simple numerical examples of learning feed-forward neural networks are presented here. One is the Spam E-mail classification using the machine learning database by Asuncion and Newman (2007). Another one is the following classification problem using "$\sin x$." The data structure here is given by

$$ y = \begin{cases} 1 & \text{if } x_2 \geq \sin x_1 \\ 0 & \text{otherwise} \end{cases} , $$

where (x_1, x_2) $(\in \mathbb{R}^2)$ and y $(\in \{0, 1\})$ are the variables of the input nodes and output node, respectively. The samples used in this experiment are generated by a pair of random variables on the domain $[0, 2\pi] \times [0, 1]$. Table 12.1 shows the settings of the two numerical experiments and their results. The number of hidden nodes and iterations for training were appropriately determined. Training the neural networks with a small number of training observations appears to give a suitable prediction for test observations. Training the neural networks and the number of hidden nodes needs to be carefully considered to avoid overfitting of the training observations. The approach may be generalized by considering some weakly-learned neural networks such as those introduced by Breiman (1996) and Schapire (1990).

12.3.5 Related Works

There are many kinds of weight parameters satisfying the given teacher-data constraint. We focus on a simple approach to neural networks and restrict ourselves to retrieving a suitable weight parameter satisfying the teacher-data constraint. The estimation problem has been fundamentally treated as a non-regular statistical model by Watanabe (2001).

As an advanced learning method based on information geometry (Amari and Nagaoka, 2000), the natural gradient method was proposed by Amari (1998b), and the natural gradient descent algorithm for training multilayer perceptrons was presented by Amari (1998a). In another theoretical work, Amari (2007) presented the minimum α-divergence-based approach to an integrated predictor consisting of an ensemble of weakly learned predictors.

TABLE 12.1
Parameters of numerical experiments and results.

Parameters	Spam E-mail	Sin function
Total number of instances	4601	2601
Number of training observations	500	1000
Number of test observations	4101	1501
Number of input attributes	57	2
Number of hidden nodes	7	5
Number of output attributes	1	1
Training ratio (ϵ_1 and ϵ_2)	0.01	
Number of iterations for training	10000 or Average error < 0.01	
Correct answers for unknown samples	85.49 %	82.95 %

12.4 Fuzzy Theory

12.4.1 Fundamental Elements of Fuzzy Sets

The concept of "fuzzy sets" was proposed by Zadeh (1965). This presents a non-probabilistic theoretical framework for treating uncertainty in information sciences. The fundamental elements of fuzzy set theory are introduced next.

Let $X = \{x_1, \cdots, x_n\}$ be the universal set. Define the function μ_A from X to the closed interval $[0, 1]$; it denotes the degree of belongingness to the set A. Such a set A with function μ_A is called "fuzzy set A," and μ_A is called a membership function. Let $F(X)$ be all the fuzzy sets of X. If the membership function $\mu_A(x)$ is 0 or 1 for all $x \in X$, A is called a "crisp set." Let $P(X)$ be the family of all the crisp sets of X. Therefore, a crisp set is treated as a normal subset of X.

Concerning the operations between two fuzzy sets as the mapping from $F(X) \times F(X)$ to $F(X)$, the following conventions and definitions are used. For two fuzzy sets A and B, the intersection $A \cap B$ is the fuzzy set with the membership function $(\min\{\mu_A(x_1), \mu_B(x_1)\}, \cdots, \min\{\mu_A(x_n), \mu_B(x_n)\})$, the summation $A \cup B$ is the fuzzy set with the membership function $(\max\{\mu_A(x_1), \mu_B(x_1)\}, \cdots, \max\{\mu_A(x_n), \mu_B(x_n)\})$, and the complement A^c is the fuzzy set with the membership function $(1 - \mu_A(x_1), \cdots, 1 - \mu_A(x_n))$.

If two fuzzy sets A and B are crisp, the sets obtained by the above operations are also crisp, and these are the normal set operations. Also $[a]$ is the fuzzy set of X satisfying $\mu_{[a]}(x) = a$ ($a \in [0, 1]$) for all $x \in X$.

12.4.2   Measures of Fuzzy Sets

Inspired by the formulation of the Shannon entropy, a complexity measure for a fuzzy set called "fuzzy entropy" was introduced by De Luca and Termini (1972) as

$$H_{\text{fuzzy}}^{\text{LT}}(A) = \sum_{x \in X} S(\mu_A(x)),$$

where $A \in F(X)$ and $S(u) = -u \log u - (1-u) \log(1-u)$.

In respect of a generalized fuzzy entropy E_{fuzzy}, the conditions (a1)–(a4) below are called the four De Luca-Termini axioms.

(a1) $E_{\text{fuzzy}}(A) = 0$ if and only if A is a non-fuzzy set.

(a2) $E_{\text{fuzzy}}(A)$ is maximum if and only if $\mu_A(x) = \mu_{[0.5]}(x)$.

(a3) $E_{\text{fuzzy}}(A) \leq E_{\text{fuzzy}}(B)$ (i.e., A is less fuzzy than B), if $\mu_A(x) \leq \mu_B(x)$ for $x \in \{\tilde{x} | \mu_B(\tilde{x}) \leq 0.5\}$ and $\mu_A(x) \geq \mu_B(x)$ for $x \in \{\tilde{x} | \mu_B(\tilde{x}) \geq 0.5\}$.

(a4) $E_{\text{fuzzy}}(A) = E_{\text{fuzzy}}(A^c)$.

Based on the property that $A \cap A^c \neq \phi$ for a fuzzy set A, Yager (1979) proposed, based on the L_p-distance, the fuzzy entropy

$$Y_p(A) = 1 - \frac{L_p(A, A^c)}{n^{\frac{1}{p}}},$$

where $L_p(A, B) = \{\sum_{i=1}^{n} |\mu_A(x_i) - \mu_B(x_i)|^p\}^{1/p}$, $p \geq 1$, and $L_p(C, C^c)$ is equal to $n^{\frac{1}{p}}$ for any crisp set C. In addition, $L_p(A, B)$ can be used as the distance between two fuzzy sets A and B. Kaufman (1975) proposed the normalized fuzzy entropy

$$K_p(A) = \frac{2}{n^{\frac{1}{p}}} L_p(A, \bar{A}),$$

where $\bar{A}$ is the nearest non-fuzzy set with the membership function

$$\mu_{\bar{A}}(x) = \begin{cases} 1, & \text{if } \mu_A(x) \geq 0.5, \\ 0, & \text{if } \mu_A(x) \leq 0.5. \end{cases}$$

Note that $\bar{A}$ is not uniquely determined for a fuzzy set A. However, if $\mu_A(x) = 0.5$, $|\mu_A(x) - \mu_{\bar{A}}(x)|$ is equal to 0.5 for $\mu_{\bar{A}}(x) = 0$ and is the same for $\mu_{\bar{A}}(x) = 1$. Therefore, the value of $K_p(A)$ is uniquely determined for a given fuzzy set A.

Based on the Kullback–Leibler divergence, the following divergence measure between two fuzzy sets was introduced by Bhandari and Pal (1993).

$$D_{\text{fuzzy}}^{S_1}(A, B) = \frac{1}{n} \sum_{i=1}^{n} [S_1(\mu_A(x_i), \mu_B(x_i)) + S_1(\mu_B(x_i), \mu_A(x_i))], \quad (12.3)$$

where

$$S_1(u, v) = u \log \left(\frac{u}{v} \right) + (1 - u) \log \left(\frac{1 - u}{1 - v} \right),$$

$A, B \in F(X)$, and $0 < \mu_A(x) < 1, 0 < \mu_B(x) < 1$ for all $x \in X$. The inequality $D_{\text{fuzzy}}^{S_1}(A, B) \geq 0$ is satisfied and the equality is established if and only if A is equal to B. Therefore, this measure is called a "fuzzy divergence," and it satisfies all the metric properties except for the triangle inequality. In addition, some properties of the fuzzy divergence $D_{\text{fuzzy}}^{S_1}(A, B)$ are presented in Bhandari and Pal (1993), e.g., the equality $D_{\text{fuzzy}}^{S_1}(A \cup B, A \cap B) = D_{\text{fuzzy}}^{S_1}(A, B)$ is shown. Moreover, the authors introduced the modified fuzzy divergence $D_{\text{fuzzy}}^{S_2}(A, B)$, where

$$S_2(u, v) = u \log \left(\frac{1 + u}{1 + v} \right) + (1 - u) \log \left(\frac{2 - u}{2 - v} \right).$$

This divergence measure can be used for two fuzzy sets A and B with membership functions $\mu_A(x) \in [0, 1]$ and $\mu_B(x) \in [0, 1]$, respectively. This modified divergence is finite for any two membership functions, but the fuzzy divergence $D_{\text{fuzzy}}^{S_2}(A, B)$ is not a Kullback–Leibler-based divergence.

Pal and Pal (1991) developed an exponential-type fuzzy entropy as

$$H_{\text{fuzzy}}^{\text{ex}}(A) = \sum_{i=1}^{n} h(\mu_A(x)),$$

where $h(u) = u \exp(1 - u) + (1 - u) \exp(u) - 1$. This fuzzy entropy is based on an exponential-type statistical entropy $\{\sum_{i=1}^{n} p_i \exp(1 - p_i)\} - 1$ for a discrete probability distribution $(p_1, \cdots, p_n)$. Using the exponential ratio $\exp(\mu_A) / \exp(\mu_B)$ and $H_{\text{fuzzy}}^{\text{ex}}(A)$, Fan and Xie (1999) introduced the following discrimination from A to B.

$$I_{\text{fuzzy}}(A, B) = \sum_{i=1}^{n} \{1 - (1 - \mu_A(x_i)) \exp(\mu_A(x_i) - \mu_B(x_i))$$
$$- \mu_A(x_i) \exp(\mu_B(x_i) - \mu_A(x_i))\}.$$

They also defined the fuzzy divergence

$$D_{\text{fuzzy}}^{\text{FX}}(A, B) = I_{\text{fuzzy}}(A, B) + I_{\text{fuzzy}}(B, A).$$

This divergence satisfies the fundamental inequality $D_{\text{fuzzy}}^{\text{FX}}(A, B) \geq 0$, where equality is obtained if and only if $A = B$ (i.e., $\mu_A = \mu_B$).

12.4.3 Generalized Fuzzy Divergence

To define a statistical-distance-based fuzzy divergence, a well formulated information divergence is needed. The f-divergence of (Csiszár, 1963) between two probability functions p, q on X, given by

$$D_f(p\|q) = \sum_{x \in X} q(x) f\left(\frac{p(x)}{q(x)}\right),$$

is such a well defined measure where $f(u)$ is a convex function on $(0, \infty)$, is strictly convex at $u = 1$, and $f(1) = 0$. Various effective divergence measures are obtained by choosing different convex functions. The detailed properties of f-divergences have been discussed in Chapter 11. By combining the f-divergence and the fuzzy divergence of Bhandari and Pal (1993), the "fuzzy f-divergence"

$$D_{\text{fuzzy}}^f(A, B) = \frac{1}{n} \sum_{i=1}^{n} [F_f(\mu_A(x_i), \mu_B(x_i)) + F_f(\mu_B(x_i), \mu_A(x_i))],$$

is naturally obtained, where

$$F_f(u, v) = v f\left(\frac{u}{v}\right) + (1 - v) f\left(\frac{1 - u}{1 - v}\right).$$

We use $f^*(u) = uf(1/u)$, $f(0) = \lim_{t \to 0} f(t)$, $f^*(0) = \lim_{t \to 0} tf(1/t)$, and $0f(0/0) = 0$. When $f(u) = u \log u$ or $f(u) = -\log u$, $D_{\text{fuzzy}}^f(A, B)$ gives the fuzzy divergence $D_{\text{fuzzy}}^{S_1}(A, B)$. If there exists x satisfying $\mu_A(x) = 0$ or 1, or $\mu_B(x) = 0$ or 1, the fuzzy divergence $D_{\text{fuzzy}}^{S_1}(A, B)$ is infinite. For this purpose the modified divergence $D_{\text{fuzzy}}^{S_2}$ was introduced; however, some of the finer statistical properties based on the Kullback–Leibler divergence are then lost.

    To obtain a class of finite fuzzy f-divergences, it is important to first construct a finite f-divergence between two probability distributions. The upper bound of an f-divergence was presented by Vajda (1972) in terms of $f(0)$ and $f^*(0)$. If $f(0)$ and $f^*(0)$ are finite, the upper bound of the f-divergence between p_{θ_1} and p_{θ_2} (where $p_{\theta_i} = (\theta_i, 1 - \theta_i)$, $0 \le \theta_i \le 1$, $i = 1, 2$) is presented by the following inequality

$$D_f(p_{\theta_1}, p_{\theta_2}) \le f(0) + f^*(0),$$

where equality is obtained if and only if $p_{\theta_1} \perp p_{\theta_2}$. Therefore, based on the f-divergence with a convex function f satisfying $f(0) + f^*(0) < \infty$, the upper bound of the fuzzy f-divergence is obtained as

$$D_{\text{fuzzy}}^f(A, B) \le 2(f(0) + f^*(0)), \tag{12.4}$$

where equality is obtained if and only if $\mu_A(x) = 1$ for all $x \in \{y | \mu_B(y) = $

$0, y \in X\}$ and $\mu_B(x) = 1$ for for all $x \in \{y | \mu_A(y) = 0, y \in X\}$. For example, both of the following convex functions generate a finite fuzzy f-divergence

$$f_\alpha(t) = \frac{1}{\alpha(1-\alpha)}(1 - t^\alpha),$$

$$\bar{f}_\alpha(t) = \frac{1}{\alpha(1-\alpha)}(1 - t^{1-\alpha}),$$

where $0 < \alpha < 1$. Using the symmetry of D^f_{fuzzy} and Equation (12.4), we have

$$D^{f_\alpha}_{\text{fuzzy}}(A, B) = D^{\bar{f}_\alpha}_{\text{fuzzy}}(A, B) \leq \frac{2}{\alpha(1-\alpha)}.$$

The relationships between f_α (and $\bar{f}_\alpha$) and the fuzzy divergence given in (12.3) are presented as follows:

$$\lim_{\alpha \to 0} D^{f_\alpha}_{\text{fuzzy}}(A, B) = \lim_{\alpha \to 1} D^{\bar{f}_\alpha}_{\text{fuzzy}}(A, B) = D^{S_1}_{\text{fuzzy}}(A, B).$$

The above generalization of the fuzzy divergence measures was obtained by an information-theoretic approach. Various applications using fuzzy divergences are available in practice, and as such the measures play an important role in many scientific problems of different disciplines.

12.5   Phase Retrieval

12.5.1   Diffractive Imaging

The technology for material imaging is progressing at a rapid pace. For achieving atomic scale resolution, microscopes using various light sources, such as X-rays, lasers, and electron beams, have been developed. By using the Fourier-intensity measurement based on Fourier phase retrieval, lensless imaging is progressing toward wavelength resolution. Figure 12.3 shows the relationship between the target material (ρ) and its diffraction pattern (I_{obs}). A beam of light source strikes an object and diffracts into many specific directions. This phenomenon has been widely used for learning the atomic-scale arrangements of periodic materials in crystallography. A charge-coupled device (CCD) has often been used as a detector for measuring the diffraction intensities. The quantity I_{obs} is the observed intensity, and it corresponds to the Fourier intensity of ρ. The objective is to retrieve the missing Fourier phase in the observation of the diffraction wave."

An iterative Fourier phase retrieval method using intensity measurements for recovering the phase has been applied in a diverse range of fields, including electron microscopy, astronomy, crystallography, and synchrotron X-ray.

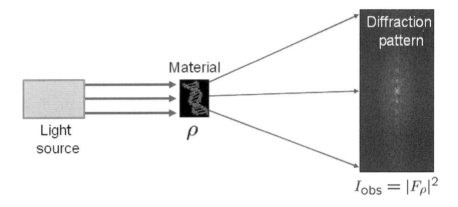

$$I_{obs} = |F_\rho|^2$$

FIGURE 12.3
Relationship between target material ρ and its diffraction pattern I_{obs} (Fourier intensity of ρ).

The Gerchberg–Saxton algorithm was presented based on iterative transformations between the object domain and the Fourier domain (Gerchberg and Saxton, 1972). The process of replacing a prior object function obtained under the Fourier constraints has been performed, including the error reduction (ER) method, input and output (IO) method, and hybrid input and output (HIO) method. Fienup (1982) presented a comparison of these methods, clarifying in particular the relationship between the ER and the steepest descent method; this was an important achievement in research regarding phase retrieval methods (Fienup, 1982). With regard to the uniqueness of the phase problem, the oversampling ratio has been discussed in Miao et al. (1998) and the mathematical analysis of the problem has been presented in Bruck and Sodin (1979) and Barakat and Newsam (1984). Recently, based on these iterative algorithms, advanced experiments of nano-scale imaging have progressed through the use of a synchrotron X-ray having a coherent source and a transmission electron microscope equipped with a field emission gun (Zuo et al., 2003). The phase-retrieved imaging for materials is often called diffractive imaging.

The maximum entropy method (MEM) used in the crystallography field by Collins (1982), Sakata and Sato (1990), Podjarny et al. (1988), Piro (1983), Wei (1985), and Gull et al. (1987) does not provide a clear representation of an iterative algorithm. On the basis of an information-theoretic approach to phase retrieval, Shioya and Gohara (2008) presented the relationship between the MEM for crystallography and the phase retrieval algorithms and introduced an iterative MEM-type phase retrieval algorithm for diffractive imaging. In the following, section, an iterative MEM-type algorithm as a statistical-distance-based phase retrieval method is introduced.

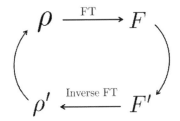

FIGURE 12.4
Gerchberg–Saxton iterative diagram of phase retrieval.

12.5.2   Algorithms for Phase Retrieval

Phase retrieval is represented as a correlative diagram of the Fourier and inverse Fourier transforms between an object domain and the Fourier domain. Figure 12.4 shows the correlative diagram called the Gerchberg–Saxton diagram. Fienup's reconstruction of the phase process using the intensity measurements of both domains is as follows (Fienup, 1982): (i) Fourier transform the prior object ρ into F; (ii) replace F with F', whose amplitude is given by the experiment in the Fourier domain (i.e., the phase of F' is the same as that of F, and the replaced amplitude is the constraint in the Fourier domain); (iii) inverse Fourier transform F' into ρ'; and (iv) replace ρ' with ρ using the constraints in the object domain. Although the setting of the constraints in the object domain is very important for phase retrieval experiments, we focus on the mathematical framework of iterative algorithms.

In practical computation, we assume that both domains are 2D discrete spaces with squared arrays. Using the discrete Fourier transform, the nth iteration is given by the following set of equations:

$$F_n(\mathbf{k}) = \sum_{\mathbf{r}} \rho_n(\mathbf{r}) \exp(-i2\pi \mathbf{k} \cdot \mathbf{r}/N)$$

and

$$\rho'_n(\mathbf{r}) = \frac{1}{N^2} \sum_{\mathbf{k}} F'_n(\mathbf{k}) \exp(2\pi i \mathbf{k} \cdot \mathbf{r}/N),$$

where $\mathbf{r} = (r_1, r_2)$, $\mathbf{k} = (k_1, k_2)$ and r_1, r_2, k_1, and $k_2 = 0, \ldots, N-1$. Also ρ_n is the object function of the nth iteration, and F_n is the Fourier transform of ρ_n. $F'_n(\mathbf{k}) = |F_{\mathrm{obs}}(\mathbf{k})| \exp\{i\psi_n(\mathbf{k})\}$, $\psi_n(\mathbf{k})$ is the phase of $F_n(\mathbf{k})$, and $|F_{\mathrm{obs}}(\mathbf{k})|$ is the amplitude of the observed intensity measurement I_{obs} in the Fourier domain. The object $\rho'_n(\mathbf{r})$ is estimated by the inverse Fourier transform of $F'_n(\mathbf{k})$. We restrict ourselves to the problem of recovering the phase in the Fourier domain using a single intensity measurement; moreover, we suppose that an object function is a real nonnegative function. The ER is then described as

$$\rho_{n+1}(\mathbf{r}) = \begin{cases} \rho_n'(\mathbf{r}) & \mathbf{r} \notin D_{\text{voc}} \\ 0 & \mathbf{r} \in D_{\text{voc}} \end{cases},$$

where D_{voc} is the set of points at which ρ_n' violates the object-domain constraints. As a modified algorithm of the ER, the hybrid input output (HIO) algorithm (Fienup, 1982) was presented as

$$\rho_{n+1}(\mathbf{r}) = \begin{cases} \rho_n'(\mathbf{r}) & \mathbf{r} \notin D_{\text{voc}} \\ \rho_n(\mathbf{r}) - \beta\rho_n'(\mathbf{r}) & \mathbf{r} \in D_{\text{voc}} \end{cases},$$

where $0 < \beta < 1$.

12.5.3   Statistical-Distance-Based Phase Retrieval Algorithm

We introduce the integrated measure between two nonnegative real functions as

$$I(\rho\|\tau) = \sum_{\mathbf{r}} \rho(\mathbf{r}) \log \frac{\rho(\mathbf{r})}{\tau(\mathbf{r})} + \sum_{\mathbf{r}} \tau(\mathbf{r}) - \sum_{\mathbf{r}} \rho(\mathbf{r}).$$

The Fourier intensity is observed, but the phase is not obtained in this problem. That is, the information of the phase is unknown, so the Fourier constraints using the structure factors are given as

$$E(|F_{\text{cal}}|, |F_{\text{obs}}|) = \frac{1}{N^2} \sum_{\mathbf{k}} |\,|F_{\text{cal}}(\mathbf{k})| - |F_{\text{obs}}(\mathbf{k})|\,|^2,$$

where $F_{\text{cal}}(\mathbf{k}) = |F_{\text{cal}}(\mathbf{k})| \exp[i\psi(\mathbf{k})]$ and F_{obs} is conventionally given by using an amplitude of observed intensity measurement I_{obs} and a phase of F_{cal}, i.e., $F_{\text{obs}}(\mathbf{k}) = |F_{\text{obs}}(\mathbf{k})| \exp[i\psi(\mathbf{k})]$. Using the property of the least squares of the Fourier domain for phase retrieval (Fienup, 1982), we have

$$\frac{\partial E(|F_{\text{cal}}|, |F_{\text{obs}}|)}{\partial \rho(\mathbf{r})} = \frac{2}{N^2}(\rho(\mathbf{r}) - \rho'(\mathbf{r})).$$

Using Lagrange's method for the minimum of $I(\rho\|\tau) + \frac{\lambda}{2}E(|F_{\text{cal}}|, |F_{\text{obs}}|)$, we obtain

$$\rho(\mathbf{r}) = \exp[\log \tau(\mathbf{r}) + \frac{\lambda}{N^2}(\rho'(\mathbf{r}) - \rho(\mathbf{r}))]. \tag{12.5}$$

We use the assumption that ρ is close to τ (the norm $|\rho - \tau|$ is sufficiently small). We then derive an update rule for the object function with respect to Equation (12.5) by using the following settings: $\rho(\mathbf{r})$ and $\tau(\mathbf{r})$ on the right side of the equation refer to the prior object $\rho_n(\mathbf{r})$, and $\rho(\mathbf{r})$ on the left side of the equation is an updated object $\rho_{n+1}(\mathbf{r})$; $\psi(\mathbf{k})$ ($= \psi_n(\mathbf{k})$) is the phase of $F_n(\mathbf{k})$ given by the Fourier transform of $\rho_n(\mathbf{r})$ for the nth iteration. Then we have

$$\rho_{n+1}(\mathbf{r}) = \rho_n(\mathbf{r}) \exp[\xi(\rho_n'(\mathbf{r}) - \rho_n(\mathbf{r}))],$$

where $\rho_n'(\mathbf{r})$ is the object obtained by the inverse Fourier transform of $|F_{\text{obs}}| \exp(i\psi_n)$ and ξ is a positive constant.

FIGURE 12.5
(a) 2D target object; (b) Fourier intensity of Figure (a) in the log scale; (c) Fourier intensity contaminated with Poisson noise in the log scale; (d) Object-domain constraint used to obtain target image by phase retrieval. The white and black regions are computational support and no-object domain, respectively.

12.5.4 Numerical Example

In this section, a simple numerical example using the HIO-ER-based algorithm and the MEM-type phase retrieval algorithm is presented. Figure 12.5 (a) is a 2D target object $\rho_{\text{target}}(\mathbf{r})$ on the discrete square array domain 256×256, where $\mathbf{r} = (r_1, r_2) \in 256 \times 256$; $\rho_{\text{target}}(\mathbf{r}) = 255$ if $\mathbf{r}$ is in the white area of (a), and $\rho_{\text{target}}(\mathbf{r}) = 0$ if $\mathbf{r}$ is in the black area of (a). Figure 12.5 (b) is the Fourier intensity function $|F_{\text{target}}(\mathbf{k})|^2$ obtained by the discrete Fourier transform on the Fourier domain 256×256, where $\mathbf{k} = (k_1, k_2) \in 256 \times 256$. The Poisson noise is the contamination in the observation on measuring the diffraction pattern. In Figure 12.5 (c), $F_{\text{obs}}^{\text{poisson}}(\mathbf{k})$ is the Fourier intensity contaminated by the Poisson noise due to the following equation.

$$\text{Poisson}\left\{ c^2 \, |F_{\text{target}}|^2 \right\} \sim c^2 |F_{\text{obs}}^{\text{poisson}}|^2,$$

where c is a positive coefficient based on the total count. The objective is to estimate the missing Fourier phase $\phi(\mathbf{k})$ (where $F_{\text{target}}(\mathbf{k}) = |F_{\text{target}}(\mathbf{k})| \exp(i\phi(\mathbf{k}))$) from the observation $|F_{\text{obs}}^{\text{poisson}}|$ and certain object-domain constraints. In addition, the estimated object is obtained by the inverse Fourier transform of the observed intensity and the estimated Fourier phase. Figure 12.5 (d) shows the object-domain constraint used in the numerical example in this section; the white area represents the computational support, and the black area represents the no-object domain.

We use two types of algorithms for this problem setting. In the HIO-ER algorithm, 500 iterations each of the HIO algorithm and ER algorithm are used. This method is often used for retrieving the missing Fourier phase. In the HIO-MEM algorithm, 500 iterations each of the HIO algorithm and iterative MEM algorithm are used. The total number of iterations for the

R-factor

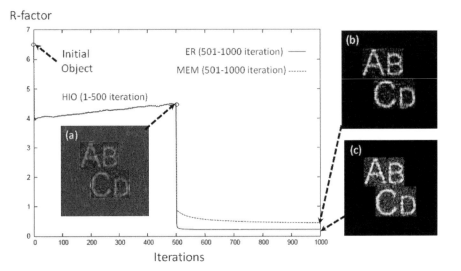

FIGURE 12.6

Graph presenting the change in the R-factor for the phase retrieval process using HIO-ER and HIO-MEM algorithms. The HIO-ER algorithm consists of HIO ($0{\to}500$ iterations) and ER ($501{\to}1000$ iterations). The HIO-MEM algorithm consists of HIO ($0{\to}500$ iterations) and iterative MEM ($501{\to}1000$ iterations). (a) Object obtained by HIO (500 iterations). (b) Estimated object by iterative MEM algorithms with 500 iterations using Fig. (a) as initial object. (c) Estimated object by the ER algorithms with 500 iterations using Fig. (a) as initial object.

phase retrieval is 1000 in both cases. The parameter β in the HIO algorithm is 0.5, and the parameter ξ in the iterative MEM algorithm is scheduled by the number of iterations, i.e., $\xi = \xi(n)$, where n is the number of iterations for phase retrieval. The following "R-factor" is often used as the criterion for phase retrieval.

$$R(|F_{\text{cal}}|, |F_{\text{obs}}|) = \frac{\sum_{\mathbf{k}} ||F_{\text{cal}}(\mathbf{k})| - |F_{\text{obs}}(\mathbf{k})||}{\sum_{\mathbf{k}} |F_{\text{obs}}(\mathbf{k})|},$$

where $|F_{\text{obs}}|^2$ is a given intensity measurement for the problem setting of phase retrieval and $|F_{\text{cal}}|^2$ is the Fourier intensity obtained by an estimated object.

The graph in Fig. 12.6 presents the changing of the R-factor in the phase retrieval process using the HIO (from the start to 500 iterations) and the ER or the MEM (from 501 to 1000 iterations). Figure 12.6 (a) is the object obtained by the HIO with 500 iterations. We use this as the initial object for the ER and MEM algorithms. Figure 12.6 (b) is the object estimated by the iterative MEM algorithms with 500 iterations, and Fig. 12.6 (c) is the estimated object by the ER algorithms with 500 iterations. Many noise particles are found in the object estimated by the ER algorithm. Such noise particles are almost completely eliminated in the object estimated by the iterative MEM algorithm (Fig. 12.6 (b)). This effectiveness was presented as an one-dimensional example in Shioya and Gohara (2008). By using an appropriate setting for scheduling the parameter ξ in the iterative MEM algorithm, the MEM algorithm turns out to be more effective than the ER algorithm in the case where the Fourier intensity is contaminated with Poisson noise.

12.6 Summary

The minimum distance method is often effective in various engineering and scientific problems. As a fundamental ingredient of this method we have reviewed the general content of information divergences. Such a foundation is important for applying the method to various scientific problems.

Feed-forward neural networks and the corresponding machine-learning architecture were introduced in connection with a statistical algorithm based on minimum divergence estimation. The fuzzy set theory was introduced as a non-probabilistic framework, and some fuzzy entropies and fuzzy divergences were considered. A generalized fuzzy divergence based on the Csiszár f-divergence was also discussed. As an application in the materials science domain, the Fourier phase retrieval for structure analysis from an information-theoretic approach was presented. In particular, an iterative algorithm for phase retrieval based on the maximum entropy method was described.

The discussions of this chapter, together with those of the previous ones, show that the minimum distance methods have great potential in many different branches of science and engineering. We feel that in the future these methods are destined to assume a more important role in data analysis and model building.

Bibliography

Agostinelli, C. (2002a). Robust model selection in regression via weighted likelihood methodology. *Statistics & Probability Letters 56*, 289–300.

Agostinelli, C. (2002b). Robust stepwise regression. *Journal of Applied Statistics 29*, 825–840.

Agostinelli, C. (2003). Robust time series estimation via weighted likelihood. In R. Dutter, P. Filzmoser, P. Rousseeuw, and U. Gather (Eds.), *Developments in Robust Statistics, International Conference on Robust Statistics 2001*. Physica Verlag.

Agostinelli, C. and M. Markatou (1998). A one step robust estimator for regression based on the weighted likelihood reweighting scheme. *Statistics & Probability Letters 37*, 341–350.

Agostinelli, C. and M. Markatou (2001). Tests of hypotheses based on the weighted likelihood methodology. *Statistica Sinica 11*, 499–514.

Agresti, A. (1984). *Analysis of Ordinal Categorical Data*. New York, USA: Wiley.

Aherne, F. J., N. A. Thacker, and P. I. Rockett (1997). The Bhattacharyya metric as an absolute similarity measure for frequency coded data. *Kybernetika 32*, 1–7.

Aldrich, J. (1997). R. A. Fisher and the making of maximum likelihood 1912-1922. *Statistical Science 12*, 162–176.

Alexandridis, R. (2005). *Minimum Disparity Inference for Discrete Ranked Set Sampling Data*. Ph. D. thesis, Ohio State University, USA.

Ali, S. M. and S. D. Silvey (1966). A general class of coefficients of divergence of one distribution from another. *Journal of the Royal Statistical Society B 28*, 131–142.

Alin, A. (2007). A note on penalized power-divergence test statistics. *International Journal of Computational and Mathematical Sciences (WASET) 1*, 209–215.

Alin, A. (2008). Comparison of ordinary and penalized power-divergence test statistics for small and moderate samples in three-way contingency tables via simulation. *Communications in Statistics: Simulation and Computation 37*, 1593–1602.

Alin, A. and S. Kurt (2008). Ordinary and penalized minimum power-divergence estimators in two-way contingency tables. *Computational Statistics 23*, 455–468.

Altaye, M., A. Donner, and N. Klar (2001). Inference procedures for assessing interobserver agreement among multiple raters. *Biometrics 57*, 584–588.

Amari, S. (1967). Theory of adaptive pattern classifiers. *IEEE Transactions on Electron Computers EC-16*, 299–307.

Amari, S. (1998a). Complexity issues in natural gradient descent method for training multilayer perceptrons. *Neural Computation 10*, 2137–2157.

Amari, S. (1998b). Natural gradient works efficiently in learning. *Neural Computation 10*, 251–276.

Amari, S. (2007). Integration of stochastic models by minimizing α-divergence. *Neural Computation 19*, 2780–2796.

Amari, S. and H. Nagaoka (2000). Methods of information geometry. In *Translations of Mathematical Monographs, Vol. 191*. Oxford University Press.

An, H.-Z. and C. Bing (1991). A Kolmogorov–Smirnov type statistic with application to test for nonlinearity in time series. *International Statistical Review 59*, 287–307.

Anderson, T. W. (1962). On the distribution of the two-sample Cramér–von Mises criterion. *Annals of Mathematical Statistics 33*, 1148–1159.

Anderson, T. W. and D. A. Darling (1952). Asymptotic theory of certain "goodness-of-fit" criteria based on stochastic processes. *Annals of Mathematical Statistics 23*, 193–212.

Anderson, T. W. and D. A. Darling (1954). A test of goodness of fit. *Journal of the American Statistical Association 49*, 765–769.

Asuncion, A. and D. Newman (2007). UCI repository of machine learning database. http://archive.ics.uci.edu/ml/.

Banerjee, A., S. Merugu, I. S. Dhillon, and J. Ghosh (2005). Clustering with Bregman divergences. *Journal of Machine Learning Research 6*, 1705–1749.

Barakat, R. and G. Newsam (1984). Necessary conditions for a unique solution to two-dimensional phase retrieval. *Journal of Mathematical Physics 23*, 3190–3193.

Barnett, N. S., P. Cerone, and S. S. Dragomir (2002). Some new inequalities for Hermite–Hadamard divergence in information theory. In *RGMIA Research Report Correction.* Victoria University. Vol. 5, Article 8.

Barnett, N. S. and S. S. Dragomir (2002). Bounds in terms of the fourth derivative for the remainder in the corrected trapezoid formula. *Computers and Mathematics with Applications 44*, 595–605.

Barron, A. R. and T. Cover (1991). Minimum complexity density estimation. *IEEE Transactions on Information Theory 37*, 1034–1054.

Bartels, R. H., S. D. Horn, A. M. Liebetrau, and W. L. Harris (1978). A computational investigation of Conover's Kolmogorov–Smirnov test for discrete distributions. *Journal of Statistical Computation and Simulation 7*, 151–161.

Bassetti, F., A. Bodini, and E. Regazzini (2007). Consistency of minimum divergence estimators based on grouped data. *Statistics & Probability Letters 77*, 937–941.

Basu, A. (1991). *Minimum disparity estimation in the continuous case: Efficiency, distributions, robustness and algorithms.* Ph. D. thesis, The Pennsylvania State University, University Park, USA.

Basu, A. (1993). Minimum disparity estimation: Application to robust tests of hypothesis. *Technical Report* 93-10, Center for Statistical Sciences, University of Texas at Austin, Texas, USA.

Basu, A. (2002). Outlier resistant minimum divergence methods in discrete parametric models. *Sankhya B 64*, 128–140.

Basu, A. and S. Basu (1998). Penalized minimum disparity methods for multinomial models. *Statistica Sinica 8*, 841–860.

Basu, A., S. Basu, and G. Chaudhuri (1997). Robust minimum divergence procedures for the count data models. *Sankhya B 59*, 11–27.

Basu, A. and I. R. Harris (1994). Robust predictive distributions for exponential families. *Biometrika 81*, 790–794.

Basu, A., I. R. Harris, and S. Basu (1996). Tests of hypotheses in discrete models based on the penalized Hellinger distance. *Statistics & Probability Letters 27*, 367–373.

Basu, A., I. R. Harris, and S. Basu (1997). Minimum distance estimation: The approach using density based distances. In G. S. Maddala and C. R. Rao (Eds.), *Handbook of Statistics Vol. 15, Robust Inference*, pp. 21–48. New York, USA: Elsevier Science.

Basu, A., I. R. Harris, N. L. Hjort, and M. C. Jones (1998). Robust and efficient estimation by minimising a density power divergence. *Biometrika 85*, 549–559.

Basu, A. and B. G. Lindsay (1994). Minimum disparity estimation for continuous models: Efficiency, distributions and robustness. *Annals of the Institute of Statistical Mathematics 46*, 683–705.

Basu, A. and B. G. Lindsay (2004). The iteratively reweighted estimating equation in minimum distance problems. *Computational Statistics & Data Analysis 45*, 105–124.

Basu, A., S. Ray, C. Park, and S. Basu (2002). Improved power in multinomial goodness-of-fit tests. *Journal of the Royal Statistical Society D 51*, 381–393.

Basu, A. and S. Sarkar (1994a). Minimum disparity estimation in the errors-in-variables model. *Statistics & Probability Letters 20*, 69–73.

Basu, A. and S. Sarkar (1994b). On disparity based goodness-of-fit tests for multinomial models. *Statistics & Probability Letters 19*, 307–312.

Basu, A. and S. Sarkar (1994c). The trade-off between robustness and efficiency and the effect of model smoothing in minimum disparity inference. *Journal of Statistical Computation and Simulation 50*, 173–185.

Basu, A., S. Sarkar, and A. N. Vidyashankar (1997). Minimum negative exponential disparity estimation in parametric models. *Journal of Statistical Planning and Inference 58*, 349–370.

Basu, S. and A. Basu (1995). Comparison of several goodness-of-fit tests for the kappa statistic based on exact power and coverage probability. *Statistics in Medicine 14*, 347–356.

Basu, S., A. Basu, and M. C. Jones (2006). Robust and efficient parametric estimation for censored survival data. *Annals of the Institute of Statistical Mathematics 58*, 341–355.

Beaton, A. E. and J. W. Tukey (1974). The fitting of power series, meaning polynomials, illustrated on band spectroscopic data. *Technometrics 16*, 147–185.

Benson, Y. E. and Y. Y. Nikitin (1995). On conditions of Bahadur local asymptotic optimality of weighted Kolmogorov–Smirnov statistics. *Journal of Mathematical Sciences 75*, 1884–1888.

Beran, R. J. (1977). Minimum Hellinger distance estimates for parametric models. *Annals of Statistics 5*, 445–463.

Berkson, J. (1980). Minimum chi-square, not maximum likelihood! *Annals of Statistics 8*, 457–487.

Bhandari, D. and N. R. Pal (1993). Some new information measures for fuzzy sets. *Information Sciences 67*, 209–228.

Bhandari, S., A. Basu, and S. Sarkar (2006). Robust inference in parametric models using the family of generalized negative exponential disparities. *Australian and New Zealand Journal of Statistics 48*, 95–114.

Bhattacharya, B. and A. Basu (2003). Disparity based goodness-of-fit tests for and against order restrictions for multinomial models. *Journal of Nonparametric Statistics 15*, 1–10.

Bhattacharyya, A. (1943). On a measure of divergence between two statistical populations defined by their probability distributions. *Bulletin of the Calcutta Mathematical Society 35*, 99–109.

Billingsley, P. (1986). *Probability and Measure* (2nd ed.). New York, USA: John Wiley & Sons.

Birch, J. B. (1980). Some convergence properties of iterated least squares in the location model. *Communications in Statistics: Simulation and Computation 9*, 359–369.

Birch, M. W. (1964). A new proof of the Pearson–Fisher theorem. *Annals of Mathematical Statistics 35*, 817–824.

Bishop, Y. M. M., S. E. Fienberg, and P. W. Holland (1975). *Discrete Multivariate Analysis: Theory and Practice*. Cambridge, USA: The MIT Press.

Bockstaele, F. V., A. Janssens, F. Callewaert, V. Pede, F. Offner, B. Verhasselt, and Philippé (2006). Kolmogorov–Smirnov statistical test for analysis of ZAP-70 expression in B-CLL, compared with quantitative PCR and IgVH mutation status. *Cytometry Part B (Clinical Cytometry) 70B*, 302–308.

Boos, D. (1981). Minimum distance estimators for location and goodness of fit. *Journal of the American Statistical Association 76*, 663–670.

Boos, D. (1982). Minimum Anderson–Darling estimation. *Communications in Statistics: Theory and Methods 11*, 2747–2774.

Box, G. E. P. (1953). Non normality and tests on variances. *Biometrika 40*, 318–335.

Bregman, L. M. (1967). The relaxation method of finding the common point of convex sets and its application to the solution of problems in convex programming. *USSR Computational Mathematics and Mathematical Physics 7*, 200–217. Original article is in *Zh. vȳchisl. Mat. mat. Fiz.*, **7**, pp. 620–631, 1967.

Breiman, L. (1996). Bagging predictors. *Machine Learning 24*, 123–140.

Broniatowski, M. (2003). Estimation of the Kullback–Leibler divergence with application to test of hypotheses. *Mathematical Methods of Statistics 12*, 391–409.

Broniatowski, M. and A. Keziou (2006). Minimization of ϕ-divergences on sets of signed measures. *Studia Scientiarum Mathematicarum Hungarica 43*, 403–442.

Broniatowski, M. and A. Keziou (2009). Parametric estimation and tests through divergences and duality technique. *Journal of Multivariate Analysis 100*, 16–36.

Broniatowski, M. and S. Leorato (2006). An estimation method for the Neyman chi-square divergence with application to test of hypotheses. *Journal of Multivariate Analysis 97*, 1409–1436.

Brown, B. M. (1982). Cramér–von Mises distributions and permutation tests. *Biometrika 69*, 619–624.

Brown, L. D. and J. T. G. Hwang (1993). How to approximate a histogram by a normal density. *The American Statistician 47*, 251–255.

Bruck, Y. M. and L. G. Sodin (1979). On the ambiguity of the image reconstruction problem. *Optics Communications 30*, 304–308.

Burbea, J. and C. R. Rao (1982). Entropy differential metric, distance and divergence measures in probability spaces: A unified approach. *Journal of Multivariate Analysis 12*, 575–596.

Byrd, R. H. and D. A. Pyne (1979). Some results on the convergence of the iteratively reweighted least squares. In *ASA Proceedings of Statistical Computing Section*, pp. 87–90.

Campbell, L. L. (1965). A coding theory of Rényi entropy. *Information and Control 8-4*, 423–429.

Canty, M. G. (2007). *Image Analysis, Classification and Change Detection in Remote Sensing: With Algorithms for ENVI/IDL*. Boca Raton, USA: CRC Press, Taylor and Francis.

Cao, R., A. Cuevas, and R. Fraiman (1995). Minimum distance density based estimation. *Computational Statistics & Data Analysis 20*, 611–631.

Cao, R. and L. Devroye (1996). The consistency of a smoothed minimum distance estimate. *Scandinavian Journal of Statistics 23*, 405–418.

Cerone, P. and S. S. Dragomir (2005). Approximation of the integral mean divergence and f-divergence via mean results. *Mathematical and Computer Modelling 42*, 207–219.

Chakraborty, B., A. Basu, and S. Sarkar (2001). Robustification of the MLE without loss in efficiency. Unpublished manuscript.

Chapman, J. W. (1976). A comparison of the X^2, $-2\log R$, and multinomial probability criteria for significance tests when expected frequencies are small. *Journal of the American Statistical Association 71*, 854–863.

Chen, H.-S., K. Lai, and Z. Ying (2004). Goodness-of-fit tests and minimum power divergence estimators for survival data. *Statistica Sinica 14*, 231–248.

Chen, J. H. and J. D. Kalbfleisch (1994). Inverse problems in fractal construction: Hellinger distance method. *Journal of the Royal Statistical Society B 56*, 687–700.

Chen, Z. (2000). A new two-parameter lifetime distribution with bathtub shape or increasing failure rate function. *Statistics & Probability Letters 49*, 155–161.

Cheng, A.-L. and A. N. Vidyashankar (2006). Minimum Hellinger distance estimation for randomized play the winner design. *Journal of Statistical Planning and Inference 136*, 1875–1910.

Chernoff, H. (1952). A measure of asymptotic efficiency for tests of a hypothesis based on the sum of observations. *Annals of Mathematical Statistics 23*, 493–507.

Chung, K.-L. (1974). *A Course in Probability Theory*. New York, USA: Academic Press.

Coberly, W. A. and T. O. Lewis (1972). A note on a one-sided Kolmogorov–Smirnov test of fit for discrete distribution functions. *Annals of the Institute of Statistical Mathematics 24*, 183–187.

Cochran, W. G. (1952). The χ^2 test of goodness of fit. *Annals of Mathematical Statistics 23*, 315–345.

Cohen, J. (1960). A coefficient of agreement for nominal scales. *Educational and Psychological Measurement 20*, 37–46.

Collins, D. (1982). Electron density images from imperfect data by iterative entropy maximization. *Nature 298*, 49–51.

Conover, W. J. (1972). A Kolmogorov goodness of fit test for discontinuous distributions. *Journal of the American Statistical Association 67*, 591–596.

Cox, D. R. and D. V. Hinkley (1974). *Theoretical Statistics*. Chapman & Hall.

Cramér, H. (1928). On the composition of elementary errors. *Skandinavisk Aktuarietidskrift 11*, 13–74 and 141–180.

Cramér, H. (1946). *Mathematical Methods of Statistics*. Princeton, USA: Princeton University Press.

Cressie, N. and L. Pardo (2000). Minimum ϕ-divergence estimator and hierarchical testing in loglinear models. *Statistica Sinica 10*, 867–884.

Cressie, N. and L. Pardo (2002). Model checking in loglinear models using ϕ-divergences and MLEs. *Journal of Statistical Planning and Inference 103*, 437–453.

Cressie, N., L. Pardo, and M. C. Pardo (2003). Size and power considerations for testing loglinear models using ϕ-divergence test statistics. *Statistica Sinica 13*, 555–570.

Cressie, N. and T. R. C. Read (1984). Multinomial goodness-of-fit tests. *Journal of the Royal Statistical Society B 46*, 440–464.

Csiszár, I. (1963). Eine informations theoretische Ungleichung und ihre Anwendung auf den Beweis der Ergodizitat von Markoffschen Ketten. *Publ. Math. Inst. Hungar. Acad. Sci. 3*, 85–107.

Csiszár, I. (1967a). Information-type measures of difference of probability distributions and indirect observations. *Studia Scientiarum Mathematicarum Hungarica 2*, 299–318.

Csiszár, I. (1967b). On topological properties of f-divergences. *Studia Scientiarum Mathematicarum Hungarica 2*, 329–339.

Csiszár, I. (1972). A class of measures of informativity of observation channels. *Priodica Math. Hungar. 2*, 191–213.

Csiszár, I. (1975). I-divergence geometry of probability distributions and minimization problmes. *Annals of Probability 3*, 146–158.

Csiszár, I. (1991). Why least squares and maximum entropy? An axiomatic approach to inference for linear inverse problems. *Annals of Statistics 19*, 2032–2066.

Csiszár, I. (1995). Generalized projections for non-negative functions. *Acta Mathematica Hungarica 68*, 161–185.

Cutler, A. and O. I. Cordero-Braña (1996). Minimum Hellinger distance estimation for finite mixture models. *Journal of the American Statistical Association 91*, 1716–1723.

Darling, D. A. (1955). Cramér–Smirnov test in the parametric case. *Annals of Mathematical Statistics 26*, 1–20.

Darling, D. A. (1957). The Kolmogorov–Smirnov, Cramér–von Mises tests. *Annals of Mathematical Statistics 28*, 823–838.

Davies, L. and U. Gather (1993). The identification of multiple outliers. *Journal of the American Statistical Association 88*, 782–792.

De Angelis, D. and G. A. Young (1992). Smoothing the bootstrap. *International Statistical Review 60*, 45–56.

De Luca, A. and S. Termini (1972). A definition of a nonprobabilistic entropy in the setting of fuzzy sets theory. *Information and Control 20*, 301–312.

Del Pino, G. (1989). The unifying role of the iterative generalized least squares in statistical algorithms (with discussions). *Statistical Science 4*, 394–408.

Dempster, A. P., N. M. Laird, and D. B. Rubin (1977). Maximum likelihood from incomplete data via the EM algorithm. *Journal of the Royal Statistical Society B 39*, 1–38.

Devore, J. L. (1995). *Probability and Statistics for Engineering and the Sciences* (4th ed.). Duxbury Press, Inc.

Devroye, L. and L. Györfi (1985). *Nonparametric Density Estimation: The L_1 View.* Wiley.

Dey, D. K. and L. R. Birmiwal (1994). Robust Bayesian analysis using divergence measures. *Statistics & Probability Letters 20*, 287–294.

Djouadi, A., O. Snorrason, and F. Garber (1990). The quality of training-sample estimates of the Bhattacharyya coefficient. *IEEE Transactions on Pattern Analysis and Machine Intelligence 12*, 92–97.

Donner, A. and M. Eliasziw (1992). A goodness-of-fit approach to inference problems of the kappa statistic: Confidence interval construction, significance testing and sample size determination. *Statistics in Medicine 11*, 1511–1519.

Donoho, D. L. and P. J. Huber (1983). The notion of breakdown point. In P. Bickel, K. Doksum, and J. Hodges Jr. (Eds.), *A Festschrift for Erich L. Lehmann*, pp. 157–184. Belmont, USA: Wadsworth.

Donoho, D. L. and R. C. Liu (1988a). The automatic robustness of minimum distance functionals. *Annals of Statistics 16*, 552–586.

Donoho, D. L. and R. C. Liu (1988b). Pathologies of some minimum distance estimators. *Annals of Statistics 16*, 587–608.

Dragomir, S. S. and C. E. M. Pearce (2000). Selected topics on Hermite–Hadamard inequalities and applications. In *RGMIA Monographs*. Victoria University.

Drossos, C. A. and A. N. Philippou (1980). A note on minimum distance estimates. *Annals of the Institute of Statistical Mathematics 32*, 121–123.

Dupuis, D. J. and S. Morgenthaler (2002). Robust weigthed likelihood estimators with an application to bivariate extreme value problems. *The Canadian Journal of Statistics 30*, 17–36.

Durbin, J. (1975). Kolmogorov–Smirnov tests when parameters are estimated with applications to tests of exponentiality and tests on spacings. *Biometrika 62*, 5–22.

Durio, A. and E. D. Isaia (2004). On robustness to outliers of parametric L_2 estimate criterion in the case of bivariate normal mixtures: A simulation study. In M. Hubert, G. Pinson, A. Struyf, and S. V. Aelst (Eds.), *Theory and Applications of Recent Robust Methods*. Basel, Switzerland: Birkhauser Verlag.

Efron, B. (1975). Defining the curvature of a statistical problem (with applications to second order efficiency). *Annals of Statistics 3*, 1189–1242.

Efron, B. (1978). The geometry of exponential families. *Annals of Statistics 6*, 362–376.

Efron, B. (1988). Logistic regression, survival and the Kaplan–Meier curve. *Journal of the American Statistical Association 83*, 414–425.

Eguchi, S. and S. Kato (2010). Entropy and divergence associated with power function and the statistical application. *Entropy 12*, 262–274.

Eslinger, P. W. and W. A. Woodward (1991). Minimum Hellinger distance estimation for normal models. *Journal of Statistical Computation and Simulation 39*, 95–114.

Fan, J. and W. Xie (1999). Distance measure and induced fuzzy entropy. *Fuzzy Sets and Systems 104*, 305–314.

Feller, W. (1971). *An Introduction to Probability Theory and Its Applications* (2nd ed.), Volume II. New York, USA: John Wiley & Sons.

Fernholz, L. T. (1983). *Statistical Functionals*. New York, USA: Springer-Verlag.

Field, C. and B. Smith (1994). Robust estimation: A weighted maximum likelihood approach. *International Statistical Review 62*, 405–424.

Fienberg, S. E. (1980). *The Analysis of Cross Classified Categorical Data*. Cambridge, USA: The MIT Press.

Fienup, J. R. (1982). Phase retrieval algorithms: A comparison. *Applied Optics 21*, 2758–2769.

Finch, S. J., N. R. Mendell, and H. C. Thode (1989). Probabilistic measures of accuracy of a numerical search of a global maximum. *Journal of the American Statistical Association 84*, 1020–1023.

Fisher, R. A. (1912). On an absolute criterion for fitting frequency curves. *Mess. of Math. 41*, 155.

Fisher, R. A. (1922). On the mathematical foundations of theoretical statistics. *Proceedings of the Cambridge Philosophical Society 222*, 309–368.

Fisher, R. A. (1925). Theory of statistical estimation. *Proceedings of the Cambridge Philosophical Society 22*, 700–725.

Fisher, R. A. (1934). Two new properties of mathematical likelihood. *Proceedings of the Royal Society A 144*, 285–307.

Fisher, R. A. (1935). The logic of inductive inference (with discussion). *Journal of the Royal Statistical Society 98*, 39–82.

Freeman, D. H. (1987). *Applied Categorical Data Analysis*. New York, USA: Marcel Dekker.

Freeman, M. F. and J. W. Tukey (1950). Transformations related to the angular and the square root. *Annals of Mathematical Statistics 21*, 607–611.

Fujisawa, H. and S. Eguchi (2006). Robust estimation in the normal mixture model. *Journal of Statistical Planning and Inference 136*, 3989–4011.

Fujisawa, H. and S. Eguchi (2008). Robust parameter estimation against heavy contimation. *Journal of Multivariate Analysis 99*, 2053–2081.

Gauss, K. F. (1821). Theoria combinationis observationum erroribus minimis obnoxiae. *Pars prior*.

Gerchberg, R. W. and W. O. Saxton (1972). A practical algorithm for the determination of phase from image and diffraction plane pictures. *Optik 35*, 237–246.

Gleser, L. J. (1985). Exact power of goodness-of-fit tests of Kolmogorov type for discontinuous distributions. *Journal of the American Statistical Association 80*, 954–958.

Green, P. J. (1984). Iteratively reweighted least squares for maximum likelihood estimation, and some robust and resistant alternatives (with discussions). *Journal of the Royal Statistical Society B 46*, 149–192.

Grenander, U. and M. Rosenblatt (1957). *Statistical Analysis of Stationary Time Series*. New York, USA: Wiley.

Gull, S. F., A. K. Livesey, and D. S. Sivia (1987). Maximum entropy solution of a small centrosymmetric crystal structure. *Acta Crystallographica, Section A 43*, 112–117.

Hald, A. (1998). *A History of Mathematical Statistics from 1750 to 1930*. New York, USA: John Wiley.

Hald, A. (1999). On the history of maximum likelihood in relation to inverse probability and least squares. *Statistical Science 14*, 214–222.

Hall, P. and J. S. Marron (1991). Lower bounds for bandwidth selection in density estimation. *Probability Theory and Related Fields 90*, 149–173.

Hampel, F. R. (1968). *Contributions to the theory of robust estimation*. Ph. D. thesis, University of California, Berkeley, USA.

Hampel, F. R. (1971). A general qualitative definition of robustness. *Annals of Mathematical Statistics 42*, 1887–1896.

Hampel, F. R., E. Ronchetti, P. J. Rousseeuw, and W. Stahel (1986). *Robust Statistics: The Approach Based on Influence Functions*. New York, USA: John Wiley & Sons.

Hardle, W., P. Hall, and J. S. Marron (1988). How far are automatically chosen regression smoothing parameters from their optimum? *Journal of the American Statistical Association 83*, 86–95.

Harris, I. R. and A. Basu (1994). Hellinger distance as a penalized log likelihood. *Communications in Statistics: Simulation and Computation 23*, 1097–1113.

Harris, I. R. and A. Basu (1997). A generalized divergence measure. *Technical Report*, Applied Statistics Unit, Indian Statistical Institute, Calcutta 700 035, India.

Haynam, G. E., Z. Govindarajulu, and G. C. Leone (1962). Tables of the cumulative non-central chi-square distribution. In H. L. Harter and D. B. Owen (Eds.), *Case Statistical Laboratory, Publication No. 104*. Chicago, USA: Markham. Part of the tables have been published in *Selected Tables in Mathematical Statistics Vol. 1. 1970*.

He, X., D. G. Simpson, and S. L. Portnoy (1990). Breakdown robustness of tests. *Journal of the American Statistical Association 85*, 446–452.

Healy, M. J. R. (1979). Outliers in clinical chemistry quality-control schemes. *Clinical Chemistry 25*, 675–677.

Heathcote, C. and M. Silvapulle (1981). Minimum mean squared estimation of location and scale parameters under misspecifications of the model. *Biometrika 68*, 501–514.

Hellinger, E. D. (1909). Neue begründung der theorie der quadratischen formen von unendlichen vielen veränderlichen. *Journal für Reine & Angewandte Mathematik 136*, 210–271.

Hettmansperger, T. P., I. Hueter, and J. Husler (1994). Minimum distance estimators. *Journal of Statistical Planning and Inference 41*, 291–302.

Hjort, N. L. (1994). Minimum L_2 and robust Kullback–Leibler estimation. In P. Lachout and J. A. Višek (Eds.), *Proceedings of the 12th Prague Conference on Information Theory, Statistical Decision Functions and Random Processes*, Prague, pp. 102–105. Czech Republic Academy of Sciences.

Hoeffding, W. (1965). Asymptotically optimal tests for multinomial distributions. *Annals of Mathematical Statistics 36*, 369–408.

Holland, P. W. and R. E. Welsch (1977). Robust regression using iteratively reweighted least squares. *Communications in Statistics: Theory and Methods 6*, 813–827.

Holst, L. (1972). Asymptotic normality and efficiency for certain goodness-of-fit tests. *Biometrika 59*, 137–145.

Hong, C. and Y. Kim (2001). Automatic selection of the tuning parameter in the minimum density power divergence estimation. *Journal of the Korean Statistical Association 30*, 453–465.

Huber, P. J. (1964). Robust estimation of a location parameter. *Annals of Mathematical Statistics 35*, 73–101.

Huber, P. J. (1981). *Robust Statistics*. John Wiley & Sons.

Jaynes, E. T. (1968). Prior probabilities. *Transactions on Systems Science and Cybernetics SSC-4*, 227–241.

Jeffreys, H. (1948). *Theory of Probability* (2nd ed.). Oxford Clarendon Press.

Jennrich, R. I. (1969). Asymptotic properties of non-linear least square estimators. *Annals of Mathematical Statistics 40*, 633–643.

Jeong, D. and S. Sarkar (2000). Negative exponential disparity based family of goodness-of-fit tests for multinomial models. *Journal of Statistical Computation and Simulation 65*, 43–61.

Jimenez, R. and Y. Shao (2001). On robustness and efficiency of minimum divergence estimators. *Test 10*, 241–248.

Jones, L. K. and C. L. Byrne (1990). General entropy criteria for inverse problems, with applications to data compression, pattern classification, and cluster analysis. *IEEE Transactions on Information Theory 36*, 23–30.

Jones, M. C. and N. L. Hjort (1994). Comment on "How to approximate a histogram by a normal density" by Brown and Hwang (1993). *American Statistician 48*, 353–354.

Jones, M. C., N. L. Hjort, I. R. Harris, and A. Basu (2001). A comparison of related density based minimum divergence estimators. *Biometrika 88*, 865–873.

Juárez, S. and W. R. Schucany (2004). Robust and efficient estimation for the generalized Pareto distribution. *Extremes 7*, 237–251.

Juárez, S. and W. R. Schucany (2006). A note on the asymptotic distribution of the minimum density power divergence estimator. In J. Rojo (Ed.), *Optimality, the Second Erich Lehmann Symposium*, pp. 334–339. Institute of Mathematical Statistics.

Kafka, P., F. Österreicher, and I. Vincze (1991). On powers of f-divergence defining a distance. *Studia Scientiarum Mathematicarum Hungarica 26*, 415–422.

Kailath, T. (1967). The divergence and Bhattacharyya distance measures in signal selection. *IEEE Transactions on Communication Technology 15*, 52–60.

Kaplan, E. L. and P. Meier (1958). Nonparametric estimation from incomplete observations. *Journal of the American Statistical Association 53*, 457–481.

Karagrigoriou, A. and K. Mattheou (2009). Measures of divergence in model selection. In C. H. Skiadas (Ed.), *Advances in Data Analysis: Theory and Applications to Reliability and Inference, Data Mining, Bioinformatics, Lifetime Data and Neural Networks*, pp. 51–65. Boston, USA: Birkhauser.

Karagrigoriou, A. and K. Mattheou (2010). On distributional properties and goodness-of-fit tests. *Communications in Statistics: Theory and Methods 39*, 472–482.

Karlis, D. and E. Xekalaki (1998). Minumum Hellinger distance estimation for Poisson mixtures. *Computational Statistics & Data Analysis 29*, 81–103.

Karlis, D. and E. Xekalaki (2001). Robust inference for finite population mixtures. *Journal of Statistical Planning and Inference 93*, 93–115.

Karunamuni, R. J. and J. Wu (2009). Minimum Hellinger distance estimation in a nonparametric mixture model. *Journal of Statistical Planning and Inference 139*, 1118–1133.

Kaufman, A. (1975). *Introduction to the Theory of Fuzzy Subsets, Vol. I*. New York, USA: Adademic.

Kiefer, J. (1959). K-sample analogous of the Kolmogorov–Smirnov and Cramér–von Mises tests. *Annals of Mathematical Statistics 30*, 420–447.

Kiefer, J. and J. Wolfowitz (1956). Consistency of the maximum likelihood estimator in the presence of infinitely many incidental parameters. *Annals of Mathematical Statistics 27*, 886–906.

Kirmani, S. N. U. A. (1971). Some limiting properties of Matusita's measure of distance. *Annals of the Institute of Statistical Mathematics 23*, 157–162.

Koehler, K. J. and K. Larntz (1980). An empirical investigation of goodness-of-fit statistics for sparse multinomials. *Journal of the American Statistical Association 75*, 336–344.

Kolmogorov, A. N. (1933). Sulla determinazione empirica di una legge di distribuzione. *Giornale dell' Instituto Italiano degli Attuari 4*, 83–91.

Kosko, B. (1986). Fuzzy entropy and conditioning. *Information Sciences 40*, 165–174.

Kraft, C. (1955). Some conditions for consistency and uniform consistency of statistical procedures. *University of California Publications in Statistics 2*, 125–141.

Kraus, D. (2004). Goodness-of-fit inference for the Cox–Aalen additive multiplicative regression model. *Statistics & Probability Letters 70*, 285–298.

Kullback, S. (1967). The two concepts of information. *Journal of the American Statistical Association 62*, 685–686.

Kullback, S. (1970). Minimum discrimination information estimation and application. In *Invited paper presented to Sixteenth Conference on the Design of Experiments in Army Research, Development and Testing*, Ft. Lee, VA, USA, pp. 1–38. Logistics Management Center.

Kullback, S. and R. A. Leibler (1951). On information and sufficiency. *Annals of Mathematical Statistics 22*, 79–86.

Kumar, P. and S. Chhina (2005). A symmetric information divergence measures of the Csiszár's f-divergence class and its bounds. *Computers and Mathematics with Applications 49*, 575–588.

Landgrebe, D. A. (2003). *Signal Theory Methods in Multispectral Remote Sensing*. Hoboken, USA: John Wiley and Sons.

Larntz, K. (1978). Small sample comparisons of exact levels for chi-squared goodness-of-fit statistics. *Journal of the American Statistical Association 73*, 253–263.

Le Cam, L. (1953). On some asymptotic properties of maximum likelihood estimates and related Bayes estimates. In *University of California Publications in Statistics*, Volume 1, pp. 277–330.

Lee, C. C. (1987). Chi-squared tests for and against an order restriction on multinomial parameters. *Journal of the American Statistical Association 82*, 611–618.

Lee, S. and O. Na (2005). Test for parameter change based on the estimator minimizing density-based divergence measures. *Annals of the Institute of Statistical Mathematics 57*, 553–573.

Lee, S., T. N. Sriram, and X. Wei (2006). Fixed-width confidence interval based on a minimum Hellinger distance estimator. *Journal of Statistical Planning and Inference 136*, 4276–4292.

Lehmann, E. L. (1983). *Theory of Point Estimation*. John Wiley & Sons.

Lehmann, E. L. (1999). *Elements of Large-Sample Theory*. Springer.

Lehmann, E. L. and J. P. Romano (2008). *Testing Statistical Hypothesis* (3rd ed.). Springer.

Leise, F. and I. Vajda (1987). *Convex Statistical Distances*. Teubner, Leipzig.

Leise, F. and I. Vajda (2006). On divergence and information in statistics and information theory. *IEEE Transactions on Information Theory 52*, 4394–4412.

Lenth, R. V. and P. J. Green (1987). Consistency of deviance based M-estimators. *Journal of the Royal Statistical Society B 49*, 326–330.

Lilliefors, H. (1967). On the Kolmogorov–Smirnov test for normality with mean and variance unknown. *Journal of the American Statistical Association 62*, 399–402.

Lin, J. (1991). Divergence measures based on the Shannon entropy. *IEEE Transactions on Information Theory 37*, 145–151.

Lin, N. and X. He (2006). Robust and efficient estimation under data grouping. *Biometrika 93*, 99–112.

Lindsay, B. G. (1994). Efficiency versus robustness: The case for minimum Hellinger distance and related methods. *Annals of Statistics 22*, 1081–1114.

Lindsay, B. G. and A. Basu (2010). Contamination envelopes for statistical distances with applications to power breakdown. *Technical Report* BIRU/2010/5, Indian Statistical Institute, Kolkata, India.

Loeve, M. (1977). *Probability Theory, Volume I* (4th ed.). Springer.

Lu, Z., Y. V. Hui, and A. H. Lee (2003). Minimum Hellinger distance estimation for finite mixtures of Poisson regression models and its applications. *Biometrics 59*, 1016–1026.

Lubischew, A. (1962). On the use of discriminant functions in taxonomy. *Biometrics 18*, 455–477.

Mandal, A. and A. Basu (2010a). Minimum distance estimation: Improved efficiency through inlier modification. *Technical Report* BIRU/2010/3, Indian Statistical Institute, Kolkata, India.

Mandal, A. and A. Basu (2010b). Multinomial goodness-of-fit testing and inlier modification. *Technical Report* BIRU/2010/4, Indian Statistical Institute, Kolkata, India.

Mandal, A., A. Basu, and L. Pardo (2010). Minimum Hellinger distance inference and the empty cell penalty: Asymptotic results. *Sankhya A 72*, 376–406.

Mandal, A., S. Bhandari, and A. Basu (2011). Minimum disparity estimation based on combined disparities: Asymptotic results. *Journal of Statistical Planning and Inference 141*, 701–710.

Mansuy, R. (2005). An interpretation and some generalizations of the Anderson-Darling statistics in terms of squared Bessel bridges. *Statistics & Probability Letters 72*, 171–177.

Mantalos, P., K. Mattheou, and A. Karagrigoriou (2010a). An improved divergence information criterion for the determination of the order of an AR process. *Communications in Statistics: Simulation and Computation 39*, 1–15.

Mantalos, P., K. Mattheou, and A. Karagrigoriou (2010b). Forecasting ARMA models: A comparative study of information criteria focusing on MDIC. *Journal of Statistical Computation and Simulation 80*, 61–73.

Mantalos, P., K. Mattheou, and A. Karagrigoriou (2010c). Vector autoregressive order selection and forecasting via the modified divergence information criterion. *International Journal of Computational Economics and Econometrics 1*, 254–277.

Margolin, B. H. and W. Maurer (1976). Tests of the Kolmogorov–Smirnov type for exponential data with unknown scale, and related problems. *Biometrika 63*, 149–160.

Markatou, M. (1996). Robust statistical inference: Weighted likelihood or usual M-estimation. *Communications in Statistics: Theory and Methods 25*, 2597–2613.

Markatou, M. (1999). Weighting games in robust linear regression. *Journal of Multivariate Analysis 70*, 118–135.

Markatou, M. (2000). Mixture models, robustness and the weighted likelihood methodology. *Biometrics 56*, 483–486.

Markatou, M. (2001). A closer look at weighted likelihood in the context of mixtures. In C. A. Charalambides, M. V. Koutras, and N. Balakrishnan (Eds.), *Probability and Statistical Models with Applications*, pp. 447–467. Chapman and Hall/CRC.

Markatou, M., A. Basu, and B. Lindsay (1996). Weighted likelihood estimation equations: The continuous case. *Technical Report* 323, Department of Statistics, Stanford University.

Markatou, M., A. Basu, and B. Lindsay (1997). Weighted likelihood estimating equations: The discrete case with applications to logistic regression. *Journal of Statistical Planning and Inference 57*, 215–232.

Markatou, M., A. Basu, and B. G. Lindsay (1998). Weighted likelihood equations with bootstrap root search. *Journal of the American Statistical Association 93*, 740–750.

Marron, J. S. (1989). Comments on a data based bandwidth selector. *Computational Statistics & Data Analysis 8*, 155–170.

Martin, N. and L. Pardo (2008). New families of estimators and test statistics in log-linear models. *Journal of Multivariate Analysis 99*, 1590–1609.

Massey, Jr, F. J. (1951). The Kolmogorov–Smirnov test of goodness of fit. *Journal of the American Statistical Association 46*, 68–78.

Mattheou, K. and A. Karagrigoriou (2009). On new developments in divergence statistics. *Journal of Mathematical Sciences 163*, 227–237.

Mattheou, K. and A. Karagrigoriou (2010). A new family of divergence measures for tests of fit. *Australian and New Zealand Journal of Statististics 52*, 187–200.

Mattheou, K., S. Lee, and A. Karagrigoriou (2009). A model selection criterion based on the BHHJ measure of divergence. *Journal of Statistical Planning and Inference 139*, 228–235.

Matusita, K. (1954). On the estimation by the minimum distance method. *Annals of the Institute of Statistical Mathematics 5*, 59–65.

Matusita, K. (1955). Decision rules, based on the distance for problems of fit, two samples, and estimation. *Annals of Mathematical Statistics 26*, 631–640.

Matusita, K. (1964). Distance and decision rules. *Annals of the Institute of Statistical Mathematics 16*, 305–320.

McCullagh, P. and J. A. Nelder (1989). *Generalized Linear Models* (2nd ed.). Chapman & Hall.

Menéndez, M. L., D. Morales, L. Pardo, and M. Salićru (1995). Asymptotic behaviour and statistical applications of divergence measures in multinomial populations: A unified study. *Statistical Papers 36*, 1–29.

Menéndez, M. L., D. Morales, L. Pardo, and I. Vajda (2001). Minimum disparity estimators based on grouped data. *Annals of the Institute of Statistical Mathematics 53*, 277–288.

Menendez, M. L., L. Pardo, and M. C. Pardo (2008). Preliminary test estimators and phi-divergence measures in generalized linear models with binary data. *Journal of Multivariate Analysis 99*, 2265–2284.

Miao, J., D. Sayre, and H. N. Chapman (1998). Phase retrieval from the magnitude of the Fourier transforms of nonperiodic objects. *Journal of the Optical Society of America A 15*, 1662–1669.

Millar, P. W. (1981). Robust estimation via minimum distance methods. *Probability Theory and Related Fields 55*, 73–89.

Minami, M. and S. Eguchi (2002). Robust blind source separation by beta divergence. *Neural Computation 14*, 1859–1886.

Miyamura, M. and Y. Kano (2006). Robust Gaussian graphical modelling. *Journal of Multivariate Analysis 97*, 1525–1550.

Mollah, M. N. H., M. Minami, and S. Eguchi (2006). Exploring latent structure of mixture ICA models by the minimum beta-divergence method. *Neural Computation 18*, 166–190.

Mollah, M. N. H., M. Minami, and S. Eguchi (2007). Robust prewhitening for ICA by minimizing beta-divergence and its application to fast ICA. *Neural Processing Letters 25*, 91–110.

Mollah, M. N. H., N. Sultana, M. Minami, and S. Eguchi (2010). Robust extraction of local structures by the minimum beta-divergence method. *Neural Networks 23*, 226–238.

Moore, D. S. and M. C. Spruill (1975). Unified large-sample theory of general chi-squared statistics for tests of fit. *Annals of Statistics 3*, 599–616.

Murata, N., T. Takenouchi, and T. Kanamori (2004). Information geometry of U-boost and Bregman divergence. *Neural Computation 16*, 1437–1481.

Nagaoka, H. and S. Amari (1982). Differential geometry of smooth families of probability distributions. *METR 82-7*. University of Tokyo.

Neyman, J. (1949). Contribution to the theory of the χ^2 test. In *Proceedings of the First Berkeley Symposium on Mathematical Statistics and Probability*, pp. 239–273. University of California Press.

Neyman, J. and E. S. Pearson (1928). On the use and interpretation of certain test criteria for purposes of statistical inference. *Biometrika Series A, 20*, 175–240.

Noether, G. E. (1963). Note on the Kolmogorov statistics in the discrete case. *Metrika 7*, 115–116.

Ortega, J. M. (1990). *Numerical Analysis: A Second Course*. Philadelphia, USA: Society for Industrial and Applied Mathematics.

Ozturk, O. (1994). *Minimum distance estimation*. Ph. D. thesis, The Pennsylvania State University, University Park, USA.

Öztürk, O. and T. P. Hettmansperger (1997). Generalized Cramer–von Mises distance estimators. *Biometrika 84*, 283–294.

Öztürk, O., T. P. Hettmansperger, and J. Husler (1999). Minimum distance and non-parametric dispersion functions. In S. Ghosh (Ed.), *Asymptotics, Nonparametrics, and Time Series*, pp. 511–531. New York, USA: Marcel Dekker.

Pak, R. J. (1996). Minimum Hellinger distance estimation in simple linear regression models: Distribution and efficiency. *Statistics & Probability Letters 27*, 263–269.

Pak, R. J. and A. Basu (1998). Minimum disparity estimation in linear regression models: Distribution and efficiency. *Annals of the Institute of Statistical Mathematics 50*, 503–521.

Pal, N. R. and S. K. Pal (1991). Entropy, a new definition and its applications. *IEEE Transactions on Systems, Man, and Cybernetics 21*, 1260–1270.

Pardo, J. A. and M. C. Pardo (1999). Small-sample comparisons for the Rukhin goodness-of-fit-statistics. *Statistical Papers 40*, 159–174.

Pardo, L. (2006). *Statistical Inference based on Divergences*. CRC/Chapman-Hall.

Pardo, L. and N. Martin (2009). Homogeneity/heterogeneity hypothses for standardized mortality ratios based on minimum power-divergence estimators. *Biometrical Journal 51*, 819–836.

Pardo, L. and M. C. Pardo (2003). Minimum power-divergence estimator in three-way contingency tables. *Journal of Statistical Computation and Simulation 73*, 819–831.

Pardo, M. C. (1998). Improving the accuracy of goodness-of-fit tests based on Rao's divergence with small sample size. *Computational Statistics & Data Analysis 28*, 339–351.

Pardo, M. C. (1999). On Burbea–Rao divergence based goodness-of-fit tests for multinomial models. *Journal of Multivariate Analysis 69*, 65–87.

Park, C. and A. Basu (2000). Mininum disparity inference based on trigonometric disparities. *Technical Report* ASD/2000/23, Applied Statistics Division, Indian Statistical Institute, Calcutta, India.

Park, C. and A. Basu (2003). The generalized Kullback–Leibler divergence and robust inference. *Journal of Statistical Computation and Simulation 73*, 311–332.

Park, C. and A. Basu (2004). Minimum disparity estimation: Asymptotic normality and breakdown point results. *Bulletin of Informatics and Cybernetics 36*, 19–33. Special Issue in Honor of Professor Takashi Yanagawa.

Park, C., A. Basu, and S. Basu (1995). Robust minimum distance inference based on combined distances. *Communications in Statistics: Simulation and Computation 24*, 653–673.

Park, C., A. Basu, and I. R. Harris (2001). Tests of hypotheses in multiple samples based on penalized disparities. *Journal of the Korean Statistical Society 30*, 347–366.

Park, C., A. Basu, and B. G. Lindsay (2002). The residual adjustment function and weighted likelihood: A graphical interpretation of robustness of minimum disparity estimators. *Computational Statistics & Data Analysis 39*, 21–33.

Park, C., I. R. Harris, and A. Basu (1997). Robust predictive distributions based on the penalized blended weight Hellinger distance. *Communications in Statistics: Simulation and Computation 26*, 21–33.

Parr, W. C. (1981). Minimum distance method: A bibliography. *Communications in Statistics: Theory and Methods 10*, 1205–1224.

Parr, W. C. and T. De Wet (1981). Minimum CVM-norm parameter estimation. *Communications in Statistics: Theory and Methods 10*, 1149–1166.

Parr, W. C. and W. R. Schucany (1980). Minimum distance and robust estimation. *Journal of the American Statistical Association 75*, 616–624.

Parr, W. C. and W. R. Schucany (1982). Minimum distance estimation and components of goodness-of-fit statistics. *Journal of the Royal Statistical Society B 44*, 178–189.

Parzen, E. (1962). On estimation of a probability density function. *Annals of Mathematical Statistics 33*, 1065–1076.

Patra, R., A. Mandal, and A. Basu (2008). Minimum Hellinger distance estimation with inlier modification. *Sankhya B 70*, 310–322.

Pearson, E. S. and H. O. Hartley (1972). *Biometrika Tables for Statisticians.* Cambridge University Press.

Pearson, K. (1900). On a criterion that a given system of deviations from the probable in the case of a correlated system of variables is such that it can reasonably be supposed to have arisen from random sampling. *Philosophical Magazine 50*, 157–175.

Peng, F. and D. Dey (1995). Bayesian analysis of outlier problems using divergence measures. *Canadian Journal of Statistics 23*, 199–213.

Petitt, A. N. and M. A. Stevens (1977). Kolmogorov–Smirnov goodness-of-fit statistics for discrete and grouped data. *Technometrics 19*, 205–210.

Pečarić, J. E. (1992). Notes on convex functions. In W. Walter (Ed.), *General Inequalities 6, International Series of Numerical Mathematics*, Volume 103. Brikhäuser Basel.

Pečarić, J. E., F. Proschan, and Y. L. Tong (1992). *Convex Functions, Partial Orderings, and Statistical Applications.* Academic Press.

Piro, O. E. (1983). Information theory and the 'phase problem' in crystallography. *Acta Crystallographica, Section A 39*, 61–68.

Pitman, E. J. G. (1979). *Some Basic Theory for Statical Inference.* London, United Kingdom: Chapman and Hall.

Podjarny, A., D. Moras, J. Navaza, and P. Alizari (1988). Low-resolution phase extension and refinement by maximum entropy. *Acta Crystallographica, Section A 44*, 545–551.

Qi, F., S.-L. Xu, and L. Debnath (1999). A new proof of monotonicity for extended mean values. *International Journal of Mathematics and Mathematical Sciences 22*, 417–421.

Ralston, A. and P. Rabinowitz (1978). *A First Course in Numerical Analysis.* New York, USA: McGraw-Hill.

Rao, C. R. (1948). Large sample tests of statistical hypotheses concerning several parameters with applications of problems of estimation. *Proceedings of the Cambridge Philosophical Society 44*, 50–57.

Rao, C. R. (1961). Asymptotic efficiency and limiting information. In *Proceedings of Fourth Berkeley Symposium on Mathematical Statistics and Probability, Volume I*, pp. 531–546. University of California Press.

Rao, C. R. (1962). Efficient estimates and optimum inference procedures in large samples (with discussion). *Journal of the Royal Statistical Society B 24*, 46–72.

Rao, C. R. (1963). Criteria of estimation in large samples. *Sankhya A 25*, 189–206.

Rao, C. R. (1973). *Linear Statistical Inference and Its Applications*. New York, USA: John Wiley & Sons.

Rao, C. R., B. K. Sinha, and K. Subramanyam (1983). Third order efficiency of the maximum likelihood estimator in the multinomial distribution. *Statistics & Decisions 1*, 1–16.

Rassokhin, D. N. and D. K. Agrafiotis (2000). Kolmogorov–Smirnov statistic and its application in library design. *Jounal of Molecular Graphics and Modelling 18*, 268–382.

Read, C. B. (1993). Freeman–Tukey chi-squared goodness-of-fit statistics. *Statistics & Probability Letters 18*, 271–278.

Read, T. R. C. (1984a). Closer asymptotic approximations for the distributions of the power divergence goodness of fit statistics. *Annals of the Institute of Statistical Mathematics 36*, 59–69.

Read, T. R. C. (1984b). Small sample comparisons for the power divergence goodness-of-fit statistics. *Journal of the American Statistical Association 79*, 929–935.

Read, T. R. C. and N. Cressie (1988). *Goodness-of-Fit Statistics for Discrete Multivariate Data*. New York, USA: Springer-Verlag.

Rényi, A. (1960). Some fundamental questions of information theory. *MTA III. Osztályának Közleményei 10*, 251–282.

Rényi, A. (1961). On measures of entropy and information. In *Proceedings of Fourth Berkeley Symposium on Mathematical Statistics and Probability, Volume I*, pp. 547–561. University of California Press.

Roberts, A. W. and D. E. Varberg (1973). *Convex Functions*. New York, USA: Academic Press.

Royden, H. L. (1988). *Real Analysis*. New York, USA: Macmillan.

Ruckstuhl, A. F. and A. E. Welsh (2001). Robust fitting of the binomial model. *Annals of Statistics 29*, 1117–1136.

Rumelhart, D., G. Hinton, and R. Williams (1986). Learning representations by back-propagating errors. *Nature 323*, 533–536.

Sahler, W. (1970). Estimation by minimum discrepancy methods. *Metrika 16*, 85–106.

Sakata, M. and M. Sato (1990). Accurate structure analysis by the maximum-entropy method. *Acta Crystallographica, Section A 46*, 263–270.

Sarkar, S. and A. Basu (1995). On disparity based robust tests for two discrete populations. *Sankhya B 57*, 353–364.

Sarkar, S., C. Kim, and A. Basu (1999). Tests for homogeneity of variances using robust weighted likelihood estimates. *Biometrical Journal 41*, 857–871.

Sarkar, S., K. Song, and D. Jeong (1998). Penalizing the negative exponential disparity in discrete models. *The Korean Communications in Statistics 5*, 517–529.

Savage, L. J. (1976). On rereading R. A. Fisher. *Annals of Statistics 4*, 441–500.

Schapire, R. E. (1990). The strength of weak learnability. *Machine Learning 5*, 197–227.

Schmid, P. (1958). On the Kolmogorov and Smirnov limit theorems for discontinuous distribution functions. *Annals of Mathematical Statistics 29*, 1011–1027.

Scholz, F. W. and M. A. Stephens (1987). K-sample Anderson–Darling tests. *Journal of the American Statistical Association 82*, 918–924.

Scott, D. W. (2001). Parametric statistical modeling by minimum integrated square error. *Technometrics 43*, 274–285.

Serfling, R. J. (1980). *Approximation Theorems of Mathematical Statistics*. New York, USA: John Wiley & Sons.

Shannon, C. E. (1948). A mathematical theory of communication. *Bell System Technical Journal 27*, 379–423, 623–656.

Shapiro, S. S. and M. B. Wilk (1965). An analysis of variance test for normality (complete samples). *Biometrika 52*, 591– 611.

Sharma, B. D. and D. P. Mittal (1977). New non-additive measures of relative information. *Journal of Combinatory Information and Systems Science 2*, 122–133.

Shen, S. and I. R. Harris (2004). The minimum L_2 distance estimator for Poisson mixture models. *Technical Report* TR-326, Department of Statistical Sciences, Southern Methodist University, Dallas, USA.

Shin, D. W., A. Basu, and S. Sarkar (1995). Comparisons of the blended weight Hellinger distance based goodness-of-fit test statistics. *Sankhya B 57*, 365–376.

Shin, D. W., A. Basu, and S. Sarkar (1996). Small sample comparisons for the blended weight chi-square goodness-of-fit test statistics. *Communications in Statistics: Theory and Methods 25*, 211–226.

Shioya, H. and T. Da-Te (1995). A generalization of Lin divergence and the derivation of a new information divergence. *Electronics and Communications in Japan, Part III: Fundamental Electronic Science 78*, 34–40.

Shioya, H. and K. Gohara (2006). Generalized phase retrieval algorithm based on information measures. *Optics Communications 266*, 88–93.

Shioya, H. and K. Gohara (2008). Maximum entropy method for diffractive imaging. *Journal of the Optical Society of America A 25*, 2846–2850.

Shorack, G. R. and J. A. Wellner (1986). *Empirical Processes with Applications to Statistics*. New York, USA: Wiley.

Shützenberger, M. B. (1954). Contribution aux applications statistiques de la théories de l'information. *Publications de l'Institut de Statistique de l' Universit de Paris 3*, 3–117.

Silverman, B. W. (1986). *Density Estimation for Statistics and Data Analysis*. London, United Kingdom: Chapman & Hall.

Simpson, D. G. (1987). Minimum Hellinger distance estimation for the analysis of count data. *Journal of the American Statistical Association 82*, 802–807.

Simpson, D. G. (1989a). Choosing a discrepancy for minimum distance estimation: Multinomial models with infinitely many cells. *Technical Report*, Department of Statistics, University of Illinois, Champaign, USA.

Simpson, D. G. (1989b). Hellinger deviance test: Efficiency, breakdown points, and examples. *Journal of the American Statistical Association 84*, 107–113.

Siotani, M. and Y. Fujikoshi (1984). Asymptotic approximations for the distributions of multinomial goodness-of-fit tests. *Hiroshima Mathematical Journal 14*, 115–124.

Smirnov, N. V. (1936). Sur la distribution de ω^2. *C. R. Acad. Sci. Paris 202*, 449–452.

Smirnov, N. V. (1937). On the distribution of the ω^2 distribution of von Mises. *Rec. Math. 2*, 973–993.

Smirnov, N. V. (1939). Estimation of deviation between empirical distribution functions in two independent samples (in Russian). *Bulletin of Moscow University 2*, 3–16.

Sriram, T. N. and A. N. Vidyashankar (2000). Minimum Hellinger distance estimation for supercritical Galton–Watson processes. *Statistics & Probability Letters 50*, 331–342.

Stather, C. R. (1981). *Robust Statistical Inference using Hellinger Distance Methods*. Ph. D. thesis, LaTrobe University, Melbourne, Australia.

Stephens, M. A. (1974). EDF statistics for goodness of fit and some comparisons. *Journal of the American Statistical Association 69*, 730–737.

Stephens, M. A. (1976). Asymptotic results for goodness-of-fit statistics with unknown parameters. *Annals of Statistics 4*, 357–369.

Stephens, M. A. (1977). Goodness of fit for the extreme value distribution. *Biometrika 64*, 583–588.

Stephens, M. A. (1979). Tests of fit for the logistic distribution based on the empirical distribution function. *Biometrika 66*, 591–595.

Stigler, S. M. (1977). Do robust estimators work with real data? *Annals of Statistics 5*, 1055–1098.

Stigler, S. M. (1986). *The History of Statistics: The Measurement of Uncertainty before 1900.* Harvard University Press.

Stigler, S. M. (1999). *Statistics on the Table: The History of Statistical Concepts and Methods.* Harvard University Press.

Stute, W. (1995). The central limit theorem under random censorship. *Annals of Statistics 23*, 422–439.

Stute, W. and J. L. Wang (1993). A strong law under random censorship. *Annals of Statistics 21*, 1591–1607.

Takada, T. (2009). Simulated minimum Hellinger distance estimation of stochastic volatility models. *Computational Statistics & Data Analysis 53*, 2390–2403.

Tamura, R. N. and D. D. Boos (1986). Minimum Hellinger distance estimation for multivariate location and covariance. *Journal of the American Statistical Association 81*, 223–229.

Thas, O. and J. P. Ottoy (2003). Some generalizations of the Anderson–Darling statistic. *Statistics & Probability Letters 64*, 255–261.

Toma, A. (2008). Minimum Hellinger distance estimators for multivariate distributions from the Johnson system. *Journal of Statistical Planning and Inference 138*, 803–816.

Toma, A. (2009). Optimal robust M-estimators using divergences. *Statistics & Probability Letters 79*, 1–5.

Toma, A. and M. Broniatowski (2010). Dual divergence estimators and tests: Robustness results. *Journal of Multivariate Analysis 102*, 20–36.

Toma, A. and S. Leoni-Aubin (2010). Robust tests based on dual divergence estimators and saddlepoint approximations. *Journal of Multivariate Analysis 101*, 1143–1155.

Tsallis, C. (1988). Possible generalization of Boltzmann–Gibbs statistics. *Journal of Statistical Physics 52*, 479–487.

Tsallis, C. (2009). Computational applications of nonextensive statistical mechanics. *Journal of Computational and Applied Mathematics 227*, 51–58.

Tukey, J. W. (1960). A survey of sampling from contaminated distributions. In I. Olkin, S. G. Ghurye, W. Hoeffding, W. G. Madow, and H. B. Mann (Eds.), *Contributions to Probability and Statistics: Essays in Honor of Harold Hotelling*, pp. 448–485. Stanford, USA: Stanford University Press.

Uchida, M. (2007). Information theoretic aspects of fairness criteria in network resource allocation problems. In *Value Tools 2007: Proceedings of the 2nd international conference on Performance evaluation methodologies and tools*, ICST, Brussels, Belgium, pp. 1–9. ICST (Institute for Computer Sciences, Social-Informatics and Telecommunications Engineering).

Uchida, M. and H. Shioya (2005). An extended formula for divergence measures using invariance. *Electronics and Communications in Japan, Part III: Fundamental Electronic Science 88*, 35–42.

Ullah, A. (1991). Entropy, divergence and distance measures with applications to econometrics. *Journal of Statistical Planning and Inference 49*, 137–162.

Vajda, I. (1972). On the f-divergence and singularity of probability measures. *Periodica Math. Hunger. 2*, 223–234.

Vajda, I. (1989). *Theory of Statistical Inference and Information*. Dordrecht: Kluwer Academic.

Victoria-Feser, M. and E. Ronchetti (1997). Robust estimation for grouped data. *Journal of the American Statistical Association 92*, 333–340.

von Mises, R. (1931). *Wahrscheinlichkeitsrechnung*. Liepzig-Wien.

von Mises, R. (1936). Les dois de probabilite pour les fonctions statistiques. *Ann. Inst. H. Poincare B. 6*, 185–212.

von Mises, R. (1937). Sur les fonctions statistiques. In *Soc. Math. de France, Conference de la Reunion Internat. des Math*, Paris, France.

von Mises, R. (1947). On the asymptotic distribution of differentiable statistical functions. *Annals of Mathematical Statistics 18*, 309–348.

Vonta, F. and A. Karagrigoriou (2010). Generalized measures of divergence in survival analysis and reliability. *Journal of Applied Probability 47*, 216–234.

Wakimoto, K., Y. Odaka, and L. Kang (1987). Testing the goodness-of-fit of the multinomial distribution based on graphical representation. *Computational Statistics & Data Analysis 5*, 137–147.

Wald, A. (1943). Tests of statistical hypotheses concerning several parameters when the number of observations is large. *Transactions of the American Mathematical Society 54*, 426–482.

Wand, M. P. and M. C. Jones (1995). *Kernel Smoothing*. London, United Kingdom: Chapman and Hall.

Wang, J. L. (1995). M-estimators for censored data: Strong consistency. *Scandinavian Journal of Statistics 22*, 197–206.

Wang, J. L. (1999). Asymptotic properties of M-estimators based on estimating equations and censored data. *Scandinavian Journal of Statistics 26*, 297–318.

Warwick, J. and M. C. Jones (2005). Choosing a robustness tuning parameter. *Journal of Statistical Computation and Simulation 75*, 581–588.

Watanabe, S. (2001). Algebraic analysis for nonidentifiable learning machines. *Neural Computation 13*, 899–933.

Weber, M., L. Leemis, and R. Kincaid (2006). Minimum Kolmogorov–Smirnov test statistic parameter estimates. *Journal of Statistical Computation and Simulation 76*, 195–206.

Wei, W. (1985). Application of the maximum entropy method to electron density determination. *Journal of Applied Crystallography 18*, 442–445.

Welch, W. J. (1987). Rerandomizing the median in matched-pairs designs. *Biometrika 74*, 609–614.

West, E. N. and O. Kempthorne (1972). A comparison of the Chi^2 and likelihood ratio tests for composite alternatives. *Journal of Statistical Computation and Simulation 1*, 1–33.

Wiens, D. P. (1987). Robust weighted Cramér–von Mises estimators of location, with minimax variance in ϵ-contamination neighborhoods. *Canadian Journal of Statistics 15*, 269–278.

Wilks, S. S. (1938). The large sample distribution of the likelihood ratio for testing composite hypothesis. *Annals of Mathematical Statistics 9*, 60–62.

Windham, M. P. (1995). Robustifying model fitting. *Journal of the Royal Statistical Society B 57*, 599–609.

Wolfowitz, J. (1952). Consistent estimator of the parameters of a linear structural relation. *Skandinavisk Aktuarietidskrift 35*, 132–151.

Wolfowitz, J. (1953). Estimation by the minimum distance method. *Annals of the Institute of Statistical Mathematics 5*, 9–23.

Wolfowitz, J. (1954). Estimation by the minimum distance method in non-parametric difference equations. *Annals of Mathematical Statistics 25*, 203–217.

Wolfowitz, J. (1957). The minimum distance method. *Annals of Mathematical Statistics 28*, 75–88.

Wood, C. L. and M. M. Altavela (1978). Large sample results for Kolmogorov–Smirnov statistics for discrete distributions. *Biometrika 65*, 235–239.

Woodruff, R. C., J. M. Mason, R. Valencia, and A. Zimmering (1984). Chemical mutagenesis testing in drosophila — I: Comparison of positive and negative control data for sex-linked recessive lethal mutations and reciprocal translocations in three laboratories. *Environmental Mutagenesis 6*, 189–202.

Woodward, W. A., W. C. Parr, W. R. Schucany, and H. Lindsay (1984). A comparison of minimum distance and maximum likelihood estimation of a mixture proportion. *Journal of the American Statistical Association 79*, 590–598.

Woodward, W. A., P. Whitney, and P. W. Eslinger (1995). Minimum Hellinger distance estimation of mixture proportions. *Journal of Statistical Planning and Inference 48*, 303–319.

Wu, J. (2007). *Minimum Hellinger distance estimation in semiparametric models*. Ph. D. thesis, University of Alberta, Edmonton, Canada.

Wu, J. and R. J. Karunamuni (2009). On minimum Hellinger distance estimation. *Canadian Journal of Statistics 37*, 514–533.

Wu, J., R. J. Karunamuni, and B. Zhang (2010). Minimum Hellinger distance estimation in a two-sample semiparametric model. *Journal of Multivariate Analysis 101*, 1102–1122.

Yager, R. R. (1979). On the measure of fuzziness and negation, Part I: Membership in the unit interval. *International Journal of General Systems 5*, 221–229.

Yan, H. (2001). *Generalized minimum penalized Hellinger distance estimation and generalized penalized Hellinger deviance testing for discrete generalized linear models*. Ph. D. thesis, Utah State University, Logan, USA.

Yang, S. (1991). Minimum Hellinger distance estimation of parameter in the random censorship model. *Annals of Statistics 19*, 579–602.

Yarnold, J. K. (1972). Asymptotic approximations for the probability that a sum of lattice random vectors lie in a convex set. *Annals of Mathematical Statistics 43*, 1566–1580.

Ying, Z. (1992). Minimum Hellinger-type distance estimation for censored data. *Annals of Statistics 20*, 1361–1390.

Zadeh, L. (1965). Fuzzy sets. *Information and Control 8*, 338–353.

Zografos, K., K. Ferentinos, and T. Papaioannou (1990). ϕ-divergence statistics: Sampling properties and multinomial goodness of fit divergence tests. *Communications in Statistics: Theory and Methods 19*, 1785–1802.

Zuo, J. M., I. Vartanyants, M. Gao, R. Zhang, and L. A. Nagahara (2003). Atomic resolution imaging of a carbon nanotube from diffraction intensities. *Science 300*, 1419–1421.

Index

Printed and bound by CPI Group (UK) Ltd, Croydon, CR0 4YY

24/10/2024

01778493-0006